AGRICULTURE AND INTERNATIONAL TRADE
Law, Policy and the WTO

This book is dedicated to the memory of

Louis Lorvellec

AGRICULTURE AND INTERNATIONAL TRADE
Law, Policy and the WTO

Edited by

Michael N. Cardwell
University of Leeds, UK

Margaret Rosso Grossman
University of Illinois, USA

and

Christopher P. Rodgers
University of Newcastle upon Tyne, UK

CABI Publishing

CABI is a trading name of CAB International

CABI Head Office
Nosworthy Way
Wallingford
Oxfordshire OX10 8DE
UK

Tel: +44 (0)1491 832111
Fax: +44 (0)1491 833508
Email: cabi@cabi.org
Web site: www.cabi.org

CABI North American Office
875 Massachusetts Avenue
7th Floor
Cambridge, MA 02139
USA

Tel: +1 617 395 4056
Fax: +1 617 354 6875
Email: cabi-nao@cabi.org

A catalogue record for this book is available from the British Library, London, UK.

Library of Congress Cataloging-in-Publication Data
Agriculture and international trade : law, policy and the WTO / edited by Michael N. Cardwell, Margaret R. Grossman and Christopher P. Rodgers.
p. cm.
Includes bibliographical references and index.
ISBN 0-85199-663-9 (alk. paper)
1. Agricultural laws and legislation. 2. World Trade Organization. 3. Free trade. 4. Nontariff trade barriers—Law and legislation. 5. Tariff on farm produce. I. Cardwell, Michael. II. Grossman, Margaret R. III. Rodgers, C.P. (Christopher P.) IV. Title.
K3870.A944 2003
341.7'5471—dc21

2003004769

ISBN-13: 978-085199-663-9
ISBN-10: 0-85199-663-9

First published 2003
Transferred to print on demand 2010

Printed and bound in the UK by
CPI Antony Rowe, Chippenham and Eastbourne.

Contents

Contributors

David E. Adelman *is an Associate Professor at the James E. Rogers College of Law, University of Arizona, 12001 E. Speedway, PO Box 210176, Tucson, AZ 85721-0176, USA.*

Michael N. Cardwell *is a Senior Lecturer in the School of Law, University of Leeds, Leeds LS2 9JT, UK.*

Theodore A. Feitshans *is a Lecturer/Extension Specialist, Department of Agricultural and Resource Economics, North Carolina State University, Room 3340, Nelson Hall/Campus Box 8109, Raleigh, NC 27695-8109, USA.*

Duncan A. French *is a Lecturer in Law, in the Department of Law, University of Sheffield, Sheffield S10 1FL, UK.*

Wyn Grant *is a Professor of Politics, Department of Politics and International Studies, University of Warwick, Coventry CV4 7AL, UK.*

Margaret R. Grossman *is a Professor of Agricultural Law, Department of Agricultural and Consumer Economics, University of Illinois, 333 Mumford Hall, 1301 W. Gregory Drive, Urbana, IL 61801-3605, USA.*

Christopher Hilson *is a Reader in Law, University of Reading, School of Law, Old Whiteknights House, Whiteknights, PO Box 217, Reading RG6 6AH, UK.*

John Linarellii *is an Associate Professor of Law, University of La Verne College of Law, 320 East D Street, Ontario, CA 91764, USA.*

Christopher P. Rodgers *is a Professor of Law, Newcastle Law School, University of Newcastle, 21–24 Windsor Terrace, Newcastle Upon Tyne NE1 7RU, UK.*

Daniel A. Sumner *is Frank H. Buck, Jr Professor in the Department of Agricultural and Resource Economics, University of California, Davis, CA 95616, USA, and he is also the Director of the University of California Agricultural Issues Centre (website: www.aic.ucdavis.edu).*

Norman W. Thorson *is Judge Harry A. Spencer Professor of Law at the University of Nebraska, College of Law, Lincoln, NE 68583-090, USA.*

Preface

This book has its origins in a symposium on International Trade Law held at the National Agricultural Library of the US Department of Agriculture in Washington, DC, in May 2001. It takes as its focus the legal and policy issues for international agricultural trade arising in the World Trade Organization (WTO) Millennium Round. The agricultural negotiations are not only complex and multifaceted but have proved to be one of the most bitterly contested and difficult spheres of that Round. Achieving a satisfactory conclusion in 2005, as envisaged in the Doha Declaration, is likely to prove difficult. The emergence of issues such as agricultural biotechnology, intellectual property rights and environmental protection as key policy areas where consensus is now sought means that a large range of contentious factors will have to be accommodated within the framework of international trade law. Some of these arise in the context of the 1994 Uruguay Round Agreement on Agriculture itself, but many issues of direct relevance to agriculture also arise in the context of the WTO Agreement on Technical Barriers to Trade, the Agreement on Sanitary and Phytosanitary Measures and the Agreement on Trade in Intellectual Property Rights.

The breadth of the issues arising in the WTO agriculture negotiations is, the editors hope, reflected in the contributions to this book. The essays address four areas of principal concern. The first is the emerging concept of 'multifunctionality', and the extent to which WTO rules should enable signatories to configure agricultural support towards animal welfare, food quality and other non-agricultural measures that support joint products with food production. A second area of major concern is trade in intellectual property rights, and the extent to which rights in new varieties of agricultural product should be subjected to special disciplines following the Millennium Round. The advent of biotechnology and its application to the development of new agricultural products is a third area that has generated significant controversy in current public debate. And, finally, the integration of environmental protection requirements into agricultural policy is an area that gives rise to a number of issues of importance for the WTO negotiations, not least the extent to which support arrangements can be decoupled from production and reconfigured to target environmentally beneficial farming systems. The contributions in this collection seek to address these issues from the perspectives of two of the principal protagonists in the Millennium Round negotiations – the European Community and the USA – each of whom has a distinctive perspective on many of the issues they raise.

The majority of the contributions approach these issues from a legal perspective, and their focus is on the international legal order for agricultural trade and related issues of domestic law and policy reform. A true understanding of the role and operation of the WTO agreements requires an interdisciplinary approach, however. This is a forum where there are many interfaces between law, economics and political science. For this reason the collection also includes contributions on the politics of international agricultural trade (see Chapter 3 by Wyn Grant) and on the economics of subsidy programmes and international trade (see Chapter 4 by Daniel Sumner). It is hoped that an interdisciplinary approach will enable the reader to appreciate not only the difficult legal issues arising in this context, but also the interaction of the legal order for international trade with the process of domestic policy reform and the key economic issues arising from agricultural support programmes. If in some modest way it enables the reader to appreciate the complexities of the interactions between law, economics and politics in this contentious area, then it will have achieved its purpose.

Finally, I must, on behalf of all three editors, add my thanks to the contributors for their efficiency and patience. Editing multi-author texts is not always an easy process. In this case, however, it has been both enjoyable and relatively painless. All the contributors responded efficiently and speedily to the collective hectoring and pressure exerted by the editors, with the consequence that script was delivered on time and without problems. This made the job of the editors immeasurably easier, and my own job in collating the script and preparing the final text for the publishers relatively straightforward. We are also grateful to the University of Arkansas for their continued support in bringing this project to fruition. As the reader will note from the Foreword, this volume is dedicated to the memory of the late Professor Louis Lorvellec. Louis was a great ambassador for the subject, lecturing widely on agricultural law and policy around the world. His many colleagues and friends, of whom I consider myself greatly privileged to have been one, remember him with enormous affection. To those who knew him he was not only a scholar of great insight and imagination, but also a constant friend and thoughtful colleague. It is hoped that this book will be a fitting tribute to his memory.

Professor Christopher Rodgers
Aberystwyth

July 2003

Foreword

In May 2001, the National Center for Agricultural Law Research and Information at the University of Arkansas School of Law and the United States Department of Agriculture National Agriculture Library sponsored an International Symposium on International Trade Law in Washington, DC. This book reflects the ideas and observations advanced at this Symposium. Its inspiration, however, can only be ascribed to the person to whom this book is dedicated, Louis Lorvellec, Professor of Law at the University of Nantes, France.

Professor Lorvellec died shortly before the International Symposium was convened. This book carries forward his passion to educate and his scholarly rigour. All who have made this book an exemplary examination of the complex and often confusing law and policy of international agricultural trade were motivated by their respect and admiration for him. At the time of his death, Professor Lorvellec was the President of the Union Mondiale des Agraristes Universitaires (Worldwide Union of Agricultural Law Professors and Researchers), and this book is a fitting tribute to a teacher and scholar who knew that agriculture and agricultural law must be approached and understood from an international and interdisciplinary perspective.

This book is a book of perspectives. It explores its subject from many vantage points and from the differing perspectives of law, economics and politics. After all, this is what a book about agricultural trade should do, for views about what should or should not be the legal and economic framework for agricultural trade are nearly as diverse as the agricultural products traded. As a book of perspectives, it is intended for anyone who desires to explore the legal and economic framework of international agricultural trade. It also seeks to address the many unsettled issues confronting the establishment of a comprehensive and equitable framework for the international trading system as it relates to agriculture. That this book should carry its readers on a journey of exploration is entirely fitting, for Professor Lorvellec enjoyed reminding his audiences that he taught and wrote in Nantes, the city where Jules Verne was born.

On behalf of the National Center for Agricultural Law Research and Information, I hope readers will find this book a useful interdisciplinary introduction to the law and policy of international agricultural trade and to the issues and controversies it now faces. We are confident that our pride in our sponsorship of this book will be matched by the value that readers find in it.

I am required to add that the contents of this book are based upon work supported by the United States Department of Agriculture, under Agreement No. 59-08201-9-115, and any opinions, findings, conclusions or recommendations expressed in this publication are those of the contributors and do not necessarily reflect the view of the United States Department of Agriculture.

Finally, it is with sadness that I must record the death of Norm Thorson, one of the contributors, shortly before this book went to press. The contributors would wish to join me, on behalf of the National Center, in extending our deepest condolences to his family and expressing our gratitude for the contribution he made, despite failing health, to this book project.

Christopher R. Kelley
Associate Professor of Law and Faculty Director
National Center for Agricultural Law Research and Information
University of Arkansas School of Law
Fayetteville, Arkansas, USA

10 February 2003

The WTO, International Trade and Agricultural Policy Reform

1

Christopher Rodgers and Michael Cardwell

Introduction

The Uruguay Round of negotiations establishing the World Trade Organization (WTO) was formally concluded with the Marrakesh Protocol of 15 April 1994. In relation to agriculture, the new WTO order introduced important new disciplines for regulating international trade in agricultural products, with major implications for the domestic support arrangements and external trade policies of the signatories. The Uruguay Round achieved several new agreements of relevance to agriculture. These included not only a specific Agreement on Agriculture (URAA) with bespoke rules for international agricultural trade to replace the more general provisions of the General Agreement on Tariffs and Trade (GATT) previously applicable, but also the Agreement on the Application of Sanitary and Phytosanitary Measures (SPS Agreement) and the Agreement on Technical Barriers to Trade (TBT Agreement). Together, these represented a major change of regulatory framework for agricultural trade. The market and subsidy disciplines in the URAA, and the rules implementing them, are considered in more detail by Grossman in Chapter 2 below.

While undoubtedly a significant step, the conclusion of the URAA was but one stage in a process of ongoing trade reform. 'Recognizing that the long-term objective of substantial progressive reductions in support and protection resulting in fundamental reform is an ongoing process', the signatories committed themselves in Article 20 of the URAA to initiate negotiations for continuing the process 1 year before the end of the implementation period for the URAA reforms (1995–2000), taking into account:

(eds M.N. Cardwell, M.R. Grossman and C.P. Rodgers)

1. The experience to that date from implementing the reduction commitments.
2. The effects of the reduction commitments on world trade in agriculture.
3. Non-trade concerns, special and differential treatment for developing countries and the objective of establishing a fair and market-oriented agricultural trading system.
4. What further commitments are necessary to achieve these long-term objectives.

The current agriculture negotiations enjoyed their own timetable under the URAA and commenced prior to the formal launch of the Millennium Round. Under the terms of the negotiating programme agreed in the Doha Declaration (WTO, 2001c), they are scheduled to conclude not later than 1 January 2005. The contributions in this book are intended to consider, from an interdisciplinary position, the problems and challenges of further reform in the agricultural trading system and proposed reforms to the legal basis for international trade in agriculture currently set out in the URAA, SPS Agreement and TBT Agreement.

Domestic Reform and the WTO: the Uruguay Round

The URAA disciplines are grouped under three heads: market access, domestic support and export subsidies. The market access provisions in Article 4 of the URAA committed Members to a process of tariffication, or conversion of all non-tariff barriers into tariffs, and then, using 1986–1988 as the base period, their average reduction by 36% over 6 years in the case of developed countries.

The provisions on domestic support were predicated upon the adoption of the 'aggregate measure of support' (AMS) as a basis for calculating domestic support for farmers in contracting states and measuring the reduction commitments undertaken in the URAA. AMS was defined by Article 1(a) to mean 'the annual level of support, expressed in monetary terms, provided for an agricultural product in favour of the producers of the basic agricultural product or non-product specific support provided in favour of agricultural producers in general'. This was amplified by detailed rules in Annex 3, using 1986–1988 as the base period. The URAA committed Members, in the case of developed countries, to a reduction of 20% in the aggregate base AMS over 6 years. So-called green-box subsidies (defined in Annex 2 to the URAA) were excluded from the calculation of the AMS and the reduction commitments, and there was a *de minimis* exclusion for product-specific domestic support that did not exceed 5% of the Member's total value of production of that product in a given year or for non-product-specific support that did not exceed 5% of the value of the Member's total agricultural production in a given year (Art. 6.4). In the case of developing countries, the threshold was raised to 10%. A further exemption was included for certain direct payments made under production-limiting programmes, the so-called blue box (Art. 6.5). The availability of the green-box exemption has proved a particularly important factor, both in terms of domestic policy development within the European Community and elsewhere and in terms of its significance as a focal point for dispute in the Millennium Round negotiations. This aspect is considered further below. Those subsidies in the amber box, on the other hand, were deemed to distort trade and were thus subject to inclusion in the calculation of the AMS and, accordingly, reduction.

The URAA also committed Members, in the case of developed countries, to reduce export subsidies by 21% by volume and 36% by value from a 1986–1990 base.

The negotiating process leading to the adoption on the URAA in this form was both lengthy and tortuous. A detailed examination of the negotiating process is beyond the scope of this work (as to which see Paarlberg, 1997; Coleman and Tangermann, 1999; and, especially, Fennell, 1997, pp. 378–397). Some of the key issues that proved problematic merit discussion here, however, as they remain closely linked to potential pressure points in the Millennium Round negotiations. The Uruguay Round was intended to last from 1987 to 1990, with a mid-term review of progress at the end of 1988. In the event, due to a number of fundamental difficulties, the final agreements were not signed until April 1994. The European Community favoured an approach to domestic support that focused on its economic impact and therefore proposed a definition of AMS that included all those factors which have an impact upon the production decisions of farmers. This was initially contested by the USA, which opposed the use of AMS to quantify support commitments. Such a basis would have had adverse implications for US internal policy. In particular, deficiency payments would fall within the Community's proposed definition of AMS (Fennell, 1997, p. 382). The Community, for its part, was strongly opposed to moves to outlaw export subsidies, which it saw as integral to its policy of differential pricing for products marketed internally to the Community and for external export.

Final negotiations intended to settle the outstanding differences failed in December 1990. The impasse was eventually broken when Dunkel, the Director-General of GATT, presented the parties with a set of proposals in December 1991 that ultimately formed the basis of the URAA. Crucially, this included the proposal for a reduction of 20% in all government subsidies over a 7-year period, calculated by reference to a 1986–1988 base and using the AMS as, in principle, the measure of calculation. Together with his other proposals on market access and export subsidies, the Dunkel draft also formed the basis of the Blair House Accord, subsequently reached between the USA and the Community in November 1992, which opened the door to conclusion of the Uruguay Round. Under this Accord agreement was also reached on the 'Peace Clause', which would provide wide-ranging immunity from challenge for domestic support and export subsidies until 31 December 2003 (Fennell, 1997, p. 388).

Prior to the introduction of the Agenda 2000 reform process, the principal changes effected to the Common Agricultural Policy (CAP) of the European Community had been the MacSharry reforms of 1992. An aspect of the Blair House Accord of particular concern in the Community was its compatibility with the MacSharry reforms. Their principal feature was a switch in the weight of agricultural support away from payments to support prices towards 'decoupled' payments to producers, alongside moves to promote the multifunctional role of agriculture in rural areas. In the cereals sector, for example, farmers saw the intervention price fall, in return for compensatory payments based on the area of land under cereal cultivation and a compulsory set-aside requirement. The introduction of the 'Agri-environment Regulation' required Member States to submit agri-environmental programmes to promote traditional and environmentally friendly farming practices (Council Regulation (EEC) 2078/92 (OJ 1992, L215/85)). There were other 'accompanying measures' providing support for early retirement for farmers and support for forestry.

Because implementation of the MacSharry reforms preceded the conclusion of the Uruguay Round, a lively debate has ensued as to the relationship between the internal policy reform process within the Community and the negotiations that eventually produced the URAA. This debate is important not only for historical reasons, but also because the later Agenda 2000 reform process in Europe was (as discussed below) expressly premised upon the package of MacSharry reforms. It represented a deepening and extension of those reforms, while the proposed introduction under the Mid-term Review of the decoupled single farm payment (SFP) has been regarded as marking their logical conclusion.

Commentators differ over the interrelationship between the negotiations in the Uruguay Round and the process of domestic policy change in Europe. Some have posited a theory of interdependence between international, domestic and Community policy change (Coleman and Tangermann, 1999). Proponents of this view see the European Commission as having adopted an international entrepreneurial and leadership role, with the result that the MacSharry reforms were significantly shaped by the proposals and outcomes of the Uruguay Round. The Commission assumed this role, it is argued, because of the deadlock between protectionist and liberalizing states within the Community itself. Others dismiss such a viewpoint, however, arguing that budgetary constraints within the Community and the need to curb overproduction of key commodities were the principal factors driving change (see, for example, Paarlberg, 1997).

Whatever the precise relationship between the Uruguay Round and the MacSharry reforms, however, there can be little doubt that the URAA has had a major effect on subsequent policy reform in the European Community. It imposed, not least, a number of inbuilt 'brakes', which have limited proposals for further domestic reform of the CAP. Indeed, the Agenda 2000 reforms have been heavily conditioned by the shape of the URAA and, in particular, by the latitude it gave to signatories to switch support to direct payments to producers. Blue-box and green-box exemptions from reduction commitments have been much exploited, commencing with the payments made to producers under the MacSharry reforms in compensation for the loss of income flowing from the cuts made in intervention prices. As a consequence, it has been argued that 'the movement away from explicit export subsidies to compensatory payments in a variety of forms may imply little more than a shallow and convenient change of appearance' (Scott, 1996, p. 177). According to its critics, the development of Community policy since 1994 demonstrates this characteristic amply.

Domestic Reform and the WTO: the Millennium Round

Following the experience of the Uruguay Round, the Community approached the Millennium Round with a clear resolve to entrench its negotiating position in advance. The Agenda 2000 reforms were central to this strategy. Thus, at their launch, it was unequivocally stated in the Agenda 2000 document itself that the proposed measures would 'enhance the Union's negotiating stance in the New Round' (CEC, 1997, Part One, III, 2). Since the issue of that document, the importance of world-trade considerations in driving reform has been consistently reiterated. For example, in the 1998 *Explanatory Memorandum*, which accompanied proposals for eight regulations, the opportunity was taken to declare that 'it must be made quite clear to all that the reform to be adopted will outline the limits of what the Union is able to agree to in the forthcoming international negotiations' (CEC, 1998, para. 1).

Such sentiments were expressed even more forcefully as the commencement of the agriculture negotiations approached, by which point the Community had already concluded its first tranche of reform at the Berlin Summit of 24 and 25 March 1999. Indeed, in the view of the European Council, 'the decisions adopted regarding the reform of the CAP within the framework of Agenda 2000 will constitute essential elements in defining the Commission's negotiating mandate for the future multilateral trade negotiations in the WTO' (CEC, 1999, 1.12, 24). This policy was subsequently confirmed by Commissioner Fischler. Thus, he announced in September 1999 that 'the Agenda 2000 reforms will form the core of our negotiating position' (Fischler, 1999b); and, a month later, that: 'it must be clear that the European model of agriculture, which is based on competitive, multifunctional and sustainable farming throughout Europe is not for negotiation' (Fischler, 1999c). Indeed, he openly articulated a determination not to repeat the experience of the Uruguay Round, stating that, in contrast to that Round, 'we have concluded our reform before the trade talks start, and thus don't run the risk of running into the sorts of difficulties we faced last time' (Fischler, 1999a). Undoubtedly, the domestic subsidies in the green and blue box are key issues in the negotiations. Much will depend upon the Community's ability to defend a reformed CAP regime that exploits the ability to reposition domestic support into these categories, so maximizing exemption from domestic support reduction commitments. Another key issue will be multifunctionality. As commentators have pointed out, the USA and Cairns Group will argue that the concept has the potential to allow the Community to circumvent disciplines imposed on domestic support, being rather a convenient justification for meeting other objectives, including protectionism. If a country is tightly bound in terms of the amber box, with little leeway to increase subsidies within URAA commitments, it is likely to support the multifunctional agenda (Bohman *et al.*, 1999; Landau, 2001). These issues will be difficult to resolve.

In the case of the USA, the cycle of reform has, to a considerable extent, been dictated by the passage of Farm Bills; and, in consequence, there would seem to be less of a direct chronological link with the agriculture negotiations. Thus, the Federal Agriculture Improvement and Reform Act of 1996 (FAIR Act) was enacted 2 years after the conclusion of the Uruguay Round and 3 years before the commencement of the Millennium Round, while the Farm Security and Rural Investment Act of 2002 (FSRI Act) was enacted during the second phase of the agriculture negotiations under the Millennium Round. That said, the linkage between domestic reform and world trade commitments has been openly acknowledged. This may be illustrated by the fact that, shortly after the enactment of the FSRI Act, the USA issued the *US Proposal for Global Agricultural Trade Reform* (United States, 2002); and, significantly, the FSRI Act was described at the launch of that initiative as the 'first leg' of a three-legged stool upon which the national negotiating strategy rested (United States Department of Agriculture (USDA), 2002g).

At the same time there may be detected links between Community and US domestic reform. In any event, when presenting domestic reform, policy-makers have not been slow to reflect differences from (and, on occasion, similarities to) measures enacted on the other side of the Atlantic. Perhaps most notably, the Agenda 2000 reforms were predicated upon perceived distinctions between the European model of agriculture and the model found both in the USA and elsewhere. Again in the words of Commissioner Fischler:

> [u]nlike that of many of our negotiating partners, EU agriculture is highly diversified, and farming in Europe also performs a range of additional tasks. It supports and safeguards our unique countryside and a stable

environment. As a result of the high population density in Europe, we must produce these services in addition to actual farm produce itself. We cannot afford to confine nature and the environment to reservations.

(Fischler, 1999b)

In practice, this linkage is readily apparent in the coincidence of reform that occurred in 2002, with the FSRI Act being enacted in May and the *Mid-term Review of the Common Agricultural Policy* being issued shortly afterwards in July (CEC, 2002b). Indeed, the issue of the document was delayed by 1 month to take preliminary account of the changes effected by the FSRI Act; and, in the world-trade context, there was unequivocal recognition of parallel reforms on promulgation of the *US Proposal for Global Agricultural Trade Reform* (USDA, 2002g). In light of the fact that the USA and the Community remain, respectively, the largest and second largest exporters of agricultural produce in the world, such a state of affairs would seem inevitable.

The United States FSRI Act of 2002

The FSRI Act was signed into law by the President on 13 May 2002 (USDA, 2002a). Covering 2002–2007 crops, it was admitted to be both complex and a compromise (USDA, 2002a,b). However, its compatibility with world-trade commitments was fiercely asserted (USDA, 2002e). Further, such compatibility was considered especially meritorious in that the USA was constrained by a far lower ceiling on its URAA AMS than either Japan or the Community ($19.1 billion per annum as opposed to, respectively, $31 billion and $62 billion per annum) (USDA, 2002c).

The response from the Community was, for the most part, negative, reversing a more conciliatory approach evident in the early stages of the passage of the FSRI Act through Congress. For example, in March 2001 Commissioner Fischler had urged that the Community and the USA should 'compare notes and keep in touch on our respective plans and try to move our support systems closer to each others rather than more apart' (Fischler, 2001). In contrast, shortly before the signing into law of the FSRI Act, he alleged that the USA had 'flunked' farm policy reform (European Community, 2002a). Not least, it was argued that comparisons between ceilings on the URAA AMS were meaningless: the Community was operating well within its ceiling, with a downward trend, while, as a result of the FSRI Act, the ceiling for the USA might even be exceeded (European Community, 2002d).

In this context, six aspects of the FSRI Act may be highlighted: first, the overall budgetary outlay; secondly, the extension of direct payments, together with the option to update base areas; thirdly, the introduction of new counter-cyclical payments; fourthly, the 'circuit-breaker'; fifthly, conservation; and, sixthly, labelling requirements.

According to the Secretary of Agriculture, 'the Farm Bill continues with roughly the same amount of support as we've been providing our farm sector over the past four years'. In making this comparison, it was vital to take account of not just support provided by the FAIR Act, but also emergency supplemental support of approximately \$7.5 billion per annum over the previous 4 years. Consequently, '[t]he new Farm Bill provides roughly \$7.4 billion each year in new spending for farm programs' (USDA, 2002c). That said, the consolidation of emergency support which this represents has caused considerable concern for the Community. Thus, it has been calculated that overall budgetary outlay will increase by \$51.7 billion to some \$296.5 billion over the 6-year period of the FSRI Act. The same analysis singled out the expansion in commodity spending under Title I for particular criticism, but it may also be noted that expenditure on conservation programmes was projected to expand by an even greater percentage, from \$2 billion per annum under the FAIR Act to \$4 billion by 2007 under the FSRI Act (CEC, 2002c, p. 159). Indeed, the Secretary of Agriculture also placed great emphasis on these largest ever increases in conservation expenditure (USDA, 2002b); but it may be noted that, as also in the Community following the Mid-term Review, such expenditure remains dwarfed by other forms of payment.

Secondly, under the FAIR Act, production flexibility contracts provided direct payments irrespective of crop prices (Kelley, 1999); and, while similar fixed direct payments are made under the FSRI Act, significant changes may be identified. In particular, coverage has been extended to include soybeans (which have seen a substantially increased acreage since 1996); the rate of payment, unlike under the FAIR Act, is not degressive; and farmers enjoy the option to update the reference period for base areas to 1998–2001 (payments being made on 85% of such base areas).

Thirdly, and perhaps most controversially, the FSRI Act has introduced new counter-cyclical payments, related to price and reminiscent of the deficiency payments system in the USA prior to the FAIR Act. These apply where the 'effective price' is less than the 'target price'. For the purposes of such calculation, the 'effective price' is the fixed direct payment plus the higher of the national average market price received by producers during the marketing year or the national loan rate for the commodity (under the marketing assistance loan programme). Moreover, in regard to counter-cyclical payments, farmers enjoy the option to update not only their base areas, but also their reference yields.

As indicated, the introduction of new counter-cyclical payments has proved highly controversial. In the view of the US administration, they confer the major advantage of stability. It is also hoped that they will contribute towards obviating the need for emergency supplemental assistance (USDA, 2002d). In contrast, the Community has characterized counter-cyclical payments as 'a guarantee' that farmers will continue to receive the additional funding hitherto provided by emergency supplemental assistance. Further, in support of its criticism, the Community has made reference to an estimate that fixed direct payments and counter-cyclical payments will together account for in the region of $11–11.5 billion per annum (European Community, 2002b).

Fourthly, considerable importance may, therefore, be attached to the implementation of the 'circuit-breaker'. Under this provision, the Secretary of Agriculture may make adjustments to the maximum extent practicable to prevent the USA breaching WTO commitments (including the $19.1 billion per annum ceiling on its URAA AMS). While in theory introducing a virtually fail-safe mechanism, the circuit-breaker has been treated with much circumspection. For example, the Community has questioned the political acceptability of requiring farmers to reimburse any excess payments (European Community, 2002b).

Fifthly, it may be reiterated that conservation programmes have been accorded significantly increased funding; and this aspect has attracted favourable comment from the Community. In addition to expansion of, for example, the Conservation Reserve Program (from 36.4 million to 39.2 million acres) and the Environmental Quality Incentives Program (with a focus on livestock production), new initiatives include the Conservation Security Program and Grasslands Reserve Program. Expenditure on these two new initiatives over the 6-year period of the FSRI Act is to total, respectively, $2 billion and $254 million.

Sixthly, country-of-origin labelling is to be introduced for meat, fruit and vegetables, fish and groundnuts. While initially voluntary, this is to become mandatory in 2004. To qualify as a US product, the commodity must have been born, raised and processed in the USA. This gives rise to interesting questions where, for example, animals are imported from Canada for fattening, but there would seem to be a clear intention to adhere to a US label as opposed to a broader North American label (USDA, 2002f). At the same time, it is of note that such requirements have marked similarities to those governing beef labelling as introduced by the Community under Regulation (EC) 1760/2000 of the European Parliament and of the Council (OJ 2000, L204/1). As a result, there may be considerable identity of interest between the Community and the USA in this aspect of the world-trade negotiations.

The CAP Mid-term Review

If a linkage between the FSRI Act and the ongoing WTO negotiations is implicit, in the case of the Mid-term Review it is quite explicit. The expressly stated objectives of the review are not only to simplify the operation of the CAP and facilitate the process of eastward enlargement of the European Union, but also to 'help to better defend the CAP in the WTO' (CEC, 2002b, p. 3). The policy document issued in July 2002 envisaged that all direct payments under, *inter alia*, the common organizations of the market in cereals, beef and veal and sheepmeat and goatmeat would be decoupled from production as from 1 January 2004. Other sectors would be integrated as and when their reform was completed. However, decoupling to this extent met with considerable opposition from some Member States (*Agra Europe*, 2003, pp. EP/1-EP/3). Agreement was reached in Luxembourg on 26 June 2003, with a considerable element of compromise to accommodate both delayed implementation and numerous derogations (European Community, 2003a). In principle, the introduction of the decoupled single farm payment (SFP) has been postponed until 1 January 2005. However, Member States enjoy the option to further postpone its introduction until 1 January 2007. Other derogations include regional implementation and 'partial implementation'. For example, Member States may opt to retain up to 25% of arable-area payments as support coupled to production.

The SFP is based upon payments received under the relevant support schemes over a 2000–2002 reference period. Each farmer will be granted a payment entitlement per hectare; and these entitlements may be leased or sold. In the case of lease, but not in the case of sale, a land transaction will be necessary.

Payment of the SFP will be conditional on observance of statutory management requirements relating to: public, animal and plant health; the environment; and animal welfare. In addition, farmers will be obliged to maintain land in good agricultural and environmental condition. Member States will enjoy discretion, at national or regional level, to define minimum requirements for such 'good agricultural and environmental condition'; but, in undertaking this task, will be obliged to comply with a Community framework in order to prevent distortions of competition.

According to the European Commission, the principal advantage this scheme offers in WTO terms will be its green-box compatibility, 'which will help secure these payments in an international context' (CEC, 2002b, p. 19). The detailed provisions have, therefore, been structured with close reference to the existing green-box criteria and are clearly intended to strengthen the Community's position in any renegotiation of the green box in the Millennium Round. These aspects will be considered further below, when discussing proposals to redefine and widen the green box. In general terms, however, there are a number of problems to which the Mid-term Review gives rise, and some of these are linked to potential pressure points in the WTO negotiations. A fundamental issue is the link between agricultural support and production, and the extent to which the Mid-term Review breaks that linkage. To qualify as decoupled income support for the purposes of the URAA, no production shall be required (Annex 2, para. 6(e)). The reforms do not, as such, render receipt of the

SFP conditional upon production. However, as indicated, farmers will be obliged to maintain land in good agricultural and environmental condition; and the Community framework includes standards for crop rotations where applicable and minimum livestock stocking rates or/and appropriate regimes. Accordingly, the continuation of farming would seem necessary in order to achieve the desired result. This is an issue which will be considered further in the context of the green box.

It may be noted that the absence of any general principle rendering receipt of the SFP conditional on production was itself a driving force behind the imposition of the requirement to maintain land in good agricultural and environmental condition. Fears had been expressed that otherwise land abandonment would ensue, giving rise to environmental problems. Indeed, the requirement was considered 'a necessary complement to decoupling' (CEC, 2003a, p.10).

The imperative of avoiding land abandonment was also cited as the reason for authorization of partial implementation (European Community, 2003a). However, where Member States opt that, to the extent permissible, support should remain coupled to production, it will not be easy to argue that green-box exemption should be available. Blue-box exemption may be available; but account will need to be taken of the widespread hostility in the Millennium Round to the continuation of the blue box on anything like its present scale.

The Developing Agenda

Non-trade concerns

The agenda for the Millennium Round has already reflected the growing importance attached by numerous Members to non-trade concerns. Debate has extended well beyond the issues of market access, domestic support and export subsidies, the three central pillars of the URAA, with considerable heat generated by differing views on the weight that should be attached to, for example, the protection of the environment, food safety, food quality, food security, rural development and animal welfare.

The expanding debate is also inextricably linked with the importance attached by the Community to the recognition of the 'multifunctional' role of agriculture. While precise definition is difficult, there would none the less seem to be a general consensus that key elements of multifunctional agriculture are the joint production of commodity and non-commodity outputs and the fact that some of these non-commodity outputs may be characterized as externalities or public goods (Organisation for Economic Co-operation and Development (OECD), 2001a). The European model of agriculture, as implemented under the Agenda 2000 CAP reform, has been regarded as a 'flagship' for the multifunctionality concept. Thus, at the Berlin Summit it was stated that 'reform will ensure that agriculture is multifunctional, sustainable, competitive and spread throughout Europe, including regions with specific problems' (CEC, 1999, 1.11, 20). Moreover, the Mid-term Review has integrated non-trade concerns into the heart of the CAP. As has been seen, a wider range of conditions is to be attached to direct payments to producers, including conditions relating to the environment and to animal welfare. Moreover, from the inception of the Millennium Round, Commissioner Fischler viewed the emphasis on non-trade concerns as its defining feature, stating:

> I believe the Millennium Round will have to go beyond purely market-related issues and cover areas of social concern such as environmental protection, animal welfare, quality and food safety. We must realise that suspicion of the WTO is already running high. If people's impression that the WTO puts trade before health issues is confirmed, it will not become stronger in future, but weaker.
>
> (Fischler, 1999d)

The Community has not been alone in its advocacy of non-trade concerns. They have also proved of great importance to numerous other WTO Members and a notable development has been the creation of the grouping entitled 'The Friends of Multifunctionality'. For example, 40 countries and economies attended a Conference on Non-trade Concerns at Ullenswang, Norway, in July 2000 (WTO, 2000c). Further, it has become increasingly clear that such Members attach differing weights to differing non-trade concerns. By way of illustration, in the case of major food importers, such as Japan, food security has been accorded high status (WTO, 2000e); but for Norway vital considerations have been rural employment, agricultural landscapes and biodiversity (WTO, 2001a).

As indicated, within the world-trade legislative framework itself, Article 20 of the URAA expressly provided that non-trade concerns should be taken into account. In this regard, the Preamble specifically identified food security and the need to protect the environment. Taking advantage of this opportunity, many WTO Members have addressed non-trade concerns in their negotiating proposals, with the incidence increasing in the second phase of negotiations (commencing in March 2001). Indeed, of the ten issues recommended for consideration in the second phase, three could be described as non-trade concerns, namely, food security, food safety and rural development (WTO, 2001d). Subsequent submissions confirmed this trend, a clear illustration being provided by *Food Safety: Note by the European Communities* (WTO, 2001b). Likewise, in the Doha Declaration itself, there was an unequivocal statement that: '[w]e take note of the non-trade concerns reflected in the negotiating proposals submitted by Members and confirm that non-trade concerns will be taken into account in the negotiations as provided for in the Agreement on Agriculture' (WTO, 2001c, para. 13). The inclusion of this statement was much trumpeted by the Community (European Community, 2001). Arguably, however, it did no more than reiterate the requirements stipulated by the URAA.

At the same time, it may be highlighted that non-trade concerns have the capacity to cross the legislative boundaries of the various Uruguay Round agreements. This may be illustrated by issues of animal welfare. As stated in the Community proposal on *Animal Welfare and Trade in Agriculture*, these have the capacity to bring into consideration the SPS Agreement, the TBT Agreement, Article XX of the GATT and Article 20 of the URAA (WTO, 2000b). That said, although Article 14 did expressly provide that 'Members agree to give effect to the Agreement on the Application of Sanitary and Phytosanitary Measures', it must be recognized that the URAA remained itself largely reticent on such matters as animal welfare and food safety, despite the fact that they were already forming an ever larger constituent of agricultural policy. Accordingly, there may be doubt whether this degree of compartmentalization can be sustained.

While the Friends of Multifunctionality have throughout advocated the extension of such issues within and beyond the traditional remit of market access, domestic support and export subsidies, it must be noted that caution, or even scepticism, has been displayed by other Members. In its policy document, *Food and Agricultural Policy: Taking Stock for the New Century*, the USA did adopt a relatively broad, multifunctional approach (USDA, 2001). Further, in the *Proposal for Comprehensive Long-term Agricultural Trade Reform: Submission from the United States*, it was stated that the USA is 'committed to and supports policies that address non-trade concerns, including food security, resource conservation, rural development, and environmental protection' (WTO, 2000a). However, there was also a determination that meeting such objectives should not create new economic distortions. In particular, a consequence to be avoided was passing the cost to other countries by closing markets, or introducing unfair competition, or both. Further, the need to address non-trade concerns without 'spillover' and trade distortion has become a recurrent feature of US policy (Bohman *et al.*, 1999); and, in regard to developed countries, the subsequent *US Proposal for Global Agricultural Trade Reform* was markedly reticent on both non-trade concerns and multifunctionality more generally (United States, 2002). An even firmer line has been adopted by Cairns Group countries (ABARE, 1999).

Biotechnology

Biotechnology issues have already proved a source of major controversy in the agricultural negotiations of the Millennium Round. Their ability to generate controversy is largely matched by their novelty. They received no specific mention in the Uruguay Round agreements, a fact which is perhaps surprising given that before the conclusion of that Round the Community had already started to put in place a legislative framework for regulating biotechnology and its application to agriculture. In particular, in 1990 the Community had introduced two important legislative measures in the form of Council Directive 90/219/EEC on the contained use of genetically modified organisms (GMOs) and Council Directive 90/220/EEC on the deliberate release into the environment of GMOs (OJ 1990, L117/1 and L117/15).

What is clear is that, as between the Community and the USA, there remains a considerable gulf in terms of both the extent of genetically modified (GM) plantings and the public perception of the benefits and disadvantages of GM technology. In 2001 only about 12,000 ha of GM crops were planted in Europe, compared with some 50 million ha worldwide, with the USA accounting for a substantial proportion of such production (CEC, 2002a). Moreover, there has been no evident increase in the acceptance of GM crops by European consumers (Burton *et al.*, 2001). That said, it may be inappropriate to present Community and US attitudes and policy as polarized. The reaction of US consumers was hardly neutral when GM StarLink maize, approved only for animal feed, was discovered in human food; and the Commission has expressed a resolve to participate fully in the biotechnology revolution (CEC, 2002a).

Three aspects of relevance to the world-trade agenda may be highlighted at this juncture. First, as a general rule, genetic modification would appear to be characterized in a fundamentally different manner by policy-makers on either side of the Atlantic. Thus, in the USA, government policy has not normally required labelling of GM food products, on the basis that they do not differ from other foods in any meaningful or uniform way; or that, as a class, they present no different or greater safety concern than foods developed from traditional plant breeding (Food and Drug Administration, 1992). In contrast, the policy of the Community has been to treat genetic modification as giving rise to something different in kind and which does not occur naturally (European Community, 2002c). Accordingly, to adopt the language of world-trade discourse, there is deep disagreement as to whether GM and non-GM products are 'like products'. In this context, consideration may be given to health effects (Howse and Tuerk, 2001). It may be noted, however, that the European Court of Justice has now ruled in *Monsanto Agricoltura Italia SpA* v. *Prezidenza del Consiglio dei Ministri* (2003) that the mere presence in novel foods of residues of transgenic protein does not, such as, preclude those foods from being considered substantially equivalent to existing foods.

Secondly, against this background the Community has adopted an overtly more precautionary approach. For example, Council Directive 2001/18/EC (OJ 2002, L106/1), which replaced Council Directive 90/220/EEC, governing the deliberate release into the environment of GMOs, was expressly stated to take into account the precautionary principle, while recent case-law of the European Court of Justice has seen a deeper analysis of that principle, not least in *Pfizer Animal Health SA* v. *Council* (2002). At the same time, support for a precautionary approach may be derived from broader international initiatives. In particular, the Cartagena Protocol on Biosafety, concluded in 2000, declares itself to be enacted '[i]n accordance with the precautionary approach contained in Principle 15 of the Rio Declaration on Environment and Development'. Accordingly, although the Appellate Body of the WTO in *EC Measures Concerning Meat and Meat Products (Hormones)* (1998) took the view that, at least outside the field of international environmental law, the precautionary principle awaited authoritative formulation, its growing impact would seem undeniable. Furthermore, the issue of precaution has already been highlighted by the Community in its submissions during the Millennium Round, for example in *Food Safety: Note by the European Communities* (WTO, 2001b).

Thirdly, since the use of GM technology was not specifically addressed in the Uruguay Round agreements, it is no easy task to determine which provisions of the various WTO agreements should apply in disputes (Howse and Mavroidis, 2000). What would seem certain, however, is that recourse may be had to more than one agreement. For example, the SPS Agreement would seem applicable to disputes relating to, *inter alia*, the protection of animal or plant life or health, while the TBT Agreement would seem applicable to issues of GMO labelling. That said, the relevance of the URAA is not immediately obvious; and a challenge in the Millennium Round may be to accommodate world-trade issues flowing from GMOs within agriculture itself, the first link in the food-chain. The question became more urgent in May 2003 when the USA took initial steps under the WTO dispute settlement procedures against the community system for authorizing GMOs.

Redefining the green box

A key area in the URAA, and one that is likely to prove highly contentious, is green-box measures that are exempt from the domestic support reduction commitments. The revised arrangements for CAP support introduced under the aegis of Agenda 2000 were designed to ensure that as many support schemes as possible come within the green box as defined in the URAA. The *EC Comprehensive Negotiating Proposal* in the Millenium Round argues for the retention of both the green box and the blue box, although the Community at the outset indicated that it was prepared to discuss the detailed rules for exempt domestic support (WTO, 2000d, para. 12). If the green box is to be retained, the legal basis for exemption will require clarification and a more detailed treatment. In a number of respects, the current rules on green-box exemption lack both flexibility and clarity.

The green box rules are currently contained in Annex 2 to the URAA. To qualify for exemption under this head, domestic support must satisfy two criteria. First, it must satisfy the fundamental test, set out in paragraph 1 of Annex 2, namely, that a support scheme must have no, or at most minimal, trade-distorting effects or effects on production. Paragraph 1 further provides that support must be provided through a publicly funded government programme not involving transfers from consumers, and the support must not have the effect of providing price support for producers. Additionally, it must satisfy one of a number of policy-specific conditions set out in the remaining paragraphs of Annex 2, some of which are problematic. Before examining these, it is perhaps worth noting that the basic qualifying condition itself raises a number of problems, in particular the issue of quantification. No indication is given in the URAA as to the basis for quantifying the effects of support for the purposes of ascertaining whether the trade-distorting effects are 'minimal' or not. Similarly, no guidance is forthcoming on the determination of whether the effect of support is in fact to provide price support to producers and, if so, in what amount. These issues are linked to the more general one of quantification, which will be addressed below.

Of the specific conditions to be met in addition to the fundamental requirement set out in paragraph 1, those governing schemes which provide direct payments to farmers are likely to prove the most controversial. In particular, they potentially cover many of the decoupled CAP support arrangements. All these specific conditions are contained in paragraphs 2–13 of Annex 2 to the URAA. Three categories of green-box support would seem to present particular difficulties in this context: decoupled income support; structural-adjustment assistance provided through resource retirement programmes; and payments made under environmental programmes.

The terms of the exemption for decoupled income support, as set out in paragraph 6, are likely to prove problematic. The current green-box criteria for their exemption provide that eligibility for payments must be determined by clearly defined criteria such as income, producer status, factor use or production level in a defined and fixed base period. Moreover, the amount of annual payments in a given year must not be related to or based upon the type or volume of production undertaken by the producer in any year after the base period. Crucially, no production must be required in order to receive payments. It should be noted that, under the current rules, while payments cannot be based on the type or volume of production in any year after the base period chosen, they *can* be based on the type or volume of production in the base period itself. A key problem concerns the stipulation that no production be 'required' if a support scheme is to qualify for the green box. The intention is to break the link between payments and production. As a matter of textual analysis, however, the clause is open to conflicting legal interpretations. Does it mean that a support scheme must prohibit continued production if a producer is to qualify for payments? Or is it sufficient merely for it to remove the requirement for continued production as a mandatory prerequisite for payments? If the latter is the case, then an indirect link to production may remain, in that producers may legitimately retain the option to continue production.

The support arrangements introduced under Agenda 2000 and in the Mid-term Review clearly proceed on the basis of the second of these alternative interpretations of the green-box criteria. Further, as already noted, while there is no requirement of production for the receipt of the SFP, the proposals do require producers to maintain land in 'good agricultural and environmental condition' as a fundamental requirement for their receipt and, in practice, this may require continued production.

The Mid-term Review raises a number of other issues relevant to the renegotiation of the green-box conditions for decoupled support. The rules for the transfer of SFP entitlements have been devised to comply with green-box requirements, which currently require that entitlements for payments must be determined by clearly defined criteria, including, *inter alia*, the status of the recipient as a producer or landowner (Annex 2, para. 6(a)). Thus, the proposed Council regulation implementing the review recited that, with a view to avoiding 'speculative transfers leading to the accumulation of payment entitlements without a corresponding agricultural basis' there should be a link between payment entitlements and eligible land (CEC, 2003b). This provision was carried through into the agreed reform package.

The cross-compliance conditions to be observed by producers receiving SFPs are closely linked to the non-trade concerns underpinning the Community's negotiating position in the Millennium Round, namely, environmental-protection requirements, animal welfare and food safety. Even if the Community is successful in securing the wider acceptance within the WTO of the multifunctional role of European agriculture, the enforcement rules for securing compliance with the cross-compliance conditions applicable to the SFP and other direct payments are likely to be closely scrutinized. The Mid-term Review policy document was unspecific, commenting merely that '[i]n the case of non-respect of cross compliance requirements, direct payments should be reduced while maintaining proportionality in respect of the risk or damage concerned' (CEC, 2002b, p. 21). However, under the agreed reform package it was expressly stipulated that, in the case of intentional non-compliance, payments must be reduced by not less than 20% and total exclusion from one or several aid schemes would be permitted for 1 or more calendar years. None the less, the efficacy of the arrangements that are put in place to secure this effect is likely to be a pressure point in the WTO negotiations.

Finally, an indirect link with production will remain, in that SFP entitlements will be calculated by reference to support payments made under all relevant CAP schemes over the 2000–2002 reference period. However, this may not be problematic in WTO terms. As noted above, while payments cannot be based on production in any year after the base period chosen, the current green-box criteria in the URAA allow the calculation of payments for a decoupled scheme by reference to the type or volume of production in the base period itself (Annex 2, para. 6(b)).

Payments for what is termed structural-adjustment assistance provided through resource retirement programmes are covered by paragraph 10 of Annex 2. Set-aside was retained following the Berlin Summit as a prerequisite of the receipt of direct payments under the Arable Area Payments Scheme, with a mandatory set-aside rate fixed at 10% from the 2000/01 marketing year through to the 2006/07 marketing year (Council Regulation (EC) 1251/1999, Art. 6 (OJ 1999, L160/1)). In the Mid-term Review, the European Commission proposed an extension of the set-aside concept, with the introduction of compulsory long-term set-aside on arable land for 10 years (CEC 2002b, p. 21). This proposal perhaps reflected some nervousness as to the extent to which the existing arrangements comply with the requirements for green-box eligibility. Not least, it would have excluded the possibility of rotating the land subject to the set-aside obligation, the rotational model having been criticised by many as primarily a supply-side market-management tool, rather than structural-adjustment assistance of the kind envisaged by the URAA. In the event, however, the agreed reform package saw the option of rotation preserved.

The third category of green-box exemption which merits discussion is that concerning payments made under environmental programmes, as set out in paragraph 12 of Annex 2 to the URAA. Under this category, eligibility for payments must be determined as part of a clearly defined government environmental or conservation programme and must be dependent on the fulfilment of specific conditions, 'including conditions relating to production methods or inputs'. Further, the amount of payments must be limited to the extra costs or loss of income involved in complying with the government programme (Annex 2, para. 12(b)). Although the broad intent of the exemption is clear, no definition of a qualifying 'environmental' or 'conservation' programme is provided, an omission that could lead to dispute where the aims and objectives of support schemes are mixed. In the case of support schemes for producers in Less Favoured Areas, for example, Community support arrangements have both an environmental-protection focus and a focus on social issues and rural development. Similarly, Environmentally Sensitive Areas schemes operated by Member States under the terms of Council Regulation (EC) 1257/1999 (Rural Development Regulation) (OJ 1999, L160/80) had wider objectives than simply the promotion of traditional and environmentally friendly farming methods. Additional difficulties may be encountered as a result of the green-box criterion limiting the amount of payments to extra costs or income forgone. Many agri-environmental schemes operated under the

Rural Development Regulation have incorporated an incentive payment element to promote maximum take-up.

As already noted, the Community's negotiating position is heavily dependent on acceptance of the multifunctional nature of the European model of agriculture, and with it the importance of accommodating non-trade concerns, such as animal welfare, environmental protection and food safety. This would require the renegotiation of the green-box criteria to give a clear and workable legal definition of those categories of non-trade concern which would fall outside the AMS calculation. The *EC Comprehensive Negotiating Proposal* identified four categories of measure targeted at 'important societal goals' and for which green-box exemption should be applied: measures for the protection of the environment; measures targeted at the sustained vitality of rural areas and poverty alleviation; food security for developing countries; and animal-welfare measures (WTO 2000d, para. 13). The integration of environmental protection into the operation of agricultural support regimes in both the Community and the USA is a policy imperative of reasonably recent provenance, and is considered further in Chapter 11 by Rodgers and Chapter 12 by Adelman. If agri-environmental schemes are to be accommodated within the revised green box, however, a clear definition will be required of those environmental objectives that are to be pursued by farming activities and of those which will qualify for exemption. The *EC Comprehensive Negotiating Proposal* accepted that they must be well targeted, transparent and implemented in a way that is not more than minimally trade-distorting (WTO, 2000d, para. 16), but no thought appears to have been given to the substantive issue, namely, what is meant by 'environmental' protection in the context of the farmed environment, and the scope and objectives of schemes that will potentially qualify for green-box exemption.

This is likely to be one of the most difficult issues to resolve, given the close interrelationship between farming and the environment. As is the case with maintaining land in good agricultural and environmental condition, many environmental concerns can only be met through continued farming, albeit using farming practices adapted to give a particular environmental benefit – for example, the adoption of extensive grazing regimes aimed at recreating moorland habitats or arable cropping patterns targeted at protecting ground-nesting birds. The Community's stance is based on the premise that this situation is one where farming will produce joint products: agricultural produce and environmental services (i.e. public goods). Some sensitivity on the issue is displayed in the *Discussion Paper on Agriculture's Contribution to Environmentally and Culturally Related Non-trade Concerns* submitted by the Community in July 2000 (WTO, 2000c), with particular reference to the potentially trade-distorting effects of environmental schemes that allow continued farm production. It rejected the suggestion that agricultural produce generated by activities funded under environmental schemes should be withdrawn from the market as wasteful and cost-inefficient. However, it accepted that joint production of marketable products and environmental services should not be used to conceal distortive economic subsidies (WTO, 2000c). It proposed, therefore, that, where society requires farmers to deliver public goods in pursuit of a 'legitimate environmental or cultural objective', governments should only recompense farmers for their additional costs and income forgone, taking account of the farmers' income from selling commodities on the market. This approach suffers from at least two weaknesses. In the first place, no attempt has yet been made to define those environmental objectives which will be categorized as 'legitimate' for this purpose: is it proposed that green-box exemption will be granted not only to schemes promoting farmland biodiversity but also, for example, to those targeted on the improvement of landscape values? Additionally, the Community's stated negotiating position ignores the fact, already noted above, that the Rural Development Regulation has permitted Member States to take account not only of income foregone and additional costs when devising payment regimes for agri-environmental schemes, but also the 'need to provide an incentive' (Council Regulation (EC) 1257/1999, Art. 24 (OJ 1999 L160/80)); and this provision survived the Mid-term Review. Under the detailed rules implemented in 2002, payments must be determined on the basis of objective criteria, and a ceiling is applied limiting the incentive element to, in general, a maximum of 20% of the income forgone and the additional cost of carrying out the commitments (Commission Regulation (EC) 445/2002, Art. 19 (OJ 2002, L74/1)). It is unclear whether, were its position on green-box exemption to be accepted in the Millennium Round, the Community proposes to amend this provision. At the very least, the provision of incentive payments under environmental schemes is likely to be subjected to close scrutiny, given their potential to be trade-distorting. Whether a payment is an 'incentive' will in practice depend upon market conditions, and especially on the profitability of the production that it displaces. Environmental payments made on an incentive model may therefore be seen by some as displacing income and as, therefore, a way of maintaining an indirect link with production in the provision of agricultural support.

Quantification

Allied to these emerging issues is that of quantification. An underlying principle of the Uruguay Round was to increase transparency; and a prime example of this principle in action, in the context of market access under the URAA, was the conversion of most non-tariff barriers into tariffs capable of being bound and reduced. In contrast, many of the measures subject to the most heated debate in the Millennium Round are less susceptible to precise measurement. Thus, with the enhanced importance of the green box since the Mid-term Review, it has become of even greater importance to determine the difficult question as to whether measures have no, or at most minimal, trade-distorting effects or effects on production. In this regard, as indicated, a clear instance would be the effect of agri-environmental schemes. Likewise, quantification has already proved to be a point of controversy in the case of biotechnology. For example, the Community is to require labelling where food has been produced from a GMO, whether or not the GMO is detectable in the final product (CEC, 2001, 2002d; European Community, 2003b). In contrast, the USA has long expressed disquiet on the issue of detection procedures, with particular concern that they could be employed in a discriminatory manner (WTO, 1998).

Such difficulties have, however, to an extent been anticipated, with the OECD taking a leading role. In its 1998 Communiqué, the Committee for Agriculture at Ministerial Level laid much emphasis on the need to develop the appropriate analytical tools to monitor and evaluate developments in agricultural policies (OECD, 1998a); and the Communiqué was subsequently followed by a series of initiatives. For example, *Multifunctionality: a Framework for Policy Analysis* proposed work on the evaluation of non-market benefits (OECD, 1998b); and *Market Effects of Crop Support Measures* addressed the key question of the trade consequences of domestic support (OECD, 2001b). Parallel initiatives have also been undertaken by individual Members. For example, in 2000 the Community issued *Indicators for the Integration of Environmental Concerns into the Common Agricultural Policy* (CEC, 2000). It remains to be seen, however, whether a sufficiently transparent and robust system is developed to inspire the confidence of the broader membership of the WTO.

Conclusion

It will be appreciated that an effective evaluation of the challenges facing the agricultural trading system in the Millennium Round and beyond requires an interdisciplinary approach. The contributions to this collection are drawn from the areas of agricultural politics, law and economics. There are a number of interfaces between these areas which are both relevant and important; and, in order to develop these links, the prevailing political and economic structures are specifically explored in, respectively, Chapter 3 by Grant and Chapter 4 by Sumner. In the first place, an important interface between law and politics is revealed in that the legal order for international trade, established by the terms of the URAA, will shape the negotiating process in the Millennium Round. It will also condition the language of the discourse in which the detailed negotiation to settle pressure points and matters of dispute are conducted. A clear example would be provided by any renegotiation of legal criteria for green-box exemption from domestic support reduction commitments.

Secondly, an understanding of the interface between law and economics is important to an evaluation of the effectiveness and impact of the current URAA measures in practice. It will be apparent from the above discussion that a number of current CAP support schemes are vulnerable to the argument that they fall outside the green box. Although the decision by a signatory to exempt a subsidy is open to question before a WTO dispute settlement body, the transaction costs associated with litigating green-box cases arguably outweighs any adverse trade effects (Kennedy, 2001). Thus, although there has been no green-box litigation under the URAA, this may not be because the existing criteria are sufficiently clearly delineated (they are not), but rather because the perceived cost of litigation outweighs the economic advantages to be achieved thereby.

Many would argue that the green-box exemption criteria laid down in the URAA are too rigid and require revision, both to introduce greater flexibility and to sharpen the legal definitions in Annex 2. The Community negotiating position presupposes revision, and the challenge in the Millennium Round will be to introduce flexibility without compromising the ability of the URAA to exclude schemes that are genuinely trade-distorting. Whether measures are genuinely trade-distorting is another fundamental problem, and provides a further example of a key legal concept that is dependent for its meaning and content on economic evaluation and financial criteria (the so-called 'quantification' issue identified above).

Finally, the interface between politics and economics is of paramount importance, especially in understanding the negotiating stances taken by the parties to the Uruguay and Millennium Round negotiations. The approach adopted by the protagonists on individual issues will depend upon their evaluation of the respective economic advantages to be gained from adherence to a particular negotiating position and the long-term economic outlook for the agriculture sector. A useful illustration of this interface is provided by the Uruguay Round, where, paradoxically, because the level of protection for many commodities was actually higher in 1986–1988 than in 1995 when implementation began, the base from which the 36% agreed cut in tariffs took place was correspondingly higher. As a consequence, it is most unlikely that the tariffication agreed in the URAA in itself improved market access in any meaningful way (Fennell, 1997, p. 393). This may well explain the softening of the Community's stance against the tariffication process in the final stages of the Uruguay Round.

List of Cases

EC Measures Concerning Meat and Meat Products (Hormones) (1998), WT/DS26/AB/R and WT/DS48/AB/R.

Case C-236/01, *Monsanto Agricoltura Italia SpA* v. *Prezidenza del Consiglio dei Ministri*, judgement delivered 9 September 2003.

Case T-13/99, *Pfizer Animal Health SA* v. *Council* (Court of First Instance), judgment delivered 11 September 2002.

References

ABARE (1999) 'Multifunctionality': a pretext for protection? *ABARE Current Issues* 99(3) (August), 1–6.

Agra Europe (2003) Chirac Stalls EU farm talks as Council drags on. *Agra Europe* 2059, EP/I-EP/3.

Bohman, M., Cooper, J., Mullarkey, D., Normile, M.A., Skully, D., Vogel, S. and Young, E. (1999) *The Use and Abuse of Multifunctionality*, Economic Research Service, USDA, Washington, DC. (http://www.ers.usda.gov/briefing/wto/PDF/multifunc1119.pdf, visited on 23 January 2001).

Burton, M., Rigby, D., Young, T. and James, S. (2001) Consumer attitudes to genetically modified organisms in the UK. *European Review of Agricultural Economics* 28, 479–498.

CEC (1997) *Agenda 2000: for a Stronger and Wider Union*. COM(97)2000 (*Bulletin of the European Union, Supplement* 5/97, Brussels).

CEC (1998) *Explanatory Memorandum: the Future for European Agriculture* (http://europa.eu.int/en/comm/dg06/ag2000/agprop/mot_en.htm, visited on 3 June 1998).

CEC (1999) *Bulletin of the European Union* 3/99. Brussels.

CEC (2000) *Indicators for the Integration of Environmental Concerns into the Common Agricultural Policy*. COM(2000)20.

CEC (2001) *Proposal for a Regulation of the European Parliament and of the Council on Genetically Modified Food and Feed*. COM(2001)425.

CEC (2002a) *Life Sciences and Biotechnology – a Strategy for Europe*. COM(2002)27 (OJ 2002, C55/3).

CEC (2002b) *Mid-term Review of the Common Agricultural Policy*. COM(2002)394.

CEC (2002c) *Prospects for Agricultural Markets 2002–2009*. Brussels.

CEC (2002d) *Amended Proposal for a Regulation of the European Parliament and of the Council on Genetically Modified Food and Feed*. COM(2002)559.

CEC (2003a) *A Long-term Policy Perspective for Sustainable Agriculture*. COM(2003) 23.

CEC (2003b) *Proposal for a Council Regulation Establishing Common Rules for Direct Support Schemes under the Common Agricultural Policy and Establishing Support Schemes for Producers of Certain Crops*. 2003/0006 (CNS).

Coleman, W.D. and Tangermann, S. (1999) The 1992 CAP Reform, the Uruguay Round and the Commission: conceptualizing linked policy games. *Journal of Common Market Studies* 37, 385–405.

European Community (2001) 'New WTO Round slap in the face for isolationism', says EU Farm Commissioner Fischler. IP/01/1584, Doha, 14 November.

European Community (2002a) Fischler slams US for 'flunking farm policy reform'. IP/02/647, Brussels, 1 May.

European Community (2002b) *Questions and Answers: US Farm Bill*. MEMO/02/94, Brussels, 15 May.

European Community (2002c) *Questions and Answers on the Regulation of GMOs in the EU*. MEMO/02/160, Brussels, 1 July.
European Community (2002d) *Further Clarifications on US Farm Act*. MEMO/ 02/166, Brussels, 11 July.
European Community (2003a) EU fundamentally reforms its farm policy to accomplish sustainable farming in Europe. IP/03/898, Luxembourg, 26 June.
European Community (2003b) European legislative framework for GMOs is now in place. IP/03/1056, Brussels, 22 July.
Fennell, R. (1997) *The Common Agricultural Policy: Continuity and Change*. Clarendon Press, Oxford, 440 pp.
Fischler, F. (1999a) The Agenda 2000 agreement: 'Der Himmel über Berlin' or 'Sleepless in Seattle'?, Speech/99/101, at Brussels, 29 June.
Fischler, F. (1999b) The European model of agriculture – facing the WTO acid test, Speech/99/117, at Verona, 24 September.
Fischler, F. (1999c) WTO negotiations – agricultural aspects, Speech/99/138, at Brussels, 22 October.
Fischler, F. (1999d) WTO press conference speaking note, Speech/99/177, at Brussels, 23 November.
Fischler, F. (2001) Trends in agricultural policy: opportunities for a closer transatlantic relationship, Speech/01/148, at Brussels, 29 March.
Food and Drug Administration (1992) Statement of policy: foods derived from new plant varieties. *Federal Register* 57, 22984.
Howse, R. and Mavroidis, P.C. (2000) Europe's evolving regulatory strategy for GMOs – the issue of consistency with WTO law: of kine and brine. *Fordham International Law Journal* 24, 317–370.
Howse, R. and Tuerk, E. (2001) The WTO impact on internal regulations – a case study of the Canada–EC asbestos dispute. In: de Búrca, G. and Scott, J. (eds) *The EU and the WTO: Legal and Constitutional Issues*. Hart Publishing, Oxford, pp. 283–328.
Kelley, C.R. (1999) Recent Federal Farm Program developments. *Drake Journal of Agricultural Law* 4, 93–139.
Kennedy, K.C. (2001) Reforming farm trade in the next round of the WTO multilateral trade negotiations. *Journal of World Trade* 35, 1061–1079.
Landau, A. (2001) The agricultural negotiations in the WTO: the same old story? *Journal of Common Market Studies* 39, 913–925.
OECD (1998a) *Communiqué of the Meeting of the Committee for Agriculture at Ministerial Level*, 5–6 March (http://www.oecd.org/agr/ministerial/ commune.htm, visited on 6 March 2001).
OECD (1998b) *Multifunctionality: a Framework for Policy Analysis*. OECD Document AGR/CA(98)9, November.
OECD (2001a) *Multifunctionality: Towards an Analytical Framework*. OECD, Paris. 159pp.
OECD (2001b) *Market Effects of Crop Support Measures*. OECD, Paris. 119pp.
Paarlberg, R. (1997) Agricultural policy reform in the EC: a three level game analysis. *International Organisation* 51, 413–444.
Scott, J. (1996) Tragic triumph: agricultural trade, the Common Agricultural Policy and the Uruguay Round. In: Emiliou, N. and O'Keefe, D. (eds) *The European Union and World Trade Law: After the GATT Uruguay Round*. John Wiley & Sons, Chichester, pp. 165–180.

United States (2002) *US Proposal for Global Agricultural Trade Reform* (http://www.fas.usda.gov/itp/wto/actual.htm, visited on 3 October 2002).

USDA (2001) *Food and Agricultural Policy: Taking Stock for the New Century*. USDA, Washington, DC.

USDA (2002a) Statement by Agriculture Secretary Ann M. Veneman on the President's signing of the Farm Security and Rural Investment Act of 2002. Release 189.02, 13 May.

USDA (2002b) Transcript of Secretary of Agriculture Ann M. Veneman to USDA employees regarding Farm Bill implementation. Release 194.02, 15 May.

USDA (2002c) Statement by Secretary Ann M. Veneman regarding Farm Bill criticism. Release 201.02, 21 May.

USDA (2002d) Transcript of Agriculture Secretary Ann M. Veneman, Undersecretary for Farm and Foreign Agricultural Services J.B. Penn, Chief Economist Keith Collins, Farm Service Agency Administrator James Little. 21 May.

USDA (2002e) Transcript of Foreign Press Center briefing with J.B. Penn, Undersecretary of Agriculture for Farm and Foreign Agricultural Services on the 2002 Farm Bill and the implications for world trade. Release 207.02, 22 May.

USDA (2002f) Statement by Secretary of Agriculture Ann M. Veneman regarding country of origin labelling provisions in the new Farm Bill. Release 212.02, 29 May.

USDA (2002g) Press conference regarding the US agenda for the World Trade Organization Negotiations. Release 314.02, 25 July.

WTO (1998) *European Council Regulation No. 1139/98, Compulsory Indication of the Labeling of Certain Foodstuffs Produced from Genetically Modified Organisms: Submission by the United States*. G/TBT/W/94, 16 October.

WTO (2000a) *Proposal for Comprehensive Long-term Agricultural Trade Reform: Submission from the United States*. G/AG/NG/W/15, 23 June.

WTO (2000b) *European Communities Proposal: Animal Welfare and Trade in Agriculture*. G/AG/NG/W/19, 28 June.

WTO (2000c) *Note on Non-trade Concerns*. G/AG/NG/W/36, 22 September.

WTO (2000d) *EC Comprehensive Negotiating Proposal*. G/AG/NG/W/90, 14 December.

WTO (2000e) *Negotiating Proposal by Japan on WTO Agricultural Negotiations*. G/AG/NG/W/91, 21 December.

WTO (2001a) *WTO Agriculture Negotiations: Proposal by Norway*. G/AG/NG/W/101, 16 January.

WTO (2001b) *Food Safety: Note by the European Communities*. D(2001)(DIVERS/500186 ur), 20 July.

WTO (2001c) *Ministerial Declaration*. WT/MIN(01)/DEC/W/1, 14 November.

WTO (2001d) *Agriculture: Negotiations 2000–March 2001: Work Programme for the Second Phase*. 27 March (http://www.wto.org/english/tratop_e/agneg7_e.htm, visited on 2 April 2001).

The Uruguay Round Agreement on Agriculture and Domestic Support

2

Margaret Rosso Grossman*

Introduction

As the European Union (EU)[1] Commissioner for Agriculture, Franz Fischler, stated:

> agriculture is different . . . [A]gricultural production . . . does not take place in a controlled environment; it does not take place at a controlled pace; it does shape our countryside and, even more important it does hold a special place in human imagination and affection.
>
> (Fischler, 2000)

Agriculture is indeed 'different' and that difference has long resulted in special treatment for agriculture in international trade agreements. The Uruguay Round Agreement on Agriculture (URAA), enacted as part of the World Trade Organization (WTO) agreements in 1994, imposed significant trade disciplines on agriculture, but also recognized the differences in various Member Country agricultures and the right of Members to protect their agricultural sectors (Switzerland, 2000, p. 4).

The URAA (WTO, 1994b) made important strides in creating open markets for agricultural commodities, but, despite significant improvements, further negotiations are necessary to eliminate remaining trade restrictions and distortions. The global average tariff on agricultural goods is 62%, with regional tariffs averaging between 25% (North

* This material is based on work supported by the Cooperative State Research, Education and Extension Service, US Department of Agriculture, under Project No. ILLU-05-309.

America) and 113%. In contrast, the global average manufacturing tariff is only 5% (Gibson *et al.*, 2001, pp. 9, 15). Export subsidies and domestic supports, too, continue to distort trade. In the EU, for example, agricultural export subsidies in 1998 constituted more than 90% of world export subsidies (Burfisher, 2001, p. 19).

Post-URAA agricultural policies continue to transfer significant sums of money to farmers, especially in the EU, Japan and the USA. Estimates of support compiled by the Organization for Economic Cooperation and Development (OECD) indicate that in 1996 Japan paid out about $30,000 per farmer (over $15,000/ha); the US, about $27,000 ($161/ha); and the EU, about $17,400 ($825/ha). US transfers to farmers were about 1.3% of gross domestic product (GDP) (Josling, 1998, pp. 12–13). Support in 1999 was somewhat lower: Japan, $25,000 per farmer ($11,000/ha); the US, $22,000 ($132/ha); and the EU, $18,000 ($840/ha). Estimated support in 2001 again decreased slightly: Japan, $23,000 per farmer ($9700/ha); the US, $20,000 ($117/ha); the EU, $16,000 ($676/ha) (OECD, 2002, tables III.5, III.6). The USA has expressed concern about the high level of domestic support for agriculture, particularly in the EU. As one economist noted, 'the absolute size of the transfers to one small and declining part of the economy is still remarkable and vulnerable to both economic reform and straightforward budget-cutting pressures' (Josling, 1998, p. 12).

The URAA prescribed that renewed agricultural negotiations were to begin in 2000, and those negotiations are now under way. Because WTO Members have differing national objectives and agricultural policies, the process will require cooperation and compromise. Negotiations, now guided by the Doha Declaration (WTO, 2001a), focus on three central issues (market access, export subsidies and domestic support) and many related questions. Domestic support for agriculture has played a key role in the negotiations so far and will continue to be an important focus. After an introductory discussion of the URAA and current WTO negotiations, therefore, this chapter reviews the various categories of domestic support for agriculture under the URAA.

The Uruguay Round Agreement on Agriculture

On 15 April 1994, after more than 7 years of negotiation, 125 countries signed the Agreement Establishing the World Trade Organization at Marrakesh, Morocco. Article 1 of the WTO Agreement established the WTO, to 'provide the institutional framework for the conduct of trade relations among its Members' (WTO, 1994a). The Uruguay Round achieved several important agreements for agriculture: the new agricultural trade rules memorialized in the Agreement on Agriculture, the Agreement on the Application of Sanitary and Phytosanitary Measures, the Agreement on Technical Barriers to Trade (TBT Agreement), and the

Understanding on Rules and Procedures Governing the Settlement of Disputes.

The TBT Agreement (WTO, 1994d) helps to ensure that technical regulations and product standards do not create obstacles to trade. The TBT Agreement, which applies to agricultural products, permits technical regulations designed to meet legitimate national objectives, including 'protection of human health or safety, animal or plant life or health, or the environment'. Regulations must 'not be more trade-restrictive than necessary to fulfil a legitimate objective'. When international standards exist, WTO Members should use those standards as the basis for their technical regulations (Arts. 1, 2). The TBT Agreement does not apply to measures governed by the Agreement on the Application of Sanitary and Phytosanitary Measures (WTO, 1994c), which applies to issues of food safety, as well as animal and plant health (Art. 1) (see Thorson, Chapter 10, this volume).

Also significant, especially for issues connected with agriculture and the environment, the Preamble to the WTO Agreement recognizes the importance of environmental concerns. In establishing the WTO, the parties recognize that improved trade should be achieved:

> while allowing for the optimal use of the world's resources in accordance with the objective of sustainable development, seeking both to protect and preserve the environment and to enhance the means for doing so in a manner consistent with their [the Parties'] respective needs and concerns at different levels of economic development.[2]
>
> (WTO, 1994a, Preamble, ¶ 1)

URAA disciplines

The WTO represented a major change in discipline for agricultural trade. Under the General Agreement on Tariffs and Trade (GATT), which the WTO replaced,[3] agriculture had received special treatment. Strategic issues (food supply), economic factors (stable farm income, reasonable consumer prices), political influence and cultural values help to explain these 'exceptional arrangements' for agriculture (Delcros, 2002, pp. 219–220). Trade in agricultural products was not subject to many of the trade disciplines that governed manufactured goods. Measures prohibited in other sectors affected trade in agricultural products; for example, GATT allowed members to use import quotas and other non-tariff barriers, to subsidize exports and to subsidize production (CEC, DG Agri, 1999, pp. 5–6). Even under the WTO, agriculture has kept its special status and is now governed by a separate Agreement on Agriculture (URAA), one of the 13 Multilateral Agreements on Trade in Goods. The URAA, implemented over the 6-year period 1995–2000, began the process of subjecting agriculture to significant trade disciplines (ERS, USDA, 1998, p. 5).

The URAA is intended 'to establish a fair and market-oriented agricultural trading system,' with 'progressive reductions in agricultural support and protection', which will result in 'correcting and preventing restrictions and distortions in world agricultural markets' (URAA, Preamble, ¶¶ 2, 3). To achieve this objective, Members agreed to specific binding commitments in the areas of market access (tariffs and tariff-rate quotas (TRQs)), export competition (export subsidies and credits) and domestic support. Thus, the URAA establishes trade disciplines for agriculture that form 'three legs of the stool' (Normile, 1998, p. 28). These commitments under the URAA are to be equitable, 'having regard to non-trade concerns, including food security and the need to protect the environment' (Preamble, ¶ 6). Members also reached a separate agreement to govern sanitary and phytosanitary issues (Preamble, ¶ 4).

To facilitate market access, the URAA prohibits non-tariff barriers to trade in agricultural products (URAA, Arts. 4, 5 and Annex 5). The URAA requires conversion of non-tariff barriers into tariffs through a process called tariffication. Tariffs were to be bound, and then reduced, between 1995 and 2000, by an average of 36% from the base period 1986–1988 (Vasavada and Nimon, n.d., p. 3). Further, to ensure minimum market access, the URAA authorized a system of TRQs, or two-tiered tariffs. Under the TRQ system, countries must accept a minimum quantity of imports – by 2000, at least 5% of base-period domestic consumption – at the lower tariff. Minimum access quantities can include products imported at lower tariffs under long-standing agreements. A higher tariff can be charged for quantities above 5% of domestic consumption. Over 1300 TRQs exist, though not all are enforced (Skully, 2000, pp. 22–23).

During the same period, subsidies on exports were to be reduced significantly, in terms of both total expenditures and volume (Vasavada and Nimon, n.d., p. 3). Various types of subsidies are regulated, including direct export payments, payments financed by government action and subsidies to reduce the cost of marketing exports. The volume of subsidized exports (tons) was to be reduced by 21% (14% for developing countries) from the 1986–1990 base, and the value of export subsidies was to be reduced by 36% (24% for developing countries), from the same base. Twenty-five countries committed themselves to subsidy reduction (URAA, Arts. 8–11; ERS, USDA, 1998, pp. 21–23).

WTO Members also agreed to reductions in domestic support, which apply to all domestic-support measures in favour of agricultural producers, except those exempted under Annex 2 (the green box, discussed below) and in the URAA itself (URAA, Art. 6). Exemptions under Article 6 include *de minimis* support (¶ 4) and a category exempting certain production-limiting programmes, the so-called blue-box measures (¶ 5). Members must ensure that domestic support measures permitted under the green box continue to conform to the requirements of Annex 2

(URAA, Art. 7, ¶ 1). The WTO provisions that govern domestic support will be discussed in more detail below.

During a 9-year implementation period ending at the close of 2003, the so-called 'peace' or 'due restraint' clause operates to insulate trade in agricultural products from actions based on other WTO or GATT provisions (URAA, Arts. 1(f), 13). Under the peace clause, domestic-support measures that conform to Annex 2 (the green box) are exempt from actions and countervailing duties. Measures that conform to Article 6 (blue box and *de minimis* support) are exempt from certain actions; they can be subject to countervailing duties, but Members must show due restraint in initiating investigations. Export subsidies that conform with the URAA also receive protection (URAA, Art. 13). If the peace clause is not continued, agriculture will be subject to the general WTO Agreement on Subsidies and Countervailing Measures when the clause expires at the end of 2003 (Josling, 1998, p. 17; Ervin, 1999, pp. 76–77).

WTO agricultural negotiations

When the URAA was finalized in 1994, Members recognized trade reform as an ongoing process and agreed that negotiations for continued agricultural trade reform would begin 1 year before the end of the 6-year implementation period – that is, in 2000 (URAA, Art. 20). Among factors to be weighed in the negotiations are experience from implementing reduction commitments, the effect of those commitments on world agriculture, 'non-trade concerns, . . . the objective to establish a fair and market-oriented agricultural trading system, and other objectives and concerns mentioned in the preamble to this Agreement', as well as further commitments needed to achieve the URAA long-term objectives (URAA, Art. 20(a)-(d)).

Some would say that the URAA dealt with the easiest agricultural trade issues and that more difficult issues must still be resolved (Kerr, 2000, pp. 120, 136). Renewed negotiations will also include new trade issues that have become important since 1994 – e.g. state trading enterprises, geographical indicators, consumer information and labelling, and animal welfare (WTO, 2002a). If agricultural negotiations occur along with negotiations in other areas, a more significant reduction of agricultural trade protection could result, especially if non-agricultural groups counterbalance those that favour retained agricultural protection (Anderson *et al.*, 1999, p. 2).

The General Council of the WTO opened the renewed agricultural negotiations prescribed in URAA Article 20. The first phase of negotiations began in March 2000 and the second phase in March 2001. Each phase involved a number of meetings, accompanied by proposals and other documents submitted by Members. In November 2001, the fourth

WTO Ministerial Conference (Doha, Qatar) resulted in a Declaration with a clear statement of objectives and a timetable for negotiations (WTO, 2001a). Agriculture is now part of a single undertaking of linked negotiations, with all related negotiations to be completed by 1 January 2005. A 12-month 'modalities' work programme (March 2002–March 2003) will set targets and formulas for achieving the Doha objectives, and several meetings are scheduled for that period. Members will use the modalities document, to be completed by 31 March 2003, to make their 'comprehensive draft commitments' by September 2003, when a fifth Ministerial Conference is scheduled for Cancún, Mexico (WTO, 2002a, pp. 4, 38).

The Doha Declaration built on URAA Article 20 to provide a new mandate for negotiations (WTO, 2001a). The Declaration commits Members to a 'broad and balanced Work Programme' (Art. 11) that includes a number of trade issues. Agriculture is the focus of Articles 13 and 14 of the Declaration, which acknowledges the work and negotiating proposals submitted since negotiations began in 2000. The Declaration reconfirms a commitment to the objective of establishing 'a fair and market-oriented trading system through a programme of fundamental reform encompassing strengthened rules and specific commitments on support and protection in order to correct and prevent restrictions and distortions in world agricultural markets' (Art. 13). Article 13 continues with reference to the three pillars of the agricultural negotiations:

> Building on the work carried out to date and without prejudging the outcome of the negotiations we commit ourselves to comprehensive negotiations aimed at: substantial improvements in market access; reductions of, with a view to phasing out, all forms of export subsidies; and substantial reductions in trade-distorting domestic support.

In connection with these three central pillars, 'special and differential treatment for developing countries shall be an integral part of all elements of the negotiations' to enable developing countries to take account of development needs, 'including food security and rural development'. Further, Article 13 states, 'We take note of the non-trade concerns reflected in the negotiating proposals submitted by Members and confirm that non-trade concerns will be taken into account in the negotiations as provided for in the Agreement on Agriculture'.

Article 14 establishes the 31 March 2003 deadline for modalities and the fifth Ministerial Conference as the deadline for Member comprehensive draft schedules; the Article indicates that the agricultural negotiations will be concluded at the time of conclusion of the 'negotiating agenda as a whole'. Agriculture was also a subject of the Implementation Decision adopted at Doha to focus on developing countries' problems in implementing WTO agreements (WTO, 2001b). Among other provisions, the Decision urges Members to exercise restraint in challenging green-box measures notified by developing countries to promote rural development and ensure food security (Art. 2.1).

Domestic Support

In negotiations under the Doha Declaration, a central objective is 'substantial reductions in trade-distorting domestic support', that is, reductions going beyond those achieved by the URAA (WTO, 2001a, Art. 13). Historically, the URAA provisions that discipline domestic agricultural policies in the interest of trade were 'unprecedented' (ERS, USDA, 1998, p. 14). Indeed, the URAA itself is 'unique in the scope of rules and commitments that the WTO establishes for domestic policies because domestic agricultural policies and international trade are closely linked' (Bohman *et al.*, 1999, p. 5).

As the USDA's Economic Research Service (ERS) explained:

> Trade policies . . . refer to the set of policies designed specifically to affect trade flows and prices through use of import quotas, tariffs, and export subsidies. Domestic policies include all other agricultural policies within a country that aim to influence internal farm and rural incomes, resource use, production, consumption of agricultural products, or environmental impacts of farming.
>
> (ERS, USDA, 1998, p. 14)

But domestic agricultural policies, at least those applicable to farm production and landowners, may affect production and therefore also trade. Further, domestic policy objectives often motivate trade policies; 'by directly influencing imports and exports, trade policies can be used to facilitate domestic price and income goals'. So WTO Members were willing to discipline domestic policy in the interest of world trade in agriculture (ERS, USDA, 1998, p. 14). This willingness was influenced, in part, by already changing policies of domestic support, for example those reflected in the 1992 MacSharry Common Agricultural Policy (CAP) reform (Josling, 1998, p. 12).

Ironically, although the 'biggest conceptual breakthrough' in the URAA was this discipline of domestic policies to further trade, constraints on domestic support were rather weak and may be the part of the URAA with the least impact (Josling, 1998, pp. 4, 11; Blandford, 2001, pp. 42–44). In recent years, domestic support in many important trading countries (including the USA and the EU) has increased and continues to distort trade. By moving support to categories exempt from reduction, WTO Members have been able to meet their reduction commitments without actually reducing total support (Roberts, 2000, pp. 1–2, 4). Thus, Members have recognized domestic support as a central element for renewed negotiations.

Under the URAA, different disciplines apply to different types of domestic support. These disciplines are often referred to as 'boxes', and domestic policies generally fit into the amber box, the green box or the blue box. The amber box contains policies that distort trade and are subject to reduction; the green box contains policies with (arguably) only a minimal effect on trade; and the blue box provides an exemption for payments that would otherwise fit in the amber box. The WTO rules do not restrict, or even address, Members' policy objectives. Instead, they govern the instruments that Members use to achieve those policy objectives that affect trade. Thus, Members are free to establish policy objectives, but implementation measures that distort trade may be subject to limitations under the amber box (Bohman *et al.*, 1999, pp. 5–6).

The distinctions between amber-, green- and blue-box disciplines are summarized here. Members' negotiating positions, especially in connection with the blue and green boxes, are discussed in Chapter 5 of this book.

The amber box

Policies that support domestic agricultural prices or subsidize production can encourage overproduction and affect trade. Under the URAA, domestic programmes that 'stimulate production and trade directly' have to be cut back, while other programmes – many direct payments – 'are considered to have no direct effect' (CEC, DG Agri, 1999, p. 6). The amber box includes 'coupled' income support, which is linked directly to production through, for example, price supports, per-unit payments or input subsidies. These payments provide 'direct economic incentives to producers to increase or decrease current resource use or current production'; such incentives affect production and trade (ERS, USDA, 1998, p. 14).

Thus, the amber box contains the domestic policies that most affect trade and production. For each country, amber-box support is quantified in the aggregate measurement of support (AMS), which combines non-exempt support, under a number of policies, for all commodities to establish the value of the amber-box policies (ERS, USDA, 1998, pp. 15–16; Bohman *et al.*, 1999, p. 6, box 1). Under the URAA, the AMS is:

> the annual level of support, expressed in monetary terms, provided for an agricultural product in favour of producers of the basic agricultural product or non-product-specific support provided in favour of agricultural producers in general, other than support provided under programmes that qualify as exempt from reduction under Annex 2 [the green box] to this Agreement, which is [calculated as specified for each Member country].
>
> (URAA, Art. 1(a))

AMS is calculated for each country according to the provisions in Annex 3, URAA. The equivalent measurement of support is a substitute calculation for components for which AMS is not practicable (URAA, Arts. 1(d), 6, and Annex 4).

AMS is generally measured from the level of the base period 1986–1988. As a result of Uruguay Round negotiations, major agricultural producers and traders (then 28 countries) agreed to reduce the AMS (amber-box support) by 20% during the 6-year period 1995–2000. For example, US AMS for the base period was $23.9 billion, making the current limit, after the 20% reduction, $19.1 billion. Developing countries were to reduce their amber-box support by 13% over 10 years, and least developed countries were not to increase their support (ERS, USDA, 1998, pp. 15–16). Though the AMS was intended to be commodity-specific, negotiations at Blair House (Washington, DC) resulted in a decision to aggregate commodities in the AMS, which weakened its impact (Josling, 1998, pp. 14–15; Delcros, 2002).

Domestic support considered *de minimis* is not subject to reduction. This includes product-specific domestic support, otherwise includable in the AMS, that does not exceed 5% of a country's total value of production of an agricultural product, as well as non-product-specific domestic support that does not exceed 5% of a country's total agricultural production. For developing countries, the level of *de minimis* support is 10% (URAA, Art. 6, ¶ 4).

Categories of payments in the USA that made up the AMS for 1995 and 1996, for example, were market-price support, non-exempt direct payments (e.g. marketing loans and loan deficiency payments), other non-exempt measures (e.g. storage payments, loan interest subsidies) and non-product specific support (e.g. irrigation, grazing, crop insurance) (ERS, USDA, 1998, p. 17, table 2). The USA characterized recent emergency support to farmers as amber box, but in the *de minimis* category (Nelson, 2002).[4]

The Farm Security and Rural Investment Act of 2002 (Farm Bill, 2002) increases agricultural spending, particularly for domestic support.[5] For example, it includes, among many other provisions, a three-part safety net for commodity farmers. Marketing loan provisions are continued, and counter-cyclical payments are reintroduced, to be triggered when the effective commodity price is lower than the target price. Direct decoupled payments (green box), based on payment rate, payment acres and payment yield, are continued. Producers can update base acres and (for counter-cyclical payments) base yield. To other WTO Members, some of the Farm Bill measures (e.g. the counter-cyclical payments) have been seen as distorting production and therefore trade (Fischler, 2002). But to meet AMS requirements, the USA may plan to notify these counter-cyclical payments as non-product specific payments under the *de minimis* rule (Haniotis, 2002; Kreuzhuber and Bunyan, 2002). Analysts have suggested that US payments under the Farm Bill will remain below the current AMS limit (Hart and Babcock, 2002, with calculations based on proposed legislation). The Farm Bill (§ 1601) includes a 'circuit-breaker' intended to keep US domestic support within amber-box limits. If the Secretary of Agriculture determines that support payments will exceed the AMS limits, after reporting to Congress, the Secretary shall 'to the maximum extent practicable' adjust the amount of expenditures to keep them within AMS limits.

For the EU, market-price support has been the main component of amber-box domestic support (ERS, USDA, 1999, p. 37). Even before agreement of the URAA, the 1992 MacSharry CAP reform had reduced support prices for arable crops, which would be part of AMS, in favour of direct payments, part of the negotiated blue box (ERS, USDA, 1998, p. 19). Agenda 2000 promised to reduce amber-box support and increase blue-box payments, as it continued to shift from price support to income support (ERS, USDA, 1999, p. 37). In 2001, the EU indicated that by 2006, blue-box payments would be almost 80% of the CAP budget, with market-price support and export refunds only about 20% (in contrast to 90% in 1989–1991) (CEC, DG Agri, 2001).

The CAP Mid-term Review, mandated in Agenda 2000 and published in July 2002, proposes further reduction in amber-box support. For example, under the proposal, the intervention price for cereals would be reduced another 5% and used as a safety net for producers (CEC, 2002, pp. 13–14). In a further break between production and payments, a decoupled system of a single income payment per farm (producer, rather than product, support) would replace most other direct payments and affect most sectors. This proposed system is discussed below, in connection with the green box (CEC, 2002, pp. 19–22).

The Secretariat of the WTO prepared a background paper on domestic support, published in April 2000, with information (based on Member notifications) about total AMS for WTO Members. Thirty of the 136 WTO Members have commitments to reduce domestic support (WTO, 2000a, p. 1). ERS research published in 1998 (using 1995 AMS notifications) indicated that these countries have generally met or even exceeded their commitments, thus reducing the effect of domestic agricultural support on trade (ERS, USDA, 1998, pp. 16, 18). Though AMS reductions do not bind most WTO Members, they have helped to eliminate some support measures that most affect trade (Josling, 1998, p. 11).

WTO Members may face additional reductions in amber-box support, depending on the results of current agricultural negotiations. For example, the US Proposal for Global Agricultural Trade Reform, submitted in July 2002, would reduce trade-distorting domestic support currently under the amber and blue boxes. Under that proposal, discussed in more detail below, each Member's AMS level for non-exempt support would be a fixed percentage (5%) of agricultural production value during a base period (WTO, 2002b). A European Communities (EC) proposal, released in December 2002, would impose a 55% reduction in AMS from the URAA final bound commitment level and eliminate the *de minimis* exemption for developed countries (WTO, 2002c). Should either of these proposals be adopted, the USA and some other Members may face further reductions in non-exempt domestic support (see Hart and Babcock, 2002).

The green box

Domestic policies that fit within the green box are attractive to WTO Members because payments under those policies are not subject to reduction under the AMS and are protected from challenges under the peace clause (Josling, 1998, pp. 11–12). The green box is governed by URAA Annex 2, which sets basic requirements that apply to all exempt domestic-support measures, as well as additional policy-specific criteria and conditions. The basic requirements appear in paragraph 1:

> Domestic support measures for which exemption from the reduction commitments is claimed shall meet the fundamental requirement that they have no, or at most minimal, trade-distorting effects or effects on production. Accordingly, all measures for which exemption is claimed shall conform to the following basic criteria:
>
> (a) the support in question shall be provided through a publicly-funded government programme (including government revenue foregone) not involving transfers from consumers; and,
>
> (b) the support in question shall not have the effect of providing price support to producers;
>
> plus policy-specific criteria and conditions.
>
> (URAA, Annex 2, ¶ 1)

Green-box payments must have 'no, or at most minimal, trade-distorting effects or effects on production' (URAA, Annex 2, ¶ 1). WTO guidelines leave these 'fundamental criteria' for green-box inclusion undefined, and it is often difficult to tell whether domestic policies are 'minimally trade distorting' (ERS, USDA, 1998, pp. 14, 20; United States, 1998b, p. 5). Moreover, some green-box instruments seem to affect production by encouraging increased output. For example, crop insurance may increase production incentive by reducing risk. Some programmes are linked with production, but serve another purpose – e.g. environmental programmes (Josling, 1998, p. 14). Other programmes may increase production through higher product demand (food security, domestic food aid) or through scale of payment (e.g. large-scale payment for environment as a joint product of food production). Programmes for food aid and public funding for agricultural research may have little or no effect (Tielu and Roberts, 1998, pp. 4–5). Green-box policies 'could have significant positive effects on production if financed with a large enough total amount of government expenditure' (ERS, USDA, 1998, p. 19). Moreover, as the USA noted, '[b]ecause certain environmental and natural resources conservation green-box policies allow for small changes in production, a country may have an incentive to use domestic policy to increase its competitiveness on the world market' (United States, 1998b, p. 5).

The green box encompasses a number of types of domestic-support measures, some with policy-specific conditions established in Annex 2. For example, general services policies provide services or benefits for agriculture or the rural community, but no direct payments for producers or processors. These may include research, pest and disease control, training, extension services, inspection, marketing and promotion and infrastructure (URAA, Annex 2, ¶ 2). Public stockholding for food-security purposes and domestic food aid are appropriate green-box policies (¶¶ 3, 4). Several types of direct payments to producers fit in the green box, provided they meet the basic criteria plus specific criteria established for the various types of programmes (¶ 5). These include decoupled income support, income insurance and income safety-net programmes, natural disaster relief, producer retirement programmes, resource retirement programmes, investment aids, environmental programmes and regional assistance programmes (¶¶ 6–13). Generally, clearly defined criteria for eligibility must be established, and some payments cannot be related to level of production.

An important, but controversial, type of green-box payment is 'decoupled income support' (URAA, Annex 2, ¶ 6; see Lopez, 2000). Though the URAA does not define 'decoupled', this category requires eligibility to be determined by 'clearly defined criteria such as income, status as a producer or landowner, factor use or production level in a defined and fixed base period'. Payment amounts 'shall not be related to, or based on, the type or volume of production (including livestock units) undertaken by the producer'. Payments cannot be based on factors of production, nor can production be required to receive payments (Annex 2, ¶ 6).

In theory, decoupled support is independent of factors that affect farmers' production decisions – marginal returns and marginal costs – so that production and marketing decision are guided by market prices, rather than by support (Tielu and Roberts, 1998, p. 2). Decoupled support measures are therefore expected to distort markets less than other types of support, but distortions can still exist. For example, some programmes designed to decouple payments in connection with WTO did not fully decouple; some programmes, not fully decoupled, continue to distort markets (Tielu and Roberts, 1998, p. 1). Moreover, even decoupled support may still influence production. Direct payments to farmers raise farm incomes and wealth beyond normal market return, and farmers are likely to invest additional money in inputs and technology, thus increasing production. Lower income risk may mean lower loan costs, making more money available to enhance production. Decoupled payments may increase land values and help to keep land in farming.[6] Expectations that future support will be linked to current production may also encourage a high level of output (Young *et al.*, 2002, p. 3). In addition, decoupled payments are costly to administer and require expenditure of tax money (Tielu and Roberts, 1998, p. 3).

The URAA does not limit the amount of subsidies in the green box. Indeed, green-box support increased 54% between the base years 1986–1988 and 1995, with the USA, the EC and Japan showing most increase. In 1995, US domestic food aid was the largest single green-box item, mostly from the food-stamp programme (ERS, USDA, 1998, p. 19; WTO, 2000b, p. 219). US green-box payments include a number of general-services items: research, pest and disease control, extension and cooperative services, inspection and marketing, conservation operations. The green box also included the Food Security Commodity Reserve and domestic food aid (food stamps; the women, infants, children nutrition programme) (ERS, USDA, 1998, p. 17, table 2, based on 1995 and 1996 WTO notifications). Decoupled income support under the provisions of 1996 farm legislation – that is, production flexibility contract payments – also has fitted in the green box since 1996 (ERS, USDA, 1998, p. 19; Nelson, 2002).[7] Numerous environmental payments for soil conservation and water quality fit in the green box; these include programmes for environmental quality (Environmental Quality Incentive Program (EQIP)), farmland protection, wetland protection (Wetland Reserve), habitat conservation and others (Vasavada and Warmerdam, 1998, p. 13). Other green-box programmes include structural-adjustment programmes (the Conservation Reserve), certain farm loans and disaster relief in the form of livestock- and crop-disaster payments (with crop insurance counted in the AMS) (ERS, USDA, 1998, p. 17, table 2).

In the EC, investment aids were the main category of total green-box expenditures in 1995 and 1996. Other EC green-box programmes included general-services items (most of those listed for the USA above), domestic food aid, decoupled income support, natural-disaster relief, producer and resource retirement programmes, environmental programmes and regional programmes (WTO, 2000b, pp. 6, 183). EC green-box programmes may be expanded in the future. The Mid-term Review of the CAP has proposed the introduction of a single income payment per farm (producer, rather than product, support), based on historical references. The payment would be decoupled, with no production-specific incentives, and thus would meet green-box criteria (at least under the URAA green box). To ensure good farming practices, the payment, which would replace most direct payments, would be conditioned on cross-compliance with statutory environmental, animal-welfare and food-safety standards. As an element of cross-compliance, long-term non-rotational set-aside would achieve environmental benefits, and audits on commercial farms would ensure compliance. Under the proposal, a system of 'dynamic modulation' would make increased sums available for rural development programmes, many of which fit within the green box (CEC, 2002, pp. 19–22).

The blue box

The blue box includes direct payments under certain programmes, which might otherwise be in the amber box, that limit agricultural production (ERS, USDA, 1998, p. 15). The blue box was a 'political strategy', negotiated as part of the Blair House Agreement (United States, 1998a, p. 3), and has been termed an 'awkward bilateral deal' – originally to benefit the USA and the EU – that other WTO Members do not like (Josling, 1998, p. 14).

Direct payments to farmers fit in the blue box. These payments are excluded from a Member's AMS, and thus are not subject to the commitment to reduce domestic support, if:

> (i) such payments are based on fixed area and yields; or
> (ii) such payments are made on 85 per cent or less of the base level of production; or
> (iii) livestock payments are made on a fixed number of head.
>
> (URAA, Art. 6, ¶ 5)

As these requirements indicate, blue-box policies limit production, but are not wholly decoupled (ERS, USDA, 1999, p. 53). Payments generally compensate producers for income lost when production is reduced and may help to reduce surplus production (Bohman *et al.*, 1999, p. 6, box 1). Payments in excess of lost income, however, may encourage production and thereby influence trade (ERS, USDA, 1998, pp. 16, 18). Not all blue-box payments are effective in reducing production. For example, under set-aside programmes, farmers could set aside their least productive land, or the amount of set-aside may be low, because no WTO rules govern this (Tielu and Roberts, 1998, p. 4).

When the blue box took effect in 1995, it applied primarily to farm programmes in the USA and in the EU, though Norway, Japan and Slovenia also reported blue-box payments in that year (ERS, USDA, 1998, p. 18, table 3). US deficiency payments, under pre-1996 farm legislation, fitted in the blue box, but 1996 farm legislation ended these payments. They were replaced by decoupled green-box payments, the production flexibility contract payments, which increased US green-box support between 1995 and 1996 (ERS, USDA, 1998, p. 19).

EC blue-box payments include the area compensatory payments enacted as part of the 1992 MacSharry CAP reform. The blue box helped the EC to meet its WTO commitments to reduce domestic support, because these direct payments are not subject to AMS limits (ERS, USDA, 1999, p. 42). The area compensatory payments are not decoupled and do not qualify for the green box, so continuation of the blue box was important for the EC. CAP changes proposed in connection with the 2002 Mid-term Review may make the blue box less important (CEC, 2002), but the EC supports maintenance of the blue box (WTO, 2002c).

Domestic support in URAA negotiations

Domestic support is a key issue in the agricultural negotiations that began in 2000 and will continue until 2005. Both URAA Article 20 and the Doha Declaration point towards reductions in domestic support; Doha Article 13 expresses a commitment to 'substantial reductions in trade-distorting domestic support'. The distorting effect of domestic support depends on 'the economic incentives created by program parameters and the total amount of support provided' (Young *et al.*, 2002, p. 8). Important questions focus on whether amber-box support will be substantially reduced or totally eliminated, and whether AMS limits should be commodity-specific, instead of the present total aggregate limits. Other issues are whether the green box is flexible enough to cover important non-trade concerns and whether exempt green-box programmes actually distort trade. Another question is whether the blue box will be continued or phased out (WTO, 2002a).

Individuals and organizations have reflected on the importance of domestic support, particularly on the status and parameters of the green box and the blue box in these negotiations. In recent years, the OECD, which includes WTO Members with high levels of domestic support, has focused on agricultural trade policy. The OECD emphasized that reform must result in policies that are transparent, targeted, tailored, flexible and equitable (OECD, 2000, p. 2). Recognizing the importance of domestic policies (e.g. environment, rural development) for trade, the OECD called for the identification of policy instruments that can meet domestic policy goals 'in ways that are decoupled from production decisions and thus cause minimum or no trade distortions' (OECD, 2000, p. 6). Domestic support that is 'genuinely decoupled', Australian economists explained, ensures that 'the price that farmers receive for their output must be the world market price and marginal costs should not be affected by support payments' (ABARE, 1999, p. 6).

Others focus on specific goals of green-box programmes. For example, the World Wildlife Fund for Nature (WWF) sees the green box (perhaps with some redefinition) as a means to improve sustainable agriculture and support rural development. WWF therefore recommended continuation of subsidies that reward 'farmers for the production of agricultural goods not recognised or rewarded by markets, such as maintenance of on-farm biodiversity and prevention of off-farm environmental impacts' (WWF, 1998).

Because some green-box policies do influence production decisions and therefore affect trade, the inclusion of domestic policies in the green box will be considered carefully in the current negotiations. Indeed, as one analyst predicted before negotiations began, green-box policies will be 'a much sought-after prize by agricultural protectionist forces' (Anderson, 1998, p. 14). Modifications in green-box policies may be desirable. For example, the green box now allows payments to cover costs or income loss from compliance with government environmental programmes. Perhaps a more precise definition could allow only payments for practices that contribute positive environmental externalities; negative externalities could be reduced through application of the polluter-pays principle, instead of by paying the polluter (Anderson, 1998, p. 10). But, as others have warned, to reopen the definition of the green box during negotiations raises the risk of its expansion, perhaps to include food security or schemes to keep remote farmers farming (Josling, 1998, p. 14).

Another important decision in the current negotiations is whether the blue box, originally intended as a temporary measure, will continue,[8] or whether it can be 'emptied and locked' (Josling, 1998, p. 14). As Australian economists have noted, blue-box policies 'lock in distorted production capacity' and are not fully decoupled. Thus, opponents of the blue box suggest that the policies be decoupled or that levels of support be reduced like other forms of market-distorting (amber-box) support (ABARE, 1999, p. 6). One of the fears of those calling for further reform is that WTO Members may conclude that decoupled and blue-box payments are 'sufficient to correct trade distortions'. If these criteria are not defined carefully, they may not reduce distortion, but their existence may weaken the political will among WTO Members to make meaningful reforms (Tielu and Roberts, 1998, p. 4).

Though the blue box was negotiated to benefit both the USA and the EC, the blue box remains important only to the EC and to a few other countries (e.g. Norway). ERS analysis suggests that, even if the blue box were closed and EC compensatory payments calculated under the AMS, support reductions under Agenda 2000 would allow the EC to meet its WTO commitment (ERS, USDA, 1999, pp. 33, 42, 44). Proposed reforms articulated in the Mid-term Review of the CAP would not seem to change this conclusion, especially if direct payments to producers fit within the exempt green box, as the EC seems to suggest.

The USA and the EC are likely to play central roles in current WTO agricultural negotiations, though some assert that support provisions in the 2002 Farm Bill have reduced US credibility as a leader of farm-policy reform (Kreuzhuber and Bunyan, 2002). Both governments have strong negotiating positions in connection with domestic support. In its Proposal for Global Agricultural Trade Reform, submitted in July 2002, the USA proposed 'ambitious reforms' for export competition, market access and domestic support. Reductions in these categories over a 5-year period would eventually be followed by elimination of 'all tariffs and trade-distorting domestic support' (WTO, 2002b). The proposal espoused a simplification of domestic-support calculations, with only two categories. Exempt support would meet the green-box criteria for programmes with no, or at most minimal, trade-distorting effects or effects on production; exempt support would face no cap. Non-exempt support would include AMS and production-limiting support. This support, currently in the amber and blue boxes, would be reduced, over a 5-year period, to 5% of each WTO Member's average value of total agricultural production, measured from the base period of 1996–1998. Further reductions on a sector-specific basis could be negotiated, and developing countries could have additional exemptions for specific purposes (WTO, 2002b).

In documents submitted earlier in the negotiations, the EC committed to reduced levels of domestic support, but pleaded for continuation of the blue box,[9] as well as for recognition of a multifunctional agriculture and the associated non-trade concerns, especially in regard to the environment, rural development, food safety, consumer concerns and labelling, and animal welfare (e.g. WTO, 2000c,d). In its proposal from December 2002, the EC recommended a 55% reduction in AMS from the URAA base, continuation of the blue box and the peace clause, and elimination of the *de minimis* exemption for developed countries. Non-trade concerns, which should be addressed by special provisions, include food safety, mandatory labelling, food security, the environment, rural development and animal welfare. The EC also asked for flexibility of domestic-support limitations as part of special and differential treatment for developing countries (WTO, 2002c, pp. 5–8).

During the WTO agricultural negotiations in 2003 and beyond, WTO Members will have the opportunity to craft a new scheme for reducing agricultural support that distorts trade. As ERS researchers commented recently:

> The continuing challenge for WTO negotiations on domestic farm policy will be to obtain effective commitments to reduce agricultural trade distortions, while allowing countries flexibility to use minimally trade distorting policies to achieve their own national priorities. Part of the task facing trade negotiators is determining where and how to draw the line between benign policies and trade distorting policies.
>
> (Young *et al.*, 2002, p. 1)

WTO negotiations will result in significant decisions that must balance Member agricultural-policy measures with the interests of world agricultural trade.

Notes

1 The EU is known as European Communities (EC) in World Trade Organization (WTO) matters. Thus discussion in the WTO context may refer to the EC, rather than to the EU.

2 In addition, Article XX of the General Agreement on Tariffs and Trade (GATT) allows the adoption of measures inconsistent with general obligations under GATT, which are 'deemed to be necessary to realise environmental objectives' (Hofreither, 1998, p. 9). The ambiguous formulation of Article XX raises questions of interpretation, and this article may not be a clear 'basis for environmental management' (Ervin, 1999, p. 74).

3 Delcros (2002) gives a history of special GATT measures affecting agriculture. A revised GATT and various associated decisions (called GATT 1994) continue to apply to trade in goods (WTO, 1994a, p. 1127).

4 The Agricultural Risk Protection Act, signed in June 2000, authorized increased payments and 5 years of crop insurance funding. Commissioner Fischler criticized the more than $15 billion in supplementary support, especially in light of US statements of the need to reduce support. Fischler asserted that US support had increased 700% since 1996 and promised that the EU Commission would monitor US policy and oppose US attempts to 'misclassify these financial handouts' (EU Commissioner, 2000).

5 One informed commentator noted that 'neither tariff protection . . . nor export subsidies . . . are affected by the new farm bill' (Haniotis, 2002, p. 5).

6 An *Agra Europe* analysis focused on the effect of EU and US direct payments on production and land values.

> While both Brussels and Washington claim that compensatory payments and Production Flexibility Contract payments are production-neutral, it is clear that they are not. In both cases they have an inflationary effect on the costs of production and therefore form an important element in farmers' dependency on the state and their expectation of maintenance of support when commodity prices fall. They maintain production at levels which do not reflect the realities of the market and thus exaggerate the depression of market prices.
>
> (Agra-Europe Weekly, 2000, p. A/2)

Both EU and US aid is based on land ownership or land use, and connection of right to subsidy with land ownership or use has led to an increase in land prices after reforms both in the USA and in the EU. In the USA, high cost of land helps to lower farm incomes and lead to more need for support. In the EU, land prices reflect income flows from Brussels, not productive capacity. Support that is more production-neutral would have less effect on land prices. *Agra Europe* analysts recommended that aid for social needs be based on direct payments to individual farmers; aid for environmental or structural functions should be related to the farmer's operation and the important environmental or structural functions in that operation (Agra-Europe Weekly, 2000, p. A/2).

7 Production flexibility contract payments under the 1996 law 'have characteristics of decoupled support' (Tielu and Roberts, 1998, p. 7). They are based on 85% of the 1996 whole-farm base acreage, regardless of whether or what crops farmers plant. Thus, the contract payments decouple support payments

from prices, so farmers can respond to the market. (Some products, not included in reforms, are still market-distorting; cotton is an example.)

8 Continuation of the blue box may also mean renewal of the peace clause; otherwise, blue-box policies could be subject to Member challenge (Pruzin, 2000).

9 In calling for continuation of the blue box, the EU cited an OECD report to the effect that 1992 CAP reforms decreased the trade impact of EU support to agriculture (WTO, 2000c, citing OECD Policy Evaluation Matrix Report). But the OECD study said that the payments were 'less trade-distorting', not that they did not distort trade, according to a US trade negotiator (Yerkey and Pruzin, 2000).

References

ABARE (1999) Multilateral agricultural policy reform: an imperative for the new millennium. *ABARE Current Issues* 99(4), 1–7.

Agra-Europe Weekly (2000) Land-linked subsidies at root of farm policy woes. *Agra Europe*, 17 March, A/1–A/2.

Anderson, K. (1998) *Domestic Agricultural Policy Objectives and Trade Liberalisation: Synergies and Trade-Offs.* OECD document, COM/AGR/CA/TD/TC/WS(98)101 (October).

Anderson, K., Erwidodo and Ingco, M. (1999) Integrating agriculture into the WTO: the next phase. In: *World Bank Conference on Developing Countries and the Millennium Round*, Geneva, 19–20 September 1999. 27 pp, from Website.

Blandford, D. (2001) Are disciplines required on domestic support? *The Estey Centre Journal of International Law and Trade Policy* 2(1), 35–59.

Bohman, M., Cooper, J., Mullarkey, D., Normile, M.A., Skully, D., Vogel, S. and Young, E. (1999) The Use and Abuse of Multifunctionality. Economic Research Service, USDA (Nov.), available at www.ers.usda.gov/briefing/wto/multifunc119.pdf

Burfisher, M.E., ed. (2001) *The Road Ahead: Agricultural Policy Reform in the WTO – Summary Report.* Agricultural Economic Report No. 797, USDA, Washington, DC.

CEC (2002) *Mid-term Review of the Common Agricultural Policy.* COM(2002) 394 final (10 July). From EU Website.

CEC, Directorate-General for Agriculture (DG Agri) (1999) *EU Agriculture and the WTO: From the Uruguay Round to the Millennium Round.* Fact Sheet. From EU Website.

CEC, Directorate-General for Agriculture (DG Agri) (2001) *EU Agriculture and the WTO* (September). From EU Website.

Delcros, F. (2002) The legal status of agriculture in the World Trade Organization. *Journal of World Trade* 36(2), 219–253.

ERS, USDA (1998) *Agriculture in the WTO* (Normile, M.A., report coordinator). WRS-98-4, USDA, Washington, DC, 56 pp.

ERS, USDA (1999) *The European Union's Common Agricultural Policy: Pressures for Change* (Kelch, D.R., report coordinator). WRI-99-2, USDA, Washington, DC, 56 pp.

Ervin, D. E. (1999) Toward GATT-proofing environmental programmes for agriculture. *Journal of World Trade* 33(2), 63–82.

EU Commissioner (2000) EU Commissioner blasts new US law providing agriculture subsidies to farmers. *BNA International Trade Daily*, 22 June.

Farm Bill (2002) *Farm Security and Rural Investment Act of 2002*. Pub. L. No. 107–171, 116 Stat. 134–540, US Government Printing Office, Washington, DC.

Fischler, F. (2000) The CAP and WTO agriculture negotiations, Speech at Cairns Group Ministerial Meeting, Banff, Alberta, 11 October. http://europa.eu.int/rapid (visited 31 Oct. 2000).

Fischler, F. (2002) Hitting agriculture worldwide makes US farm bill a flunk bill. Letter to the Editor. *Financial Times*, 27 May.

Gibson, P., Wainio, J., Whitley, D. and Bohman, M. (2001) *Profiles of Tariffs in Global Agricultural Markets*. Agricultural Economic Report No. 796, USDA, Washington, DC.

Haniotis, T. (2002) The new US farm bill from an EU perspective. Available at http://europa.eu.int/comm/agriculture/external/WTO/usfarmbill/tassos.pdf

Hart, C.E. and Babcock, B.A. (2002) US farm policy and the WTO: how do they match up? *The Estey Centre Journal of International Law and Trade Policy* 3(1), 119–139.

Hofreither, M.F. (1998) *Agri-environmental Policies and Trade Issues: Selected Problems*. OECD document COM/AGR/CA/TD/TC/WS(98)133 (October). From Website.

Josling, T. (1998) *The Uruguay Round Agreement on Agriculture: A Forward Looking Assessment*. OECD document COM/AGR/CA/TD/TC/WS(98)100 (October). From Website.

Kerr, W. A. (2000) The next step will be harder: issues for the new round of agricultural negotiations at the World Trade Organization. *Journal of World Trade* 34(1), 123–140.

Kreuzhuber, G. and Bunyan, C. (2002) Further clarifications on US Farm Act, Memo/02/166. Brussels (11 July), available at europa.eu.int/comm/agriculture/external/wto/usfarmbill/memo/_en.pdf

Lopez, J.A. (2000) *Decoupling: A Conceptual Overview*. OECD paper COM/AGR/APM/TD/WP(2000)14/FINAL (21 December). From Website.

Nelson, F.J. (2002) Aligning US farm policy with world trade commitments. *Agricultural Outlook* (January–February), 12–16.

Normile, M.A. (1998) Uruguay Round Agreement on Agriculture: the record to date. *Agricultural Outlook* (December), 28–33.

OECD (2000) *Policy Brief, Agricultural Policy Reform: Developments and Prospects*, June. From Website.

OECD (2002) *Agricultural Policies in OECD Countries: Monitoring and Evaluation 2002*. OECD, Paris, 238 pp.

Pruzin, D. (2000) Cairns group officials warn EU on agriculture 'peace clause' deadline. *BNA International Trade Daily*, 29 June.

Roberts, I. (2000) Multilateral trade negotiations: what is required to reform domestic agricultural support through the WTO? *ABARE Current Issues* 2000(3), 1–7.

Skully, D. (2000) Five years of tariff-rate quotas – a status report. *Agricultural Outlook* (November), 22–25.

Switzerland (2000) Specific characteristics of agriculture and the need to treat agriculture separately within the WTO. Discussion Paper. International

Conference on Non-Trade Concerns in Agriculture, Ullensvang, Norway, 2–4 July 2000. http://odin.dep.no/ld/engelsk/multifunctionality/

Tielu, A. and Roberts, I. (1998) Farm income support: implications for gains from trade of changes in methods of support overseas. *ABARE Current Issues* 98(4), 1–8.

United States (1998a) *Domestic Support Policies.* OECD document COM/AGR/CA/TD/TC/WS(98)115 (October).

United States (1998b) *Green Box Policies and the Environment.* OECD document COM/AGR/CA/TD/TC/WS(98)118 (October).

Vasavada, U. and Nimon, W. (n.d.) Environmental Effects of Further Trade Liberalization in Agriculture. http://www.ers.usda.gov/briefing/WTO/PDF/environmentandtradeliberalization.pdf (last visited 19 April 2001).

Vasavada, U. and Warmerdam, S. (1998) Environmental policy and the WTO: unresolved questions. *Agricultural Outlook* (November), 12–14.

WTO (1994a) WTO Agreement, reprinted in *International Legal Materials* 33, 1125.

WTO (1994b) *Uruguay Round Agreement on Agriculture (URAA).* From Website.

WTO (1994c) *Agreement on the Application of Sanitary and Phytosanitary Measures.* From Website.

WTO (1994d) *Agreement on Technical Barriers to Trade.* From Website.

WTO (2000a) Secretariat, Committee on Agriculture. Domestic Support. Background Paper by the Secretariat. G/AG/NG/S/1 (13 April). From Website.

WTO (2000b) Special Session, Committee on Agriculture. Green Box Measures. G/AG/NG/S/2 (19 April). From Website.

WTO (2000c) EC: The Blue Box and Other Support Measures to Agriculture. G/AG/NG/W/17 (28 June). From Website.

WTO (2000d) EC Comprehensive Negotiating Proposal. G/AG/NG/W/90 (14 December). From Website.

WTO (2001a) Ministerial Declaration. Adopted on 14 November 2001, Doha, Qatar. WT/MIN(01)/DEC/1. From Website.

WTO (2001b) Implementation-Related Issues and Concerns. Decision of 14 November 2001. WT/MIN(01)/17. From Website.

WTO (2002a) WTO Agriculture Negotiations. The issues, and where we are now. http://www.wto.org. From Website.

WTO (2002b) United States: Proposal for Global Agricultural Trade Reform (July), available at http://www.fas.usda.gov/itp/wto/actual.htm

WTO (2002c) The EC's Proposal for Modalities in the WTO Agriculture Negotiations. Ref. 625/02 (16 Dec.), available at europa.eu.int/comm/trade/goods/agri/modalities.pdf

WWF (World Wildlife Fund for Nature) (1998) *Emerging Issues at the Interface of Domestic and International Policy: Agricultural Trade and the Environment.* OECD document COM/AGR/CA/TD/TC/WS(98)120.

Yerkey, G.G. and Pruzin, D. (2000) US blasts EU for 'resisting' US proposal to eliminate export subsidies in farm trade. *BNA International Trade Daily*, 3 July.

Young, E., Burfisher, M., Nelson, F. and Mitchell, L. (2002) Domestic Support and the WTO: Comparison of Support Among OECD Countries. USDA, ERS, available at http://cnas.tamu.edu/publications/powerpoint/papers/young.pdf.

The Politics of Agricultural Trade

3

Wyn Grant

Agricultural trade is highly politicized. It produces serious tensions in relations between the USA and the European Union (EU), as well as between the EU and Cairns Groups countries and between developed and developing countries. Disputes over agriculture prevented a speedier conclusion to the Uruguay Round. They are likely to be a prominent feature of the negotiations in the Doha Round. This is in spite of the fact that agriculture has undergone a long process of decline as a major economic activity in advanced industrial countries.

However, its political significance has not declined in proportion to the reduction in its economic significance. Indeed, with agriculture being effectively brought into the ambit of international trade negotiations for the first time in the Uruguay Round, its political displacement in world trade diplomacy has increased. Similarly, while '[a]griculture is a smaller proportion of GDP [gross domestic product] and of world trade . . . trade has tended to become more important to agriculture' (Wolfe, 1998, p. 5). Even where the share of output that is traded is small or declining, it may be an important mechanism for the disposal of surpluses and hence have an important impact on consumer prices.

The Politicization of Agricultural Trade

Why is agricultural trade so politicized? In developed countries farmers remain highly reliant on the subsidies, protection and tax reliefs provided by their governments or through the Common Agricultural Policy (CAP) of the European Community. The extent of the influence of these policies may be seen from the fact that in the EU planting and investment decisions by farmers are often influenced by developments in the CAP. Farmers have become highly reliant on subsidies and it is difficult for them to see how they could survive without them. They thus lobby vigorously for the maintenance of these subsidies.

Their lobbying activity is facilitated by the existence of specialized institutions devoted to agricultural politics, specifically departments or ministries of agriculture. They usually see it as their task to promote the interests of their clients, the farmers. It is unusual for any other industry to have its own ministry protecting its position and advancing its claims within government. Such ministries may also have a special position in international trade negotiations. For example, in the USA the Department of Agriculture (USDA) has a major role in relation to the agricultural aspects of international trade negotiations. In Britain, a Cabinet Office official noted that the Ministry of Agriculture, Fisheries and Food (MAFF) conducted its own foreign policy once Britain joined the EU: 'they were flying all over the world to negotiate in a way that the Foreign Office couldn't control. So, suddenly there was a small element of foreign policy that was run by MAFF instead of the Foreign Office' (quoted in Marsh *et al.*, 2001, p. 18).

Policy communities in agriculture

National ministries of agriculture are, however, part of a broader and relatively closed policy community concerned with agriculture. Within the influential theory of policy networks, political scientists distinguish policy communities as a particular subtype of policy network often found in agriculture. Its characteristics including stable membership, shared assumptions about policy, insulation from other policy networks and a relatively high political entry price, which makes it difficult for outsiders to participate in the policy discussion (exemplified by the complexity of the CAP). A number of consequences follow from the prevalence of policy communities in agriculture:

- Policy objectives tend to be relatively unquestioned.
- Policy change tends to be incremental.
- Members of the policy community are able to work together effectively to defend the status quo.
- Substantial policy change occurs as a result of some exogenous policy shock.

Trade policy nevertheless presents significant challenges to well-established domestic policy communities in agriculture. It brings a much broader range of actors than usual into the policy-making process, e.g.

trade ministries and heads of government as talks run into difficulty. One of the important political consequences of the Uruguay Round was that domestic subsidies that had been seen as a purely domestic (or European) issue were called into question because of their trade-distorting effects. As negotiations enter their final phase, the political costs of continuing agricultural protection have to be balanced against other policy priorities. The agricultural policy community is starting to lose its exclusive control of policy.

Defensive strategies

This does not mean that there are no strategies that agricultural policy interests can deploy to preserve their advantages. These may include:

- Ensuring that the agricultural aspect of the negotiations is in the hands of agricultural decision-makers or at least that they have a heavy input in decisions.
- Using their knowledge of the technicalities of the decision-making process to ensure that reductions in tariffs or subsidies are calculated in as favourable a way as possible, e.g. the selection of base periods.
- Finding new ways of delivering subsidies that are less explicitly trade-distorting, e.g. through environmental payments or subsidies to preserve a particular type of landscape (multifunctionality).
- Coalition building with other countries with similar interests, e.g. the attempts of the EU to form a club of 'Friends of Multifunctionality' in the run-up to the Doha Round. The ambiguity of the notion of 'multi-functionality' in relation to the environment, discussed by Grossman in Chapter 5, makes it a politically appealing way of recasting subsidies so that they are more World Trade Organization (WTO) compatible.

The attempts of agricultural interests to ensure that the essential elements of the existing package of subsidies are retained is helped by the complexity of agricultural policy, but also by the absence of effective opposition. Agriculture offers a classic example of a relatively small group of recipients being provided with concentrated benefits, while the costs are much more diffused. 'The logic of the argument is that farm policies are the function of the strong preferences of a small number of farmers interacting with the weak preferences of a large number of consumers' (Wolfe, 1998, p. 14). For example, consumers in the EU are rarely aware of how much higher their shopping bill is because the CAP leads to European prices being above world market prices for agricultural commodities. A farmer will know how large his or her subsidy cheque is. Farmers have strong incentives to form effective associations to defend their interests, whether they are principally organized at a commodity level (as in the USA) or at a more general level (as in the EU). In the case of the USA:

> [t]he farm commodity programmes that were first initiated in the United States during the 1930s have proved difficult to reform primarily because the defenders of these programs have remained better organized and more effective than the critics over the ensuing years. Numbers of voters matter, but organization . . . matters more. If it were only numbers that mattered, farmers would have lost their political power long ago
>
> (Orden *et al.*, 1999, p. 48)

Nevertheless, commentators are often puzzled by the fact that such a small economic sector receives such large benefits funded by taxpayers and by consumers. What is often overlooked is that there are a number of other economic sectors which are closely linked to the success of agriculture and which often lend it political support. These include a number of sectors that contain multinational companies, which often have sophisticated government-relations operations that are able to exert influence on decision-makers at the highest level. Among the sectors that have close links with agriculture are: seed companies; manufacturers of agricultural machinery, ranging from combine harvesters and tractors to specialized milking equipment; agrochemical and fertilizer companies; providers of specialized financial services for agriculture; veterinary drug companies; and 'first-stage' food-processing companies, such as dairy processing and sugar refining.

Sources of change

Agricultural politics does not, however, present a purely static and unchanging picture. A number of tendencies have been evident which have undermined the cohesion of the sector and presented it with new sources of opposition. There have always been distinct interests within agriculture and an important role of farm associations was to mediate between these and come up with an agreed policy package that could be negotiated with governmental authorities. However, some of the divisions within agriculture have been widening and this can give greater scope for other actors to change policy.

In the USA, the gap between agribusiness, in the sense of large-scale, corporately controlled, vertically integrated enterprises, and the more traditional 'family farm' has been widening. 'Family farmers', for example in the dairy sector, are more interested in import protection and direct subsidy payments. Agribusiness is likely to give a higher priority to export subsidies and possibly to a more level international playing-field with reduced levels of subsidy, which would offer greater opportunities to American corporate agriculture because of its technological and managerial lead. This gap is reinforced by the growing importance of 'the broader interests of upstream food investment, processing, packaging, marketing and transportation businesses of the 'agro-food' sector. Most of these businesses are pro free trade and stand to profit from greater volumes of food in commercial channels' (Balaam, 1999, p. 69). By the mid-1990s Congress was 'more sympathetic toward full-production agriculture and agribusiness interests than any previous Congress had been' (Orden *et al.*, 1999, p. 134).

The actual playing out of these competing political forces is a complex and unpredictable political process. It would be an oversimplification to see the Reagan administration's initial 'zero option' stance in the Uruguay Round as a concession to the growing influence of export-oriented agribusiness interests. Even if it was the case, more traditional forces soon reasserted themselves in a Polanyian 'double movement'. It has even been argued that some protectionist interests supported the zero option as a means of undermining the whole Uruguay Round. More straightforwardly, it is evident that 'the political clout still wielded by agricultural protectionists, farm lobbies, farm fundamentalists and other opponents of freer agricultural trade was underestimated from the beginning of the Uruguay Round' (Hillman, 1994, p. 34). Nevertheless, changes in the balance of interests as a result of a shift in agricultural structure may have given state officials more autonomy to conduct trade negotiations in terms of their own independently derived goals rather than the demands of farm organizations (Balaam, 1999, p. 61).

In the EU, the increasing number of Member States has made it more difficult to reconcile different interests and balance commodity regimes. Some commodities may be of interest to a limited number of Member States, e.g. cotton, rice, tobacco. Other commodities may be produced throughout the EU, but the domestic structure of production may vary considerably, giving rise to different interests among Member States (Grant, 1997, pp. 32–62). These difficulties are likely to be increased when candidate countries in Central and Eastern Europe join the EU. One consequence has been the weakening of Europe-wide farm organizations, such as the Comité des Organisations Profesionnelles Agricoles (COPA). 'Once a pivotal member of the supranational policy community of the EU-6, by the early 1990s COPA was one of a number of relatively important actors in a fiercely competitive policy community' (Jones and Clark, 2001, p. 97).

Environmental considerations have also been given a greater emphasis in relation to agricultural policy. Various environmental organizations have emerged to increasingly question and oppose traditional forms of agricultural policy. These organizations range from general ones, such as Friends of the Earth, concerned with the overall impact of agriculture on the environment, to those concerned with more specific issues, such as biodiversity (World Wide Fund for Nature), and even to those concerned with particular species (Birdlife International). In a broader context, these developments may be seen as part of a shift from a politics of production to a politics of collective consumption (Grant, 2000). In the immediate postwar period, food security and supply issues led to an emphasis on maximizing production, with little regard for the environmental consequences. Modern consumers are more concerned with the quality rather than the quantity of food and their definition of quality extends to how the food was produced and any external 'bads' in the form of environmental pollution or deficiencies in animal welfare.

This clash between different perspectives on how food should be produced is seen at its most stark in the debate about genetically modified (GM) seeds and crops. As Hilson and French note in Chapter 9, the potential conflict between international free trade and national environmental policies has been an issue since the early 1990s, but the potential dispute over GM products introduces a new dimension. It involves human-health questions rather than just environmental ones and it raises issues about risk and how this is managed by nation states in an international order. The issue raises fundamental questions about the relationship between trade liberalization and the regulation of risk.

From a political perspective, the EU is caught between contradictory demands:

- The desire to avoid a major trade dispute with the USA which could endanger the whole WTO disputes settlement mechanism.
- The desire to promote an important new enabling technology (biotechnology) within the EU, which could bring important economic and social benefits.
- The difficulty of supporting a technology that is strongly opposed by public opinion in northern Europe in particular.

The EU has sought to postpone difficult decisions by imposing a de facto moratorium on new approvals of GM crops, which, as Hilson and French show in Chapter 9, is of dubious legality, even if it is politically convenient. At the time of writing, the USA had held off mounting a challenge in the WTO, in part because it does not wish to provoke another major trade dispute with the EU. However, throughout 2002 there were increasing signs that the patience of the American authorities was becoming exhausted. The EU has also sought to reassure public opinion by introducing a strict regulatory regime involving labelling and traceability. This is also unlikely to be found compatible with WTO rules and is regarded with particular disfavour by the USA.

It is difficult, however, to resolve such matters by a legal ruling without intensifying political criticism of the WTO itself. Non-governmental organizations have been particularly effective at mobilizing public opposition to GM foods, no doubt helped by enhanced concerns about food safety in the wake of the bovine spongiform encephalopathy (BSE) crisis. For their part, the multinationals concerned with developing the new technology, such as Monsanto, have been inept at addressing public concerns in an effective fashion. The GM episode offers an unusual example of the adoption of a major new technology being slowed down by political opposition.

It is important, however, not to exaggerate the impact of environmental considerations on agricultural policy in general and agricultural trade policy in particular. In spite of the requirement enunciated in the 'Cardiff process' that all EU policies should integrate environmental considerations, relatively little progress has been made. Agri-environmental programmes account for a few percentage points of CAP spending and 'cross-compliance', making the receipt of subsidies conditional on meeting environmental and animal-welfare standards, remains to be implemented in 2005 following the Mid-term Review. In the USA, the Bush administration does not provide a favourable context for the pursuit of environmental goals. In any case, the environmental and human-health impacts of agriculture tend to fall between particular agencies rather than being a prime responsibility of any one of them. In terms of international trade policy, EU attempts to insert animal-welfare issues into the Doha Round have met with little success.

Nevertheless, it is not 'business as usual' for agricultural politics. One indicator is the disappearance of agricultural ministries in Britain and Germany and their replacement by departments with a much broader remit. With the disappearance of an agricultural department, the agricultural policy community loses much of its focus and potentially its cohesion. The creation of departments with 'environment' or 'consumer' in their titles signals a new set of priorities in government policies.

Agricultural politics has certain characteristics that are shared across different countries and these in turn have an influence on the conduct of agricultural trade negotiations. However, there are also important differences in the way that agricultural politics is conducted in the USA and the EU. In summary, agricultural policy-making in the USA is more fragmented than in the EU and also more politicized and less technocratic. The next section reviews these differences and their implications, after which there is a discussion of substantive changes in the goals of policy. While the EU has sought to defend the CAP as a cornerstone of the European Community, the USA seemed to be moving its agricultural policies in a more market-oriented direction with the 1996 Farm Bill, but appeared to reverse this stance with the 2002 measure.

Agricultural Trade Policy-making in the European Union

'Commercial policy', as it was originally called, has been a task of the European Community since its formation. The Commission has been able to secure a considerable amount of practical autonomy in the operation of trade policy. This has meant that it has been able to conduct policy in a relatively technocratic fashion, with an emphasis on securing deals that will protect EU interests but also maintain the international trade regime in place. This is not an accidental outcome, as the original design of the policy-making process was intended to prevent too much intrusion by domestic political considerations and protectionist forces. Politics do, however, intrude into the process as a result of demands made by Member States, while the Commission itself is not necessarily united on the appropriate course of action.

Formally the process of conducting a trade negotiation involves the Commission making a recommendation to the General Affairs Council of foreign ministers to commence a negotiation. One of the peculiarities of the institutional arrangements is that there is no council of trade ministers as such, despite the importance of the policy area, although trade ministers do sometimes join meetings of the General Affairs Council. Peterson and Bomberg (1999) argue the case for 'some intermediary (besides the Committee of Permanent Representatives: COREPER) between the technocratic Article 133 committee and the General Council of Foreign Ministers'. In the absence of such an intermediary, the Council was often unable to take decisions sufficiently quickly. 'One upshot was that EU farm ministers retained effective veto power for much of the [Uruguay] Round' (Peterson and Bomberg, 1999, p. 98).

It is thus the business of the General Affairs Council to give the Commission permission to conduct the negotiations and provide it with a negotiating mandate to guide its actions. The Commission will want as broad and unrestricted a mandate as possible, while Member States will often seek to place limitations on its freedom of action. There is often tension during the course of negotiations about whether the Commission has gone beyond its mandate. The progress of the negotiations is monitored by the Council's Article 133 (formerly 113) committee. This generally meets once a month with senior officials from national ministries to discuss general policy issues and three times a month with deputies to discuss more specific problems. 'The members of both levels serve on the Committee for an average of four to five years, giving rise to a club-like atmosphere' (Hayes-Renshaw and Wallace, 1997, p. 190). This is consistent with a pattern of politics in which the emphasis is on seeking to reach a consensus position. On the whole:

> the Committee works with, rather than against the Commission, indicating to the latter what is and what is not likely to be accepted by the ministers and what should be referred back to the Council for consideration and perhaps for new and modified negotiating mandates.
>
> (Nugent, 2001, p. 308)

Policy-making is complicated by the existence of divisions within the Commission itself. Each of the Directorate-Generals (DGs) within the

Commission services has its own mission and often its own distinctive culture. This has certainly been true of the Agriculture DG, which, at least until recently, had a strong French influence. The more rapid alternation of senior officials introduced as a result of the Kinnock reforms may start to erode these distinctive administrative cultures. However, horizontal links between different DGs remain less well developed than they would be in a bureaucracy with a longer history. Consequently, many disputes between DGs over jurisdiction and policy are referred upwards to meetings of the *chef de cabinet* or to the College of Commissioners itself.

A classic example of such a dispute occurred during the Uruguay Round in November 1991 when the then Agriculture Commissioner, Ray MacSharry, had gone to Chicago to attempt to negotiate a deal with the USA. MacSharry thought that it might be sensible to agree to a 21% cut in export volumes. The Commission President, Jacques Delors, told him on the telephone that:

> [a] figure of 21 per cent would mean cutting output by even more than the CAP reform already required – and would therefore exceed the commission's negotiating mandate. Delors said he would oppose such a deal, that the commission would vote it down, and that if it went to the Council of Ministers France would wield the 'Luxembourg Compromise' (a veto).
>
> (Grant, 1994, pp. 174–175)

MacSharry considered that Delors had undermined his negotiating position, making it more difficult for him to reach a deal with the Americans. On his return to Brussels he wrote a letter stating that he was resigning from his responsibilities in the General Agreement on Tariffs and Trade (GATT) negotiations. 'In complete disarray, the commission could not say who was responsible for negotiating on agriculture' (Grant, 1994, p. 175). Britain and Germany put substantial pressure on Delors to restart negotiations. He was obliged to allow MacSharry to return to Washington to negotiate on his terms. The result was the Blair House Accord, which agreed a 21% cut in export volumes over 6 years. The outcome was a humiliation for Delors, who felt that he had lost authority as a result. It illustrates what can be achieved by a determined Agriculture Commissioner, although it is unusual for such conflicts to be quite so dramatic or public.

The influence of Member States

Conflicts may also break out between the Commission and Member States, as is illustrated by the closing phases of the Uruguay Round. The Commission's determination to exercise its leadership capacity may go beyond what Member States find tolerable:

> [a]t Blair House the Commission was able to pull together the threads . . . despite the fact that officially it was still operating within the 'negotiating mandate' given to it by the Council in November 1990. The mandate did not even permit making separate commitments on export subsidies, market access or domestic support. Certainly, the accord went beyond the position of France, the most strongly opposed Member State.
>
> (Coleman and Tangermann, 1999, p. 400)

The French government was able to secure a reopening of the Blair House Accord and its reformulation in the Blair House 2 or Breydel Agreement. Initially isolated on the issue:

> the French government was extremely skilful and effective in deploying and making credible its threat to veto a final GATT agreement. The key to its success lay in its enlisting the support of [then German Chancellor] Kohl and the restoration of the Franco-German 'tandem' that had broken down in late 1992.
>
> (Webber, 1998, p. 589)

The French were helped by the fact that the procedural norm in Council discussions is still to reach agreement by consensus rather than voting, if at all possible.

A determined Member State that adopts a relatively intransigent position can protect its interests in agricultural trade negotiations. In practice this usually means France. The text of the accord on agriculture agreed at the start of the Doha Round commits the participants to 'comprehensive negotiations aimed at . . . reductions of, with a view to phasing out, all forms of export subsidies'. Commenting on the accord, it was stated that:

> This clause was seen by CAP traditionalists as striking a blow at the very heart of European farm policy, and France, backed by Ireland, signalled its willingness to bring down the whole Doha meeting rather than agree to it without qualification.

This impasse was resolved by inserting language 'which made clear that the objectives set out in the agenda were 'without prejudice to the outcome of the negotiations' – a formula which satisfied French pride without removing the sense of the original draft' (*Agra Europe*, 2001, p. EP/1).

The Mid-term Review

France's policy preferences also played a major part in shaping the Mid-term Review proposals for reform of the CAP initially put forward by Commissioner Fischler in the summer of 2002. It was anticipated that the final proposals to be announced in January 2003 would have to be substantially watered down to meet objections from France and other

Member States. As it was, the proposals did not go beyond discussing possible reform options in the case of the dairy sector, the most subsidized and protected sector apart from sugar beet.

The Mid-term Review offered a complex set of proposals, but they were intended to help provide the EU with a defensible position in the Doha Round. Subsidies would not be abandoned, but they would be reshaped to give a greater emphasis to their 'public good' elements. Three of the key proposals advanced were:

- Cutting or 'decoupling' the link between production and direct payments to farmers. Payments would be made on an historical basis.
- 'Dynamic modulation', cutting payments by farmers up to a 20% maximum (by 2011) and diverting the money to rural development, with a stated intention to favour poorer Member States.
- 'Cross-compliance', making payments conditional on meeting environmental, food-safety, animal-welfare and occupational-safety standards.

Of these proposals, only cross-compliance has broad support, although even there the devil is in the detail, particularly issues of Member State implementation. Decoupling attracted strong support from just three reform-oriented states (Denmark, Germany and the UK). The final settlement reached in June 2003 involved a complicated form of partial decoupling with considerable national discretion and a more limited form of dynamic modulation.

The situation was complicated by a bilateral deal reached between France and Germany at the Brussels summit in the autumn of 2002. Just as happened before the Berlin summit in 2002, Germany backed down in the face of French determination to defend the CAP, with Chancellor Schröder outmanoeuvred by President Chirac. In essence, the deal was that any reform would be delayed until 2006 in return for a reduction in the rate of increase in the farm budget after that year. The precise impact of this bilateral agreement on the Mid-term Review remained a matter of dispute, but it did contribute to making the eventual reform package more modest. One Member of the European Parliament observed that, in one 30-minute meeting, two of the pressures for reform (enlargement and pressures on the German domestic budget) had disappeared, leaving only the Doha Round itself (*European Voice*, 28 November 2002, p. 20).

Trade politics in the EU

Politics certainly inserts itself into the trade decision-making process in the EU, but it is the politics of particular national interests, in turn reflecting the influence of particular lobbies, rather than popular electoral politics. Nevertheless, Smith argues (1999, p. 280) that in the EU:

> the management of trade politics has become less technocratic and more subject to pressures for accountability and short-term responsiveness. Arguably, this has increased the constraints on the trade policy makers, and thus the possibility of inconsistency and reactiveness on their part.

It is undeniable that a wider range of actors has become involved in EU trade politics. A range of non-governmental organizations have become much more active in trade issues, stimulated in the case of agriculture by the controversy over GM crops. The general authority of the Commission has eroded over a range of policy issues. It has never really recovered from the forced resignation of the Santer Commission. Nevertheless, trade policy-making in the EU remains less politicized than in the USA, 'where politicization of trade policy is a long-established fact of life' (Smith, 1999, p. 280).

Agricultural Trade Policy-making in the United States

A far wider range of actors is involved in making trade policy in the USA than in the EU. There is a broad sharing of responsibility between the legislature and the executive, with the legislature having reasserted its traditional constitutional role in relation to trade issues. This contrasts with the very marginal role of the European Parliament in trade issues, a pattern that is generally replicated in the legislatures of the Member States. Congress itself is a notoriously atomistic institution and the fragmentation of authority is replicated within the executive branch.

The whole trade policy-making process is highly protectionist. Members of the Congress are highly attuned to demands from particular (often very small) segments of their electorates for protection or subsidization. 'In the Senate, the geographic dispersion of American agriculture still gives farm interests alone surprising political clout. There are few 'farm dependent' states, but there remain a large number of states with enough agriculture to be important' (Orden *et al.*, 1999, p. 178). One might think that urban members of the House would be indifferent to the plight of agriculture. However, given the cost of re-election, urban members of the House or Senate are happy to accept contributions from agricultural interests, who take careful note of helpful votes on issues that concern them. 'Agriculture Political Action Committees (PACs) gave a total of $15.5 million to members of Congress in 1993–4, making them the nation's third-largest source of PAC money overall' (Orden *et al.*, 1999, p. 178). The food-stamps programme, which accounts for nearly one-third of USDA's budget, has also provided a means of forging an unlikely alliance between urban members of Congress and their rural counterparts. The decision-making process is penetrated by lobbyists representing various interests, their role underwritten by successive pieces of legislation:

[i]n particular, the 'privatisation' of US trade policy under the Clinton Administration meant that American policy became increasingly responsive to whichever interests lobbied most effectively or, more cynically, contributed most to increasingly expensive US political campaigns. Little attempt was made to aggregate interests in US trade policy, as the EU still struggled to do in the name of the 'Community interest'.

(Peterson and Bomberg, 1999, p. 101)

The number of actors involved in trade policy-making in the USA means that elaborate arrangements have had to be established for policy coordination. There are no fewer than 17 federal offices and agencies involved in developing US government positions on international trade issues. The office of the US Trade Representative (USTR) seeks to coordinate policy through the Trade Policy Review Group (TPRG) and the Trade Policy Staff Committee (TPSC). The TPSC is the primary operating group, with representation at the senior civil servant level. Its work is supported by more than 60 subcommittees responsible for specialized areas and task forces that work on particular issues. If the TPSC cannot agree, or the issue is of major significance, then the TPRG (Deputy USTR/Under-Secretary) level has to become involved.

A further layer of complexity is provided by the involvement of business interests in a network of advisory committees. The Advisory Committee on Trade Policy and Negotiations, whose members are appointed by the President, has a mandate to provide overall guidance on trade issues. At the next level there are six advisory committees dealing with specific policy areas, one of which is concerned with agriculture. The role of these committees has been expanded beyond advising on trade negotiations to providing advice on the operation of trade agreements, on the development and implementation of overall US trade policy and on priorities for actions to implement such policy.

Forming part of the Executive Office of the President, USTR sees its role as to develop and coordinate US international trade policy, lead or direct international negotiations, resolve disagreements and frame issues for Presidential decision. The importance of agriculture to the USTR is underlined by the fact that, of five members of the office at deputy level, one is designated as the chief agricultural negotiator. Only one other sector of the economy has representation at this level, textiles. Nevertheless, there are limits to the expert infrastructure available to USTR in the agricultural area. USTR officials working on agriculture in Geneva are often on secondment from USDA. The relationship between USTR and USDA might be described as a functional one, where the USTR has the formal lead role but where USDA has very significant influence on the framing of policy. The Director of Agricultural Affairs at USTR has described the relationship as follows:

> [t]he role of USTR is primarily one to provide the government leadership in putting forth negotiating positions and doing the actual negotiations in the WTO. We work hand in hand with the Department of Agriculture on agricultural trade issues, we are partners together.
> (http://fas.usda.gov/itp/wto/montana/lauritsen.html, visited on 27 June 2002)

USDA's responsibilities in international trade negotiations are coordinated and directed by its Foreign Agricultural Service (FAS), which works closely with USTR. FAS has a network of agricultural diplomats and specialists stationed across the world. This service has the expertise to develop policy proposals which are then submitted to USTR. It serves as the official conduit for notifications and comments about WTO sanitary and phytosanitary issues and technical barriers to trade.

If it were a matter of sharing out responsibilities between USTR and USDA, coordination within the executive branch on agricultural trade policy might be relatively straightforward. However, a number of other agencies are also involved, not least the State Department, which styles itself as the lead department for foreign affairs. It has someone at under-secretary level who is responsible for economic, business and agricultural affairs.

Fragmentation at the executive level would not matter so much if responsibility for trade policy were not shared with the legislature. 'For most of the postwar period, the United States Congress was remarkably restrained in the exercise of its constitutional authority "to regulate commerce with foreign nations"' (Destler, 1992, p. 65). However, a combination of the erosion of authoritative leadership in Congress and increased political demands for action to combat increasingly effective foreign competition produced a more activist Congress on trade policy. Effective leadership from the executive branch became more important, but it was often unable to provide it.

The key committees in the House are Agricultural Committees, which 'jealously guard their position as initiators of legislation' (Josling, 1998, p. 57). Their stance towards agricultural policy issues is readily evident. Senator Tom Harkin, the chair of the Senate Committee, has stated that '[o]ur committee is committed to bringing strength and prosperity back to farm families and rural communities' (http://agriculture.senate.gov/, visited on 28 June 2002). Similarly, the chair of the House Agriculture Committee, Larry Combest, has stated that 'I have staked out an advocacy role for the Agriculture Committee as its chairman.' 'Iron triangles' may indeed be a reductionist simplification as a metaphor for Washington politics (Browne, 1995, p. 11). Clashes between commodity interests are common and reflected in the often asymmetrical distribution of farm subsidies. Institutional divisions of labour and fragmentation of tasks do not make it easy to produce coherent policy. Nevertheless, USDA, the Congress committees and the various interest organizations are united by a common desire to see a prosperous domestic agriculture in the USA and to advance its interests on a global stage.

The Freedom to Farm Act

Nevertheless, at one point in recent American history this coalition of pro-agricultural interests and agencies broke down and, for a while, it appeared that American agricultural policy was moving in a new and more market-oriented direction. The radicalism of the Federal Agriculture Improvement and Reform Act of 1996 (Fair Act) has sometimes been oversold. It was the result of a very specific set of economic and political circumstances and to some extent it was built on insecure foundations. Nevertheless, it was possible to claim with some justification that:

> [t]he US farmer has in effect been liberated from government controls on planting and can now react to market opportunity. It is in this respect that the FAIR Act is a fundamental change in approach to the succession of farm bills since the 1930s.
>
> (Josling, 1998, p. 59)

Josling emphasized (1998, p. 59) that '[t]he importance of the FAIR Act extends beyond the United States.' Any American farm bill is going to have an international impact because of the role of the USA as the world's leading agricultural exporter. Hence, decisions taken on farm policy in the USA can have a considerable impact on world market prices and consequently on the policy responses required elsewhere. However, the FAIR Act provided an example for other countries to follow and it also strengthened the hand of the USA in international trade negotiations. The USA had acquired the moral authority to advise other countries 'don't just do as we say, do as we do'.

One of the main measures in the FAIR Act was the replacement of deficiency payments, which boosted subsidies when market prices fell, by decoupled payments that were not tied to production and prices. The practice of set-aside or 'buying air', as it was called in the USA, was also ended. Farmers dubbed it not 'the Freedom to Farm' but 'the Freedom to Fail' Act. However, it is important to note that '[t]he FAIR Act that finally emerged from Congress in April 1996 was not as market-oriented a measure as the Freedom to Farm Act originally proposed in August 1995' (Orden *et al.*, 1999, p.169). Sceptical Europeans saw it as a recipe for subsidized expansion:

> [[t]he FAIR Act] effectively removes controls from production, while largely maintaining similar levels of subsidisation of agriculture as maintained in the 1990–95 farm bill . . . American farmers will now be in a position where they will be paid a substantial subsidy whatever they produce, while their

> output is sold on a market still protected and supported by import tariffs and export subsidisation.
>
> (*Agra Europe*, 1996b, p. P/1)

Nevertheless, the bill was a departure from existing policy. Free-market Republicans, led by Newt Gingrich, adopted a more ideological approach to traditionally consensus-based policy arenas such as agriculture. Agricultural policy became a partisan issue and 'the momentum of the budget process impelled by the zeal of the Republican freshmen overruled the protection of agricultural turf' (Josling, 1998, p. 57). No farm bill emerged from the House Agriculture Committee, an event without precedent. However, Republican control of the Congress would not in itself have secured a farm bill incorporating decoupled payments. There was a 'sudden and rapid increase in farm commodity prices. If this commodity price boom had not taken place, the Freedom to Farm Act would not have moved forward' (Orden *et al.*, 1999, p. 183). It should also be noted that the Bill reflected a shift in the balance of interests within agriculture in favour of the agribusiness end of the spectrum as distinct from 'family farms'.

> The Bill is seen as a victory for grain buyers and input suppliers who fought to reduce the government's ability to restrain acreage, and a defeat for the advocates of the traditional 'supply management' style of farm supports that have characterised US farm policies for some 50 years.
>
> (*Agra Europe*, 1996a, pp. N5–N6)

Analysts were aware that the FAIR Act represented 'a precarious victory . . . it leaves the traditional agricultural policy process establishment in place – ever ready to seek expanded benefits whenever circumstances offer the political opportunity' (Orden *et al.*, 1999, p. 234). In particular the Act did 'not repeal the permanent legislation that will automatically trigger a reversion to high price supports and supply controls if Congress fails to take further action in 2002' (Orden *et al.*, 1999, p. 169). Moreover, the Act did provide for 'supplementary assistance' payments to be made in 'emergencies'.

In 1997 the good times continued for America's farmers, but in 1998 prices for grain and many other farm products fell sharply. Farmers had also overplanted what were seen as profitable crops, creating surpluses that sent prices lower. In September 1998 President Clinton told farm representatives, 'When I signed that Farm Bill . . . I tried to make it clear that sooner or later we would have to do more to provide a safety net for hard times . . . Well, that time has arrived' (Office of the Press Secretary, 1998). Within days the President had asked Congress to provide $2.3 billion in emergency relief to farmers. In 1998 an agricultural appropriations bill signed by the President included $8.6 million in emergency assistance for farmers. However, Vice-President Gore made it clear that it was not enough. He expressed his disappointment that 'Congress did not come through with more assistance for farmers and ranchers who suffered crushing losses from this summer's drought, Hurricane Floyd, and other natural disasters'. The Vice-President argued that 'the fact that this is our second emergency farm legislation in two years is all the evidence we should need to conclude that the 1996 Freedom to Farm Bill is in need of repair'. An alternative argument might be that it showed that farmers were irredeemably wedded to subsidies and there was insufficient political opposition to keep them in check. For the administration, however, 'we can't wait until the Freedom to Farm bill expires in 2002' (Office of the Vice President, 1999).

As it turned out, emergency payments to farmers amounted to an average of $7.5 billion a year between 1996 and 2002. From an average expenditure of $8.8 million in the 1990–1997 period government payments to farmers rose to more than $24 billion in 2001. 'Close to 40% of these direct payments have taken the form of emergency assistance under three supplementary legislative packages enacted since October 1998, in response to pressure from farm interests alarmed by a continuing trend of low commodity prices' (*Agra Europe* 2002, p. A/1). Despite all this additional help, US farm exports between 1996 and 2001 fell by almost 12%, while imports rose by more than 30% (*Financial Times*, 10 May 2002). As the 2002 Farm Bill was drafted, it was evident that talk of a more market-oriented agricultural policy had been swamped by the imperative to defend key US agricultural interests.

The Farm Security and Rural Investment Act of 2002 (FSRI Act)

The FSRI Act, which covers the 2002–2006 period, 'not only reinforces the ad hoc measures introduced in the late 1990s, but also brings back into play all the market distorting measures which marked the farm programmes which preceded the 1996 FAIR Act' (*Agra Europe*, 2002, p. A/1). Neither the Democrats nor the Republicans wanted to provide their opponents with an opportunity to label them as the 'anti-farm' party in advance of the key Congressional elections in November 2002. Ayer and Swinbank (2002, p. A/2) comment:

> [t]he problem is clearly political – special interest groups gained the support of political leaders to change the original rules and did so not in the interest of the broader public, but mainly for the benefit of a few large producers. FAIR thus failed to lock-in reform.

One of the main provisions of the new legislation was the return of the counter-cyclical subsidy abandoned by the FAIR Act – the target price/deficiency-type payment. This is designed to provide additional top-up payments to farmers when commodity prices are low. However, in contrast to earlier programmes, deficiency payments are decoupled from production. They will be based on 85% of a fixed-base acreage and a fixed historical yield. Loan rates and associated subsidies were increased for key commodities. Soybeans, minor oilseeds and groundnuts were added to the list of eligible fixed-payment crops. Subsidies have been brought back for wool and honey producers.

The USA has defended the FSRI Act on the grounds that it was compatible with its WTO obligations. Should there be a chance of breaking through WTO thresholds in any year, so-called 'circuit-breakers' will come into effect. However, doubts have been expressed about whether it will be politically feasible to apply this aspect of the legislation. The USA has also claimed that the greater emphasis on channelling money into agri-environmental schemes is analogous to the growing emphasis on 'second pillar' measures within the EU.

Nevertheless, many analysts see the legislation as a repudiation of bipartisan US advocacy of agricultural trade reform. The USA will continue to press for measures such as the elimination of export subsidies within the Doha Round. However, its moral authority as a champion of liberalized agricultural trade has been severely undermined. This will make it easier for protectionist forces in the EU to defend the status quo and reduce the chances of significant progress on agricultural issues in the Doha Round.

Conclusions

Whatever the outcome of the Doha Round, the process of conducting agricultural trade negotiations is likely to remain highly politicized. Even if the Doha Round led to significant further trade liberalization, for example the phased elimination of export subsidies and the blue box, not a very likely outcome, countries and lobbies would still have much left to defend. The distinctive character of the agricultural policy arena is likely to persist. It is, of course, continuing to change, with the formation of the Agriculture Committee of the WTO providing a negotiating forum informed by a preference for more liberal solutions.

However, the WTO remains one of the weaker global governance agencies despite the way in which its opponents often characterize it. It remains more a 'Water Treading Organization' than a 'World Terror Organization'. Its secretariat can seek to facilitate agreement, but much still depends on bilateral mutual accommodations between the EU and the USA. Their stance in turn is driven to a large extent by their domestic politics. There is a stated intention to make the Doha Round 'a development round', but the underlying asymmetries of power that favour the developed world are unlikely to be easily changed.

The political process that underpins agricultural trade is in many ways a perverse one. It favours the 'haves' – rich countries, big farmers, agribusiness – over the 'have-nots'. It is a process that still prioritizes the interests of producers over consumers. The process has become more transparent, but this is yet to lead to any significant redistribution of power. The forces that make the politics of agriculture distinctive at a national, regional and global level are unlikely to change very quickly.

References

Agra Europe (1996a) Agreement on historic US Farm Bill. *Agra Europe* 1690, N/5–N/8.

Agra Europe (1996b) US Farm Bill opens way to export expansion. *Agra Europe* 1694, P/1–P/3.

Agra Europe (2001) CAP under pressure as new WTO trade round launched. *Agra Europe* 1978, EP/1–EP/2.

Agra Europe (2002) US abandons 'Freedom to Farm'. *Agra Europe* 2001, A/1–A/2.

Ayer, H. and Swinbank, A. (2002) The US Farm Bill: help or harm for CAP and WTO Reform? *Agra Europe* 2004, A/1–A/2.

Balaam, D.N. (1999) Agricultural trade policy. In: Hocking, B. and McGuire, S. (eds) *Trade Politics*. Routledge, London and New York, pp. 52–66.

Browne, W.P. (1995) *Cultivating Congress: Constituents, Issues and Interests in Agricultural Policymaking*. University of Kansas Press, Lawrence, 295 pp.

Coleman, W.D. and Tangermann, S. (1999) The 1992 CAP reform, the Uruguay Round and the Commission: conceptualizing linked policy games. *Journal of Common Market Studies* 37, 385–405.

Destler, I.M. (1992) *American Trade Politics*, 2nd edn. Institute for International Economics, Washington, DC, 453 pp.

Grant. C. (1994) *Delors: Inside the House that Jacques Built*. Nicholas Brearley, London, 305 pp.

Grant, W. (1997) *The Common Agricultural Policy*. Macmillan, Basingstoke, 244 pp.

Grant, W. (2000) *Pressure Groups and British Politics*. Macmillan, Basingstoke, 250 pp.

Hayes-Renshaw, F. and Wallace, H. (1997) *The Council of Ministers*. Macmillan, Basingstoke, 340 pp.

Hillman, J.S. (1994) The US perspective. In: Ingersent, K.A., Rayner, A.J. and Hine, A.C. (eds) *Agriculture in the Uruguay Round*. Macmillan, Basingstoke, pp. 26–54.

Jones, A. and Clark, C. (2001) *The Modalities of European Union Governance: New Institutionalist Explanations of Agri-environmental Policy*. Oxford University Press, Oxford, 233 pp.

Josling, T. (1998) *Agricultural Trade Policy: Completing the Reform*. Institute for International Economics, Washington, DC, 132 pp.

Marsh, D, Richards, D. and Smith, M.J. (2001) *Changing Patterns of Governance in the United Kingdom. Reinventing Whitehall?* Palgrave, Basingstoke, 276 pp.

Nugent, N. (2001) *The European Commission*. Palgrave, Basingstoke, 366 pp.

Office of the Press Secretary (1998) Remarks by the President to National Farmers Union. Room 450, Old Executive Office Building, 15 September. Washington, DC.

Office of the Vice President (1999) Statement by Vice President Al Gore. 22 September. Washington, DC.
Orden, D., Paarlberg, R. and Roe, T. (1999) *Policy Reform in American Agriculture: Analysis and Prognosis*. University of Chicago Press, Chicago and London, 270 pp.
Peterson, J. and Bomberg, R. (1999) *Decision-Making in the European Union*. Macmillan, Basingstoke, 336 pp.
Smith, M. (1999) The European Union. In: Hocking, B. and McGuire, S. (eds) *Trade Politics*. Routledge, London, pp. 275–289.
Webber, D. (1998) High midnight in Brussels: an analysis of the September 1993 Council meeting on the GATT Uruguay Round. *Journal of European Public Policy* 5, 578–594.
Wolfe, R. (1998) *Farm Wars: the Political Economy of Agriculture and the International Trade Regime*. Macmillan, Basingstoke, 235 pp.

4

Farm Subsidies and Agricultural Trade Policy: the Case of US Support Programmes

Daniel A. Sumner*

Introduction

Precursors to the current set of farm-subsidy programmes in the USA date back about seven decades. Substantial and thorough government programmes to subsidize and regulate agricultural commodity industries began as a part of Roosevelt's New Deal. These programmes were initially considered too intrusive and were deemed unconstitutional by the US Supreme Court, but revised programmes with the same economic implications were reinstituted and became law in the middle 1930s (Benedict, 1953; Olmstead and Sumner, 2003).

The USA has periodically renewed and reformulated legislation authorizing domestic farm-subsidy programmes since the 1930s, and most current commodity programmes are legally temporary amendments to the Farm Bill of 1949. The Farm Security and Rural Investment (FSRI) Act of 2002 replaced the Federal Agriculture Improvement and Reform (FAIR) Act of 1996, which covered the years 1996–2002. The FAIR Act replaced the Farm, Agriculture, Conservation and Trade (FACT) Act of

* Director of the University of California Agricultural Issues Center and the Frank H. Buck, Jr Professor in the Department of Agricultural and Resource Economics, University of California, Davis.

This chapter draws on some research reported in the *Australian Journal of Agricultural and Resource Economics* (2003) and presented in Korea, Italy, Australia and California. I thank Hyunok Lee, Jung Sup Choi, Giovanni Anania, Ivan Roberts, John Mullen, Bruce Gardner and Joseph Balagtas for useful suggestions on this research.

1990 (together with the farm spending portions of the Omnibus Budget Reauthorization Act of 1990), which in turn replaced the Food Security Act of 1985. Thus, over the past 50 years, the USA has implemented a series of farm-programme amendments, each scheduled to be revised 4, 5 or 6 years in the future.

US farm programmes are relatively minor economic and political issues in the USA, where the farm contribution to gross domestic product is less than 2%. But in international trade and economic development communities, US farm programmes have gained considerable attention because the USA is a large producer and exporter with the potential to influence international markets and because the USA has played a leadership role in efforts to reform such policies globally, recently through the World Trade Organisation (WTO).

This chapter reviews US farm policy and the role of farm subsidies in trade and trade negotiations of the USA. I focus especially on the current US programmes as they were revised under the FSRI Act of 2002. I shall not provide a comprehensive description or analysis of the new legislation. Instead I shall highlight the features of the Act that are most important for understanding trade policy and trade negotiations. For a useful and readable description of the FSRI Act of 2002, the reader may consult the US Department of Agriculture (USDA, 2002).

Basics of US Farm Programmes

Periodic farm bills deal with more than farm commodity subsidies. The laws contain provisions for long-term land reserves and subsidies for farm environmental improvements. They also authorize export-price subsidies, subsidies for international promotion of farm-based products, food aid, export-credit guarantee programmes and other programmes that affect imports or export directly. Farm bills include provisions for food assistance to the poor, public research and extension support, food safety and aid to rural communities for sewage treatment or electricity, among many other things. However, farm bills do not set tariff rates and do not implement trade agreements.

Core farm programmes in the USA include farm-programme payments provided for a small range of crops – grains, oilseeds (now including groundnuts) and cotton – which produce about 40% of farm cash receipts. In addition, honey and wool producers receive substantial payments relative to the size of these very small industries. A few commodities, notably sugar, groundnuts, dairy produce and frozen concentrated orange juice, have significant trade barriers. Dairy produce is supported by a complex set of marketing regulations that allows price discrimination within the USA, by trade barriers, by a small export subsidy programme and by direct payments. In addition to these specific programmes, government activities in agriculture include crop insurance subsidies, export-credit guarantees, export-promotion support, disaster aids, marketing regulations, occasional ad hoc programmes, protection from exotic pests and diseases and government-sponsored research and extension programmes.

Meats, fruits and tree nuts, vegetables and melons, ornamental crops and hay crops receive almost no programme payments and, including import barriers (with a few exceptions, such as frozen concentrated orange juice), have little support compared with the programme crops and sugar. The average producer support for these commodities, which comprise more than half of US agriculture, is less than 10% of total revenue, and this figure includes broad support such as research and extension. Using the WTO categories, amber-box support for these commodities, including the so-called trade-distorting programmes, is typically well below the WTO *de minimis* level of 5% of total revenue.

To understand current US commodity programmes it is helpful to review briefly some of the recent history (see also Olmstead and Sumner, 2003). After a period of high prices and expanded US production, world commodity prices dipped in the early 1980s and US support prices again exceeded world market prices by a wide margin. The lack of political resolve to lower high support prices in the early 1980s led to growing stockpiles of wheat, feed grains and cotton. The result was one of the largest and most expensive acreage-reduction programmes in US history, idling 20% of US cropland (77 million acres). The Food Security Act of 1985 lowered price supports to reduce the accumulation of stocks and increase American export competitiveness. For cotton and rice, a new 'marketing loan' programme replaced price-support loans so that the government no longer acquired stocks of these commodities, but instead made payments to growers whenever a specified 'adjusted world price' was below the loan rate. The 1985 Act also allocated more than $1 billion per year to direct export bonuses, mainly for wheat. A new long-term Conservation Reserve Program (CRP) paid landlords to remove erodible cropland from production for a 10-year period. In most of the years since 1986, about 36 million acres have been idled under this programme. Total annual outlays for farm programmes peaked at $26 billion in the fiscal year 1986 and direct payments peaked at $17 billion in the fiscal year 1987.

The large federal budget deficit and a push to further liberalize farm policy led to several changes in 1990 legislation. These included fewer acres eligible for deficiency payments, additional planting flexibility, lower price supports and frozen nominal target prices used to determine direct payments. Export subsidies and the CRP were continued with some reforms. The 1990 legislation replaced the price support programme for grains and oilseeds with a 'marketing loan' programme, under which payments were triggered whenever an average local market price was below the local loan rate. Because loan rates were set at between 75 and 85% of the moving average of past prices, the expectation was that this new payment scheme would trigger few payments. In fact, no payments were made under these provisions until 1998.

By 1995 the environment was ripe for further liberalization of farm commodity programmes. Budget pressures continued, additional planting flexibility was popular with some growers and marketing firms and farm prices were high. High prices meant that farmers would receive few payments under the traditional programmes. As a result, the FAIR Act of 1996 replaced payments that had been linked to market prices with new fixed 'contract' payments, which also implied much more planting flexibility (Young and Westcott, 2000).

The FAIR Act did not begin a 'phase-out' of farm subsidy programmes. It was an extension of the policy path of the previous decade, which reinforced and consolidated previous changes. The result was radical because planting requirements, land set-asides, price supports and government stockpiles were now eliminated.

However, when prices fell and remained depressed and federal budget deficits became surpluses, Congress responded with annual ad hoc legislation that raised direct payments by 50% in 1998 and doubled payments for 1999 through 2001. In addition, marketing loan programmes triggered billions of dollars of additional payments. In all, subsidies jumped from about $4.6 billion in fiscal year 1996 to $19.2 billion in fiscal year 1999 and $32.2 billion in fiscal year 2000 (USDA, 2001).

In summary, going into the debate on the 2002 Farm Bill, three farm payment programmes were making record payments to growers of programme crops. These were: (i) direct contract payments not tied to current production or prices; (ii) market-loss assistance payments not tied to current production but motivated by low commodity prices; and (iii) marketing-loan benefits tied directly to current production of a specific commodity and calculated to offset low prices for that commodity.

The WTO Context for Farm Commodity Policy in the United States

The US schedule of import-barrier reductions and related rules was set in the WTO implementing legislation in 1994. No new legislation was needed to implement the domestic-support or export-subsidy provisions of the WTO agreement. The Uruguay Round Agriculture Agreement of the WTO (URAA) has been reviewed in earlier chapters. For further discussion in the context of farm trade negotiations see Sumner and Tangermann (2003).

The main connection between recent farm legislation and the WTO relates to domestic support and the computation of aggregate measure of support (AMS) limits. The AMS indicates the amount of subsidy that is presumed to affect trade significantly and therefore is liable to be reduced according to WTO rules. Since 1995, direct support for US agriculture (in categories reportable to the WTO) has ranged from a low of about $14 billion to perhaps double that or more in 2001. Until 1997 the biggest category was green-box (or blue-box) exempt payments. Non-product-specific exempt support was a minor item during this period. For the period 1995 to 1997, the AMS was in the range of $6 billion and far below the cap of more than $20 billion for that period. The major component of the AMS was computed as the difference between dairy, sugar and groundnut price supports and the base-period border prices of these products. But this 'support' was really an artefact of trade barriers; it provided little, if any, additional benefit for the industry and did little or nothing to affect trade.

The situation changed when farm prices fell in 1998. Two sets of payments responded to low commodity prices. Marketing-loan benefits are product-specific and tied directly to the production and prices of specific commodities. They account for the rise in the AMS. Ad hoc market-loss assistance payments were not tied to the production of any specific crop, but were linked to low prices (by Congressional intent). Therefore, they have been designated non-product-specific amber support in the WTO notifications. Because the USA provides relatively little support for most commodities, even with these payments added, the total non-product-specific support remained less than 5% of the value of total farm production (approximately $10 billion) and so these payments were exempt from the AMS, which is subject to WTO reduction commitments.

Some information from the last official US notification to the WTO is presented in Table 4.1 (adapted from Nelson, 2002). The AMS in 1998 consisted mainly of price support for dairy, sugar and peanuts, plus marketing-loan benefits. Importantly, the dairy support price relative to the fixed border price accounts for about $4.3 billion of the AMS, and this policy provides almost no support in addition to that provided by the dairy trade barriers. The non-product-specific amber-box support, which was outside the AMS, included mainly ad hoc payments and crop-insurance subsidies. Finally, the green-box support included mainly decoupled payments, plus payments for the long-term land idled in the CRP and miscellaneous payments.

Current US Farm-subsidy Programmes

To consider current US farm programmes that affect global markets and WTO commitments and negotiations, we need only consider a few of the myriad topics governed by farm-subsidy legislation.

Table 4.1. US direct support for agricultural producers in 1998 notified to WTO (from Nelson, 2002).

WTO category	US programme	1998 Total ($ billion)	AMS-exempt ($ billion)	1998 AMS ($ billion)
Amber box	Dairy price support	4.33	0	4.33
Product-specific[a]	Loan Deficiency Payment and marketing loans	3.82	0.03	3.79
	Other	2.39	0.08	2.27
Amber box	Payments (0.5 Agricultural Market Transition Act)	2.81	2.81	0
Non-product-specific[a]	Crop insurance	0.75	0.75	0
	Other	1.03	1.03	0
Green-box support	AMTA	5.66	5.66	0
	CRP	1.69	1.69	0
	Disaster	1.41	1.41	0
	Environment and credit	0.35	0.35	0
Total		24.24	13.81	10.39
WTO ceiling		na	na	20.70

[a]Subsidies not in AMS ceiling if less than 5% of applicable revenue.

Conservation programmes

The CRP has idled about 36.4 million acres of cropland on 10-year contracts since the late 1980s. The FSRI Act of 2002 raises this limit to about 39 million acres. The total idled acres are equivalent to almost one-quarter of all grain acreage in the USA. The CRP tends to draw in the less productive acreage, and some cropland idled has not been planted to programme crops recently. This programme has been reported in the WTO green box because it lowers production and has environmental benefits.

The Environmental Quality Incentive Programme (EQIP) provides subsidies for farmers to implement environmentally friendly practices. The budget for EQIP increased by a factor of five to almost $6 billion over 6 years in the 2002 Act. This programme provides matching funds for farmers, especially livestock producers, to comply with environmental requirements or otherwise improve their farming operations. Thus the EQIP lowers costs for growers who would otherwise have paid for improvements with their own funds. The programme has been reported in the WTO green box because of its environmental benefits and because the production enhancing impact is expected to be small.

A new Conservation Security Program (CSP) provides annual payments for farms that use environmentally approved practices in their production operations. Because many farms already apply a number of environmentally friendly practices, this programme can be viewed as simply a direct subsidy to farmers on a per-unit basis up to a relatively small payment limit of $40,000 per farm. It is not clear how the CSP will be reported to the WTO. The total budget is relatively small (about $200 million per year) and all commodities are eligible, including livestock operations. Thus even if the programme is deemed to be in the WTO amber box, it would probably qualify as non-product-specific.

A host of other smaller conservation and environmental programmes are either renewed, amended or added to the FSRI Act of 2002, but none is large enough to have significant national implications for production.

Trade programmes and provisions

Direct trade provisions are a relatively minor part of domestic commodity policy in the USA, and these change little in the FSRI Act of 2002. The Export Enhancement Program (EEP) is authorized at the WTO maximum limits, but this authority has not been used in several years and no one expects the programme to be used in a significant way in the near future. The Dairy Export Incentive Program (DEIP) is authorized at the WTO limits. This programme is relatively small and, although it allows the export of significant amounts of dried milk selected countries, these exports have small impacts on the US domestic market and little global effect.

International food aid and export-credit subsidy programmes have not been considered export subsidies in the WTO even though they facilitate exports of commodities. The understanding on food aid is that they are exempt from WTO limits, so long as these programmes focus on humanitarian goals and do not displace commercial exports. Of course, the criteria are complex in practice and so some food aid remains controversial.

Export-credit programmes are much more problematic. For many years the US government has guaranteed repayment to private lenders who finance the export of agricultural products to selected countries. These guarantees allow buyers to acquire credit for up to 7 years at relatively low interest rates and allow US exporters to make sales that would otherwise be much more difficult. The subsidy implicit in these guarantees can be measured by the difference in interest-rate charges with and without the guarantees. Defaults are rare but have occurred in some high-profile cases, such as by Iraq in the period of the Gulf War. There is no question that credit guarantees facilitate exports, but the status of credit programmes as export subsidies is as yet unsettled in the WTO (for more specifics see WTO Secretariat, 2000, specifically para. 44).

The Market Access Program (MAP), which is heavily used by the otherwise minimally subsidized commodities, provides matching funds for industry promotions overseas. The FSRI Act of 2002 gradually doubles MAP funding to $200 million per year. The size of the programme remains small – about 0.1% of the export value of the products supported. It is not clear how effective these export programmes are, and they have not been treated as export subsidies in the WTO.

A controversial new provision of US law requires country-of-origin labelling for meat and fresh produce sold in the USA. Known by the acronym COOL, this provision required firms to certify the origin of products sold in the food-chain and to have records available for audit by federal authorities. For meat, the law requires labelling as imported meat from livestock that was born or has spent part of its life out of the USA. This programme was designed to disadvantage imports, but it is unlikely to be a major factor in discouraging trade in general and is unlikely to raise WTO issues. The importance of the import-labelling law is mainly symbolic and is a disquieting indication of the general protectionist tenor that has recently coloured much farm-policy discussion in the USA.

Finally, anticipating potential WTO problems with the new commodity subsidies, the FSRI Act of 2002 stipulates that, if the Secretary of Agriculture determines that the AMS ceiling will be exceeded in any year, the Secretary shall, to the maximum extent practicable, adjust expenditures to avoid exceeding allowable levels. This requirement is extremely difficult to implement, especially because expenditures are not known until after a crop year has concluded, when it is too late to make adjustments. Further, as discussed below, even the categories used for reporting various programmes may not be determined until long after payments have been made. This provision seems useful only to place political pressure on the Secretary and President and to claim in the WTO that with this provision the FSRI Act does not threaten WTO violations.

Crop payment programmes

Because of the complexity of commodity programmes, this section can only focus on the main provisions for major commodities, with emphasis on payments for major programme crops and dairy products. For further details, the reader may consult Becker (2002) and USDA (2002).

Under the marketing-loan programme, 'loan rates' are now used solely to determine the marketing-loan benefit rates when the market price falls below the loan rate. The programme was introduced in this form in the 1990s (in 1985 for rice and cotton), and the FSRI Act of 2002 only adjusted the loan rates. Table 4.2 shows that loan rates were raised for maize and wheat and lowered for soybeans, with no change for rice and cotton. Marketing-loan programmes are also available for other feed grains, extra-long-staple cotton, other oilseeds, groundnuts, wool, mohair, honey and field peas and lentils. Payments are applied to current production on each farm; they are clearly in the product-specific amber category and are a major component of the AMS reported by the USA to

Table 4.2. National average loan rates, direct payments and counter-cyclical (cc) target prices (from Westcott *et al.*, 2002).

Crop	FAIR Act Loan rate	FAIR Act Direct payment	FSRIA Loan rate	FSRIA Direct payment	FSRIA CC target price
Wheat, $/bu.	2.58	0.53	2.75[a]	0.52	3.92[a]
Maize, $/bu.	1.89	0.30	1.95[a]	0.28	2.63[a]
Soybeans, $/bu.	5.26	not applicable	5.00	0.44	5.80
Cotton, $/lb.	0.5192	0.0667	0.52	0.0667	0.724
Rice, $/cwt.	6.50	2.35	6.50	2.35	10.50

[a]In 2002 and 2003 the wheat loan rate is $2.80 and the target price is $3.86. The maize loan rate is $1.98 and the target price is $2.60.
bu., bushel; lb., pound; cwt., hundredweight.

the WTO. In recent years these payments have averaged about $10 billion annually.

Table 4.2 also shows direct payment rates under the FSRI Act of 2002 compared with the FAIR Act payment rates that applied in 2001. The new direct payment rates are roughly equal to the payment rates that applied in 2001. These payments were listed in the WTO green box under the FAIR Act because they are not tied directly to current production of any crop and, although the payments are tied to historical production of a specific programme crop, farmers may plant alternative crops or leave the land idle without influencing their payments. The outlays under this programme are set at about $3.8 billion annually for 6 years.

Under the new law, historical area bases may, at the farmer's option, be updated to the 1998–2001 period. Because soybeans are now eligible for these payments, some base updating would have been required to adjust for new assignments of base areas to soybeans that had been planted on the base area of other programme crops. Some restrictions continue to apply to land receiving direct payments: the payment land may not be shifted out of agriculture altogether and the land may not be used for fruits, tree nuts, wild rice or vegetables and melons. These restrictions are of little importance for most US programme cropland, but they do matter for perhaps 5% of the relevant area. For example, in California, vegetable crops, fruits and tree nuts compete with programme crops, in the north-west potatoes and wheat compete for land and in pockets in the Midwest some vegetables are grown.

The third payment programme in current legislation was designed to replace the ad hoc market-loss assistance payments that were made from 1998 to 2001. The programme parameters used in the new Counter-cyclical Program (CCP) were calibrated so that projected annual payments were approximately equal to the magnitude of ad hoc payments made in 2001 (about $5 billion). As with ad hoc payments and the direct payment programme, the CCP ties payments to historical bases rather than current production. However, in this case, if the producer elects to update the acreage base on a farm, that farm may also update the yield base to 93.5% of the average of the 1998–2001 yields. Determining the size of the counter-cyclical payment is complex. For each unit of production base, the payment is equal to the difference between 0.85 times the legislated target price, minus the larger of the national average loan rate or the national average market price and the direct payment rate. Of course, payments may not be negative. The counter-cyclical payment is designed to supplement the direct payment when the average price for the commodity is lower than the target price, but by including the loan rate in the formula, the CCP payments rate has a maximum equal to [(0.85)(target price – loan rate)] – direct payment rate.

The CCP raises two main concerns in respect of WTO domestic-support considerations. First, payment rates are tied to the market price of a specific crop, and this may in itself imply that these payments are in the amber-box category under the URAA rules. The rationale is that, for those growers considering planting the payment crop on the base land, the counter-cyclical payment provides a revenue offset when prices are low, just as the loan-rate programme does. Secondly, while the payments are not tied directly to the current production of any specific crop, updating the base in 2002 may cause growers to anticipate that bases may be updated again. This may cause growers to plant more of the payment crop than otherwise. The ad hoc market-loss assistance payments were reported by the USA as non-product-specific amber-box support. One might anticipate a similar notification for the CCP. However, tying the payments to individual market prices and updating the bases has raised concerns that these payments are more appropriately product-specific support under the URAA rules.

Dairy policy

The main elements of government support for the US dairy industry include: (i) small import tariff-rate quotas with prohibitive over-quota tariffs, which limit imports to a small share of domestic consumption; (ii) a price-discrimination scheme that raises the price and regulates the geographical movement of milk used for beverage consumption; (iii) a price-support programme under which the USDA purchases manufactured dairy products when their prices fall to a specified minimum; (iv) a small export subsidy that is applied mainly to dried milk; and (v) a new dairy-farm payment scheme introduced in the FSRI Act of 2002 (Sumner and Balagtas, 2002). The new Milk Income Loss Contract (MILC) programme will probably distribute more than $1 billion annually to dairy farms. Payments are likely to be equivalent to between 5 and 10% of total milk revenue for fully eligible dairy farms. The payments offset low milk prices; however, the programme stipulates that no payments may be made for production in excess of an annual limit of 2.4 million pounds of milk per farm. Under current milk-price projections, the average payment is likely to be about $1.00 per hundredweight or about $24,000 for a farm producing 2.4 million pounds or more.

US dairy farms range broadly in size, but most farms produce less than 2.4 million pounds, and most of the milk comes from farms that exceed this limit. Furthermore, most of the smaller farms are in the eastern half of the nation, with the larger farms in the west (Sumner and Wolf, 2002). This means that large farms will actually lose revenue, as lower prices caused by increased production on small farms more than offset the payments to large farms.

Effects of US Farm Subsidies on Global Markets

It is beyond the scope of this chapter to provide comprehensive empirical investigation of the production impacts of farm subsidies. Indeed, impacts certainly differ by commodity and specific market; measuring such impacts requires large econometric and simulation efforts for each commodity and then linkage across commodities. No such comprehensive analysis has yet been completed for current US programmes, and recent changes mean that earlier analysis is out of date.

Westcott *et al.* (2002) have provided some estimates of the production impacts of the FSRI Act of 2002 compared with previous US law for production and prices of major programme crops. Their results suggest relatively small incremental impacts of the new law. Thus, some of the public outcry about the new programme was probably exaggerated. For example, French President Jacques Chirac claimed in the wake of the signing of the 2002 Act that 'massive increases to [US] farm subsidies would hurt poor countries hardest – including those in Latin America'. Reports collected by Wiesmeyer (2002) indicate that Chirac's comment was representative of a much wider misunderstanding about the 2002 Act relative to previous law.

Westcott *et al.* (2002) have not looked directly at world market impacts and have not assessed the effects of current programmes compared with no subsidy or some other hypothetical policy. Here, I shall discuss the major impacts without assigning quantitative weight to each. Projecting the supply impact of commodity programmes is exceedingly complex given the multi-commodity nature of policies and farming enterprises, the effects of programmes on estimated supply response parameters and the effects on expectations about future programme incentives (McDonald and Sumner, 2003).

The first impact is straightforward. The marketing-loan programme raises the expected price for producers of programme crops and causes an increase in acreage planted to those crops. The magnitude of this effect depends on projected market prices. For example, many observers expect market prices to be well above loan rates in the future, as they were for maize, wheat and soybeans in the autumn of 2002. But even if the expected price is above the loan rate, by guaranteeing farmers a minimum return per unit produced, the loan programme increases US production, encourages exports and reduces potential imports.

Secondly, some economists have argued that any government payment tied to farmland increases production, even if payments are not tied to current production (Roberts and Jotzo, 2001). This means that the so-called decoupled payments also increase supply. The argument is that by shifting income to holders of a programme base, the interest rate for farm investments is lower than otherwise and commodity supply is therefore larger. Two points suggest that this impact may be small. US farmers have relatively little debt compared with their equity, and the share of wealth tied to farm payments is small. Further, recent data show that about 20% of the land value of programme crops is attributable to farm payments, but 62% of that land is owned by non-operator landlords rather than farmers (Barnard *et al.*, 2001). Young and Westcott (2000) discuss these issues in the context of FAIR Act payments and find that the production effects of direct payments with loose ties to current production are small.

Updating payment bases from the 1981–1985 base period used in previous laws adds a new production impact to farm payment programmes. Voluntary updating will cause more of the payments to flow to those producers who have grown more of the programme crop recently. However, more important for market effects and WTO considerations, updating in 2002 naturally causes growers to revise estimates of the probability of future updating. This means that, in considering what to plant, the effect on the present value of future payments becomes a larger consideration. The empirical importance of this supply response has not been investigated in practice. If growers expect current planted area and yield to have a major effect on the base used for future payments, then they will plant substantially more of the programme crop now to build a programme base for the future. Westcott *et al.* (2002) consider this effect to be a topic for future investigations. Sumner (2003) explores this impact and suggests that, with updating, counter-cyclical payments may have between one-quarter and one-third the production effects of payments tied directly to current output.

Conservation provisions also affect production and prices. The CRP will now idle about 39 million acres of cropland. Much of this land has been removed from wheat and feed grains. In total this programme may have removed more than 10% of wheat land from production and somewhat smaller shares of other grains. It has little effect on cotton or rice acreage. As noted above, the EQIP lowers the cost of production for some farms and may thus stimulate added production on some crop and livestock farms. However, these impacts are probably small and the programme has been included in WTO green-box subsidies in the past. The new CSP provides payments for farms based on their production of crops or livestock. This additional revenue will increase crop supply. The amount of funding is small, amounting to about 0.1% of farm revenue; thus, any supply effects that exist in theory are almost certainly tiny in practice.

Finally, the new dairy programme will stimulate production on those farms that produce at or below the 2.4-million-pound payment cap. By raising the marginal returns to milk production by about 10% on these farms, milk production nationwide will rise by about 4% and the price will decline by about 4% relative to the situation without these new provisions. However, global impacts are small because US prices will remain above world market levels, and import barriers mean that US markets will remain isolated.

Implications of Domestic Subsidies for WTO Compliance

I have argued elsewhere that domestic subsidies can divert attention in trade negotiations from the more important issues of tariffs and export subsidies. Countervailing duty actions in domestic law or serious prejudice actions in the WTO may be more effective remedies for trade-distorting domestic subsidies (Sumner, 2000; Sumner and Tangermann, 2002). None the less, the URAA includes complex rules on domestic support, and the main proposals for the Doha Round negotiations include similar disciplines. These raise implications for WTO compliance that are important in evaluating US policy.

Under the URAA, the first important consideration is the colour of the domestic-support 'box' in which the various subsidy payments are classified. Clearly marketing loans remain in the amber box and are included in the AMS. Depending on future market prices, these payments may fill up about half of the $19.1 billion limit that currently applies to the USA. The other major contributors to the AMS have been dairy product, sugar and groundnut price supports, which account for about one-quarter of the AMS limit. The new MILC payment for dairy producers (about $1 billion per year) will almost certainly be added to the AMS.

The next consideration is the direct-payments programme for grains and cotton. Updating area bases for direct payments has raised concerns that these payments may belong in the WTO amber box. The arguments for keeping these in the green box are that some update was required to implement the 2002 Act because soybeans were added to the programme and that the updates distort production only minimally because farmers may still leave the land idle or plant a large variety of crops. If these payments (worth about $4 billion per year) were considered amber-box, they would be declared non-product-specific and thus be 'charged' against the *de minimis* limit, which is about $10 billion. This would represent a large share of the *de minimis* limit.

The new counter-cyclical payments (which replace market-loss assistance payments) raise even more concerns about treatment under the URAA. These counter-cyclical payments allow updating of both the base area and base yield and are calculated in relation to the current market price of a specific commodity. As noted above, the USA reported market-loss assistance payments as non-product-specific amber-box support, and so it would be difficult to reverse course and declare the counter-cyclical payments to be in the green box. Counter-cyclical payments, like direct payments, are likely to be in the range of $4 billion per year, although they vary from year to year inversely with market prices. If both direct and counter-cyclical payments are included in non-product-specific amber support, the *de minimis* limit would probably be exceeded in some years (given crop insurance and a number of minor subsidies in that class). In that case, all non-product-specific support would be added to the AMS, and the AMS amber-box limit would be violated. However, because the CCP ties payments to a specific crop price and, with base updating, to lagged output of a specific crop, many analysts would suggest that these payments belong in the product-specific category. If the counter-cyclical payments were placed in the product-specific category, they would be added directly to the AMS, as are the marketing-loan benefits, and charged against the $19.1 billion limit.

The USA may have a delicate balancing act when reporting its various programmes in the product-specific and non-product-specific categories to minimize the chance of exceeding the $19.1 billion AMS limit. However, there seem to be WTO-legal ways of reporting US programmes that would reduce the prospect of exceeding the limit. When the total support for any specific commodity is less than 5% of the total revenue for that commodity, that support is not counted in the AMS. Some programmes that the USA has included in the non-product-specific category may just as easily be declared product-specific. This distinction mattered little when the non-product-specific category was well below the *de minimis* limit of 5% of total US farm revenue. But, as that category nears the limit, the USA may find it useful to reallocate support. Payments or benefits, such as a part of crop-insurance subsidy or grazing fees, may fit into the product-specific category. If used for commodities with less than 5% amber support, this reallocation would take those subsidies out of consideration and leave more room for the direct payments or perhaps the counter-cyclical payments to fit within the non-product-specific *de minimis* limit.

One way to reduce the AMS itself would be to modify the price-support policies for sugar and dairy products. For these commodities the real support comes from the import barriers. However, because the AMS is calculated as a difference between the support price and the fixed world reference price, the AMS for dairy products and sugar, in particular, is very large relative to the benefit received by producers. It would be economically simple (although perhaps politically difficult) to compensate affected producers with relatively small payments and cut the US AMS by about $4 billion.

Concluding Remarks: Implications for Current Trade Negotiations

Domestic-support programmes have several implications for current trade negotiations. In its July 2002 WTO proposal, the USA continued its call for elimination of export subsidies and reinforced its call for rapid tariff cuts. That proposal also called for cuts in trade-distorting domestic support, which would imply significant reductions for US programmes. However, given the pressure from Congress, the USA will have a limited opportunity to lower domestic support in exchange for additional market opening or lower export subsidies during the give and take of negotiations.

Furthermore, many other countries, multilateral organizations and non-governmental organizations see US domestic support as a major source of distortion in world markets. With the focus on domestic support, policies such as US, Canadian and EU dairy trade barriers or Japanese and Korean tariffs are likely to face less pressure for reduction. This is an unfortunate outcome, given that progress on reducing domestic support through trade negotiations is likely to be limited and that the most significant distortions to trade are found in border measures.

References

Barnard, C., Nehring, R., Ryan, J. and Collender, R. (2001) Higher cropland value from farm programme payments: who gains? *Agricultural Outlook* November, 26–30.

Becker, G. (2002) *Farm Support Programmes and World Trade Commitments.* US Congressional Research Service Report, updated 29 April, Washington, DC.

Benedict, M.R. (1953) *Farm Policies of the United States, 1790–1950: A Study of Their Origins and Development.* The Twentieth Century Fund, New York, 548 pp.

McDonald, J.D. and Sumner, D.A. (2003) The influence of commodity program on acreage response to market price: with an illustration concerning rice policy in the United States. *American Journal of Agricultural Economics* (in press).

Nelson, F.J. (2002) Aligning US farm policy with world trade commitments. *Agricultural Outlook* January–February, 12–16.

Olmstead, A. and Sumner, D.A. (2003) Farm policy. In: Carter, S., Gartner, S., Haines, M., Olmstead, A., Sutch, R. and Wright, G. (eds) *Historical Statistics of the United States – Millennial Edition*. Cambridge University Press, Cambridge (in press).

Roberts, I. and Jotzo, F. (2001) *2002 US Farm Bill: Support and Agricultural Trade*. Research Report 1.13, ABARE, Canberra, 98 pp.

Sumner, D.A. (2000) Domestic support and the WTO negotiations. *Australian Journal of Agricultural and Resource Economics* 44(3), 457–474.

Sumner, D.A. (2003) Implications of the US Farm Bill of 2002 for agricultural trade and trade negotiations. *Australian Journal of Agricultural and Resource Economics* 47(1), 117–140.

Sumner, D.A. and Balagtas, J.V. (2002) United States' agricultural system. In: *Encyclopedia of Dairy Sciences*. Elsevier Science, San Diego, pp. 1–6.

Sumner, D.A. and Tangermann, S. (2002) International Trade Policy and Negotiations. In: Gardner, B. and Rausser, G. (eds) *Handbook of Agricultural Economics*. North Holland Press, Amsterdam, pp. 1999–2055.

Sumner, D.A. and Wolf, C.A. (2002) Diversification, vertical integration and the regional pattern of dairy farm size. *Review of Agricultural Economics* 24(2) (Autumn/Winter), 442–457.

USDA (2002) Farm Bill 2002 website. http://www.usda.gov/farmbill. Accessed December 15.

Westcott, P.C., Young, C.E. and Price, J.M. (2002) *The 2002 Farm Act: Provisions and Implications for Commodity Markets*. Agriculture Information Bulletin AIB778, ERS, Washington, DC, 67 pp.

Wiesmeyer, J. (2002) Farm Bill opponents *still* talking and squawking. http://www.agweb.com/analysisdetail.asp?ltype=insidewashingtontoday (29 April).

WTO Secretariat (2000) *Export Credits and Related Facilities*. G/AG/NG/S/13 (26 June), Geneva, para. 44.

Young, C.E. and Westcott, P. (2000) How decoupled is US agricultural support for major crops? *American Journal of Agricultural Economics* 82(3), 762–767.

Multifunctionality and Non-trade Concerns

5

Margaret Rosso Grossman*

Multifunctionality refers to the ability of agriculture to provide goods and services valued by society, in addition to the production of marketable food and fibre. Multifunctionality, described in one recent working definition, has two key elements:

> *i)* the existence of multiple commodity and non-commodity [food and non-food] outputs that are jointly produced by agriculture; and *ii)* the fact that some of the non-commodity outputs exhibit the characteristics of externalities or public goods, with the result that markets for these goods do not exist or function poorly.
>
> (OECD, 2001, p. 13)

Multifunctionality, moreover, bears a close relationship to 'non-trade concerns' (NTCs). In the context of World Trade Organisation (WTO) negotiations, NTCs are interests that are central to the domestic policy of WTO Members. The most important NTCs linked with agriculture are protection of the environment, rural development and food security, but Members have identified additional NTCs. All countries, of course, have NTCs, and many have policies to foster the positive (and minimize the negative) externalities associated with agriculture. Countries disagree, however, about the 'jointness' of non-commodity outputs from agriculture – that is, whether agriculture necessarily produces non-commodity outputs as joint products (Normile and Bohman, 2002).

* This material is based on work supported by the Cooperative State Research, Education and Extension Service, US Department of Agriculture, under Project No. ILLU-05-309.

Further, though some non-commodity outputs may be public goods, countries also disagree about the importance of government intervention to ensure that agricultural producers (or others) provide those public goods.

The multifunctionality of agriculture and the related NTCs are significant issues in the WTO agricultural negotiations that began in 2000 and will be concluded by 2005. Early in the negotiations, Members established positions on these issues. To ensure that farmers will continue to provide non-commodity goods and services connected with agricultural production, some WTO Members assert that support (even higher tariffs or support that affects production and trade) is justified. These 'friends of multifunctionality' include the European Union (EU), where multifunctionality is at the core of the European model of agriculture, as well as Norway, Japan and others. Other WTO Members, particularly the USA and members of the Cairns Group,[1] assert that support for farmers should be decoupled from production. These countries are reluctant to use multifunctionality and the related NTCs (for example, environmental protection and rural development) to justify domestic support. They are not convinced that non-commodity benefits are joint products of agricultural production (Normile and Bohman, 2002). Instead, these nations seem to view the concept of multifunctionality as a 'fig-leaf' for domestic support (A. Burrell, Wageningen, 2001, personal communication).

The debate about multifunctionality helped to prevent agreement on a framework for continued agricultural talks in Seattle in late 1999. Some countries demanded explicit recognition of multifunctionality in the negotiations; others opposed recognition (Pruzin, 2000b).[2] For the negotiations beginning in 2000, the EU demanded that Article 20 of the Uruguay Round Agreement on Agriculture (URAA) (WTO, 1994) be the basis for further agricultural negotiations (Pruzin, 2000a). Article 20 states specifically that NTCs are to be part of the new round of agricultural negotiations. The Doha Declaration (WTO, 2001g, Art. 13), which guides negotiations that will be completed by 2005, confirms that NTCs will be taken into account. Thus NTCs and the related concept of multifunctionality continue to play important roles in WTO negotiations.

This chapter first discusses multifunctionality and the various NTCs in agriculture. The chapter then outlines WTO negotiating positions related to multifunctionality of the USA, the Cairns Group, the European Communities (EC), Norway and Japan. Finally, the chapter considers the joint-products issue and appropriate policy considerations for resolution of the debate over multifunctionality, particularly in the context of domestic support.

Multifunctionality and Non-trade Concerns in Agriculture

Multifunctionality

As WTO Members participate in renewed agricultural negotiations, multifunctionality is a contentious focus of discussion and debate. The 'friends of multifunctionality' argue strongly in favour of WTO rules that allow domestic support for the multifunctional products of agriculture. The USA and countries in the Cairns Group respect the idea of a multifunctional agriculture, but call for disciplined domestic support that does not burden other countries by affecting production and trade.

The concept of multifunctionality is not new, nor is the term. US Department of Agriculture (USDA) researchers noted that the term multifunctionality has been used since the early 1990s in connection with agricultural policy but only more recently became part of trade policy discussions.

> The basic idea is that agriculture is more than just producing and selling commodities; it also produces many intended and unintended by-products. Some by-products are 'good', such as rural employment creation; some are 'bad', such as erosion and pollution; and some are 'intangible', such as the spiritual or symbolic value of preserving our farming heritage.
>
> (Bohman *et al.*, 1999, p. 5)

Multifunctionality thus encompasses externalities, both positive and negative, from agricultural production. Expanding on the basic idea that agriculture produces services beyond food and fibre, many definitions and explanations of multifunctionality exist; most focus on positive externalities of agricultural production.

The Organization for Economic Cooperation and Development (OECD), which has played a significant role in facilitating the study and discussion of multifunctionality in agriculture, stated:

> Multifunctionality refers to the multiple goods and services provided by agriculture and the contribution that these goods and services make to the achievement of domestic non-food objectives. The distinguishing feature is that the multiple outputs are generated by one and the same activity . . . [T]he term multifunctionality favours a perspective that recognises the integrated nature of all outputs.
>
> (OECD, 1998b, pp. 5–6)

An OECD study developed a working definition of multifunctionality, quoted above. That study identified two (often interrelated) approaches to the analysis of multifunctionality. One, the 'positive' approach favoured by OECD in its analysis, interprets multifunctionality as 'a characteristic of an economic activity' like agriculture, and it focuses on 'multiple, interconnected outputs or effects'. The other, the 'normative' approach, which often appears in trade documents, focuses on the 'multiple roles assigned to agriculture' as part of its function in society. Under this approach, multifunctionality is an objective, rather than a characteristic, and has 'a value in itself' (OECD, 2001, p. 14).

The European Commission Directorate-General for Agriculture (DG Agri) provided an explanation of multifunctionality in the context of an Info-Paper, also submitted to the WTO.

> Agriculture is multifunctional because it is not limited to the sole function of producing food and fibers but it also has a number of other functions . . . Agriculture provides services which are linked to the land and are mainly of a public good character . . . Apart from its production function, agriculture encompasses other functions such as the preservation, the management and enhancement of the rural landscape, the protection of the environment, including against natural hazards, and a contribution to the viability of the rural areas. Agriculture must also be able to respond to consumer concerns for example those regarding food quality and safety.
>
> (CEC, DG Agri, 1999c, p. 1)

Multifunctionality, which implies recognition and encouragement of services that farmers provide, takes pride of place among the EU principles of rural development (CEC, DG Agri, 1999a, p. 1).

The tenor of comments about multifunctionality often reflects a value judgement. The EU statement just quoted favours multifunctionality. In contrast, economists from Australia assert, in a rather negative tone, that:

> [t]o those who wish to use 'multifunctionality' to justify agricultural protection, the term refers to any unpriced spillover benefits that are additional to the provision of food and fibre. Claimed benefits range from environmental values, rural amenities, cultural values, rural employment and rural development. In some countries food security is also emphasised.

Further, '[i]n a policy context, multifunctionality has become associated with a view that providing support to agriculture is an appropriate mechanism to enhance these spillover benefits' (ABARE, 1999, p. 1). Another statement from the Cairns Group reflects the same negative judgement: 'Discussions of multifunctionality in the international context have made one fact clear: multifunctionality is being used by some developed countries to justify their high levels of protection and other trade-distorting policies' (Hill, 2000).

Besides clear differences in attitude, practical realities help to explain the different descriptions of multifunctionality. Supply and demand for non-food services from agriculture are relevant. Because types of agriculture, as well as economic and geographical conditions, differ among nations, the non-food services supplied by agriculture also differ. Citizens in different countries expect different non-food services, and the demand for such services is subject to change (OECD, 1998b, pp. 6, 10). Different countries (or even areas within a single country) value the possible services, or output bundles, differently, and non-food services that are threatened may become more valuable. Further, in some countries, markets for non-food services have been developed, making government intervention unnecessary. Systems of property rights are relevant if, for example, access to services (e.g. landscape) on privately owned land can be restricted (OECD, 1998b, p. 6).

As a result of different value judgements, supply and demand, countries include a variety of possible non-food services within their definitions of multifunctionality. According to comments collected from OECD Member countries, these may include 'biodiversity, landscape, soil erosion, water conservation, flood prevention, rural employment, food security, cultural heritage, village development, social cohesion, urban congestion, agricultural competitiveness, income distribution and the trade balance' (OECD, 1999b, p. 4). Such an inclusive list suggests that the parameters of multifunctionality may need to be framed carefully in the context of WTO negotiations.

OECD role

In recent years, the OECD has facilitated international discussion of multifunctionality and contributed to the analytical debate. OECD Agricultural Ministers introduced the concept of multifunctionality at their March 1998 meeting. Their Ministerial Communiqué noted that:

> [b]eyond its primary function of supplying food and fibre, agricultural activity can also shape the landscape, provide environmental benefits such as land conservation, the sustainable management of renewable natural resources and the preservation of biodiversity, and contribute to the socio-economic viability of many rural areas.
>
> (OECD, 1998a, ¶ 10)

Papers submitted for the October 1998 Workshop on Emerging Trade Issues in Agriculture focused in part on aspects of multifunctionality and the related NTCs (e.g. Anderson, 1998a; Runge, 1998).

To facilitate further analysis, in November 1998, the OECD Secretariat published a discussion document, *Multifunctionality: a Framework for Policy Analysis*, and invited Member country comments (OECD, 1998b). One OECD Member, Australia, noted that this *Framework* paper was the first time a 'great international organisation' dealt with the subject of multifunctionality (OECD, 1999a, p. 8). The OECD recognized that policy reform and trade liberalization are important reasons for international interest in multifunctionality, but not the only reasons (OECD, 1999b, p. 7). Two developments have fuelled the current debate on multifunctionality: a growing demand for some non-food services, which seems to require government intervention, and an agricultural policy environment that pledges to focus on market signals, but may also affect non-food objectives. These policy developments challenge governments to 'make agriculture more market oriented and to internalise its external benefits and costs' (OECD, 1998b, p. 17).

In 1999, the OECD published written comments on the *Framework* paper submitted by Member countries (OECD, 1999a) and a summary document, *Multifunctionality: Status Report and Proposal for Further Work* (OECD, 1999b). The *Status Report* discussed the various Member comments and proposed further work in several areas: production relationships, externality and public good aspects of non-food outputs, measurement and valuation of non-food outputs and policy approaches to multifunctionality (OECD, 1999b, pp. 8–11).

The OECD continued to work towards a balanced view of multifunctionality, adding a reasoned voice to the debate. In June 2000, OECD ministers recognized:

> the multifunctional characteristics of agriculture, and the need to ensure that policies should be targeted, transparent and cost effective, maximise benefits, and avoid distorting production and trade. Food safety, food security, viability of rural areas and protection of the environment . . . are common concerns.
>
> (OECD, 2000, ¶ 23)

This comment illustrates the close analytical connection between multifunctionality and NTCs, also reflected in WTO negotiations.

In 2001, OECD published an extensive analysis, *Multifunctionality: Towards an Analytical Framework*, which focuses on the production relationships underlying multifunctionality, as well as externality and public good issues; its annexes provide further information about jointness in production of agricultural and non-agricultural outputs (OECD, 2001). In July 2001, OECD also sponsored a multifunctionality workshop to discuss analytical issues raised in the 2001 report. Numerous national and synthesis papers from the workshop provide information and technical analyses (e.g. Abler, 2001; Burrell, 2001).

Non-trade concerns

Multifunctionality is often discussed in conjunction with NTCs connected with agriculture, and the concepts involve some of the same considerations (OECD, 2001, p. 15). Indeed, one commentator asserted that the term NTCs had 'metamorphosed' into multifunctionality (Green, 2000, p. 831). When the Uruguay Round of trade negotiations was launched in 1986 by the Punta del Este Declaration, agricultural negotiations were to address 'restrictions and distortions' in world agricultural trade. No mention of NTCs occurred. Thus, URAA Article 20, which reflects WTO Members' agreement that agricultural negotiations beginning in 2000 would take into account NTCs, was an important departure from Punta del Este objectives (Mauritius, 2000, ¶¶ 7, 8). The URAA Preamble identifies two important NTCs: trade commitments should have regard for 'non-trade concerns, including food security and the need to protect the environment'. Though the Preamble mentions only these two NTCs, Members have identified rural development (or employment) as another important NTC. Some Members insist that animal welfare and food quality are also NTCs (e.g. WTO, 2000e,f). The URAA neither defines NTCs precisely nor prescribes appropriate policies for responding to the concerns. The Doha Declaration (WTO, 2001g, Art. 13) takes note of the NTCs raised in Member negotiating proposals and confirms that NTCs will be considered, as provided by the URAA. Thus, the Declaration provides no further definition or policy guidance.

WTO Members have linked multifunctionality with NTCs. For example, the Cairns Group noted that some Members use the term multifunctionality to 'describe a range of agricultural non-trade concerns' (Cairns Group, 1999). The Norwegian Minister of Agriculture, who believed that NTCs would be 'main issues' in the WTO agricultural trade negotiations, connects the two concepts: 'NTCs are embodied into the concept of multifunctionality. Agriculture may be defined as multifunctional when it has one or several roles or functions in addition to its primary role of producing food and fibre.' These functions, the Minister asserted, 'roughly correspond' to the NTCs mentioned in the URAA, that is, food security and environmental protection (Lindland, 1998, p. 1). Elsewhere Norway asserts:

> The concept of multifunctionality therefore greatly overlaps with the NTCs referred to in the Agreement on Agriculture. While the concept of multifunctionality, in our view, is preferable from an analytical point of view, for most practical purposes the concepts of multifunctionality and NTCs are to a large extent identical.
>
> (WTO, 1999m, p. 1)

Like multifunctionality, the importance of NTCs varies for different WTO Members (Lindland, 1998, p. 3).

Not all WTO Members favour special treatment for agricultural NTCs. Australia, representing the Cairns Group, asserts that:

> non-trade concerns are . . . not limited to the agricultural sector. Other sectors also have implications for a range of other societal concerns not directly related to production in those sectors. There is therefore no reason to claim, as do many rich developed countries with protectionist agricultural policies, that the special multifunctional nature of agriculture means it should be treated differently to other sectors.
>
> (Cairns Group, 1999)

NTCs are 'simply a subset of domestic policy issues that are interfacing increasingly with international policies as the globalization of the world economy proceeds' and 'a direct consequence of the lowering or outlawing of trade barriers' (Anderson, 1998a, pp. 5, 12). As barriers to trade are reduced, domestic economic and social policies become more important for competitiveness in some industries, and interest groups seek government assistance through other measures (e.g. domestic policy) (Anderson, 1998a, p. 5). But NTC may be a misnomer for these concerns. Anderson (1998a, p. 3) believes that NTCs are indeed trade-related and have economic content; he refers to them as 'so-called' NTCs.

WTO Members have the right to determine their own domestic goals and objectives; at issue is whether the means for achieving those goals are appropriate, in light of their effects on trade (Anderson, 1998a, p. 4). The WTO recognizes the importance of domestic policy objectives connected with agriculture as NTCs; therefore, 'meeting important policy objectives such as rural development, protecting the environment and ensuring food security need not be inconsistent with liberalising world trade in agriculture' (Cairns Group, 1999). The issue, for both multifunctionality and the related NTCs, is whether new trade provisions are necessary to meet domestic policy objectives. In discussions about discipline for agricultural support, however, some countries prefer to focus on the objectives of policies, rather than on their trade effects (Normile, 2001). Negotiators must decide whether existing WTO provisions (e.g. the green box) allow Members to implement policies designed to meet their national objectives and perhaps even whether farmers in developed countries should receive support at all (Agra-Europe Weekly, 2000c, p. EP/7).

Three main multifunctionality issues or NTCs

WTO Members generally agree both that NTCs are 'legitimate and valid concerns' and that the nature of agricultural production and the policies needed to sustain agriculture differ among countries. Members often demand different goods and services from agriculture, so that NTCs receive different 'weight and priority' (Norway, 2000, pp. 1, 6). Though many functions of agriculture have been emphasized among Members, the non-commodity outputs of agriculture and NTCs are generally grouped in three main categories: environment, rural development (or employment) and food security. The OECD identified these three categories as the most commonly cited concerns (OECD, 1998b, p. 6), but later acknowledged that 'inclusion of rural employment and food security in the discussion of multifunctionality has been controversial' (OECD, 2001, p. 13). USDA researchers identified environment, food security and rural development as NTCs (Bohman *et al.*, 1999). These were also the focus of discussion papers prepared for a July 2000 international conference on NTCs, attended by 40 countries (WTO, 2000m), and for a subsequent conference in May 2001, attended by 42 countries (WTO, 2001c). Though EC submissions to the WTO included animal welfare and quality food speciality as NTCs (WTO, 2000e,f) and later added food safety and mandatory labelling (WTO, 2002d), other Members have not generally embraced these issues as NTCs.

Environment

Agriculture affects the environment in both positive and negative ways, and numerous government policies can influence the interaction between agriculture and environment. Arguments for protection of NTCs related to the environment focus primarily on the positive environmental externalities from agriculture. Most important are rural landscape and biodiversity (Anderson, 1998a, p. 8), which include such amenities as open space, scenic views and wildlife habitat. Other environmental benefits may include flood control, watershed protection and recharge of groundwater. Negative externalities of agriculture, which should also be recognized, include odour, runoff of pesticides and nutrients, soil erosion, biodiversity loss, damage to habitat and other effects (Bohman *et al.*, 1999, pp. 9–10).

Rural landscape provides pleasure for urban people and others who enjoy scenic views and are reminded of an important cultural heritage of food and fibre production. In some countries, special landscape features – e.g. alpine pastures, ancient hedgerows, stone walls – may provide aesthetic pleasure for residents or tourists (Anderson, 1998a, pp. 8–9). Some people want to preserve rural landscape to ensure that it remains available for future generations (Bohman *et al.*, 1999, p. 16). Similarly, society values biodiversity in rural areas, as well as the wildlife necessary to support diverse flora and fauna.

Generally, 'markets do not exist for the environmental by-products of agriculture, which creates a rationale for government intervention'. To ensure continued production, some countries argue, governments should provide subsidies or price supports (Bohman *et al.*, 1999, p. 13). But other policies, which do not affect production or distort trade, may also be effective in ensuring the continued existence of environmental benefits that often accompany agricultural production.

Environment as a multifunctional component of agriculture raises several issues. For example, what targets for externalities (e.g. how much agricultural scenery) should governments establish, in the light of relevant costs and benefits? What government policies, or combinations of policies, are available to meet these environmental objectives (Bohman *et al.*, 1999, pp. 13–14)? Will reduced support for agriculture lead to decreased provision of environmental objectives? Should support be targeted specifically to meet society's environmental objectives, rather than provided through domestic support for farmers? Can subsidies to agriculture lead to environmental degradation, rather than preservation of environmental amenities (Anderson, 1998a, pp. 8–10)?

Rural development

Rural development as an NTC focuses on providing employment and income and on maintaining viable towns in rural areas (Bohman *et al.*, 1999, p. 19). Shrinking rural communities may result in decreased social cohesion, loss of the 'nostalgic attraction' of villages and possibly (in remote areas) military insecurity.[3] But agriculture does not always dominate rural areas, and in some countries non-farm economic opportunities are available in those areas (Anderson, 1998a, p. 11).

Agricultural policy can influence rural employment by affecting the number of existing or new farmers or the amount of labour, but support for farmers does not always increase the number of jobs (Bohman *et al.*, 1999, p. 20). In some countries, it is not clear that agricultural payments are effective in ensuring that rural villages remain economically viable. In the USA, for example, where total rural employment (e.g. manufacturing, service) is increasing, agriculture represents only a small share (6%) of rural employment. (The agricultural share of rural employment in other countries varies; e.g. Germany 2%; Norway 8%, France 11%, Japan 14%.) Moreover, the majority of US farm-programme payments go to a minority of farms, often with above-average incomes, while other farms remain poor (Bohman *et al.*, 1999, p. 19). Indeed, research in the USA indicates that, for most rural communities, where the non-farm economy has developed, farm payments play a relatively small economic role and represent only a fraction of total federal assistance (Gale, 2000, pp. 15, 17–18).

Thus, one issue connected with rural development is whether general agricultural support is an efficient mechanism. That is, would more targeted support – subsidies for rural employment or support for essential services – be more effective in protecting vulnerable areas (Anderson, 1998a, p. 12)? Public policies directed towards rural development may include structural adjustment, general services, such as training or infrastructure, and tax incentives for rural investment (Bohman *et al.*, 1999, p. 20).

Food security

Food security is also considered an NTC, one of two named in the Preamble to the URAA. Food security is achieved 'when all people, at all times, have physical and economic access to sufficient, safe and nutritious food to meet their dietary needs and food preferences for an active and healthy life' (OECD, 2001, p. 47). Countries often cite two objectives of food security as an output of agriculture. The first is national food security (adequate access by the country), which focuses on the vulnerability of the food supply to various external and natural factors. The second is household food security (access by citizens), which focuses on internal issues, including income and other factors that affect food distribution (Bohman *et al.*, 1999, pp. 16–17).

In some countries, the capacity for, or the actual existence of, domestic agricultural production is viewed as an important component of food security. The need for sufficient production in those countries justifies support to agricultural production, with the rationale that food security is an externality of agriculture or that it is a public good (Bohman *et al.*, 1999, pp. 17–18). Other countries would argue that, though food security is an NTC, it should not be analysed as part of multifunctionality. That is, unlike environment and rural development, food security focuses on the products of agriculture themselves, rather than on non-food outputs. Further, food security should be considered a joint product of agricultural trade rather than of production (Bohman *et al.*, 1999, p. 18; OECD, 1999b, p. 4)

Questions connected with food security, as with other NTCs, focus on appropriate and efficient policies to meet national objectives. Numerous policy instruments may be effective to ensure food security; these include public stockholding, food aid and emergency assistance, as well as longer-term measures, such as research and education or liberalized trade (to encourage imports) (Bohman *et al.*, 1999, pp. 18–19). Some of these measures (e.g. public holding of food stocks) fit within the green box (URAA, Annex 2).

Negotiating Positions on Multifunctionality and NTCs

Though WTO Members generally agree that NTCs play a legitimate role in Member-country agriculture, Members have different views about appropriate policy measures. Some Members believe that NTCs should be protected even if distortion of trade results, but others disagree. Some argue that domestic support should be reduced unless it meets the requirements (or perhaps even stricter requirements) for the green box; others plead for continuation of the controversial blue box. (For more information about the green and blue boxes, see Chapter 2, this volume.)

The following discussion outlines the negotiating positions of selected WTO Members. These include the USA and the Cairns Group, who are most reluctant to use multifunctionality to justify domestic support that distorts trade, and the European Community, Norway and Japan, who argue for flexible policy to accommodate multifunctionality and the associated NTCs. Though the focus of analysis is domestic support, it should be remembered that multifunctionality may also be relevant in the context of market access and tariff protection (M. Bohman, Washington, DC, 2002, personal communication).

The United States

The USA is among the WTO Members who oppose promotion of multifunctionality and protection of NTCs as justification for domestic policies that distort trade. Though the USA acknowledges the multifunctional character of agriculture, it argues for decoupled policies carefully targeted to meet NTCs in the various WTO Member countries. Research reports by USDA economists have provided helpful analysis of issues connected with multifunctionality.

WTO submissions

In submissions to the WTO, the USA articulated its position on multifunctionality in the context of domestic support. These documents express a consistent and increasingly clear stance for the negotiations. The USA asserts that WTO rules should exempt domestic support from reduction only if those policies fit within carefully defined green-box criteria ensuring that domestic support has at most a minimal effect on production and trade. The following discussion traces the US position through a series of documents submitted since 1998.

In November 1998, the USA expressed preliminary views on several negotiations mandated by the Marrakesh agreement, including the URAA Article 20 agriculture negotiations. The USA (WTO, 1998, p. 2) hoped that Members would agree on an overall objective to expand market access 'by ensuring further deep reductions in support and protection and by strengthening the rules governing trade in agriculture'. The USA recommended an ambitious target for reduction of domestic support that distorts trade and stronger rules to discipline production-related support. The green box should continue to minimize the 'link between support and production' and to exempt only support that does not distort trade (WTO, 1998, p. 3).

Similarly, in May 1999, the USA proposed that objectives for the new agricultural negotiations include further deep reductions in support and protection, 'while encouraging non-trade distorting approaches for supporting farmers and the rural sector' (WTO, 1999c, p. 1). Four communications submitted to the WTO in July 1999 focused briefly on export competition, market access, biotechnology and domestic support (WTO, 1999g,h,i,j). The one-page communication on domestic support echoed the November 1998 statement, quoted above, which emphasized discipline for production-related support. By way of background, the USA noted that '[g]overnments have the right to support farmers if they so choose. However, it is important that this support be provided in a manner that causes minimal distortions to production and trade.' Acknowledging the legitimate purpose of the green box in supporting the contributions of farmers, the communication emphasized that the green box should continue to minimize the effect of support on production through 'appropriately specified policies' (WTO, 1999j).

Following this terse communication, the USA expanded on the domestic-support issue in a communication on Trade and Sustainable Development, submitted a few days later. The USA noted that trade rules should not constrain Members from enacting 'science based measures' to protect health, safety and the environment and recognized that trade liberalization can contribute to sustainable development (WTO, 1999k, p. 2). Trade-distorting subsidies, including domestic agricultural subsidies, had led to unsustainable practices, such as overuse of farm inputs, degradation of soil and overgrazing. Asserting that elimination of agricultural subsidies could yield environmental benefits (see WTO, 1996), the USA recommended elimination of export subsidies, transition from those domestic subsidies that 'encourage degradation of natural resources and distort trade' and retention of the green box (WTO, 1999k, p. 4).

During renewed negotiations in 2000, the USA restated and expanded its position on support for agriculture in its Proposal for Comprehensive Long-Term Agricultural Trade Reform (WTO, 2000b), supplemented by a Note on Domestic Support Reform (WTO, 2000c). Adopting a softer stance on multifunctionality, the proposal states that 'the United States is . . . committed to and supports policies that address non-trade concerns, including food security,[4] resource conservation, rural development, and environmental protection' (WTO, 2000b, p. 2). Members should meet these objectives, the USA insisted, with means that do not distort trade and with 'programs targeted to the particular concern', so that closed markets and unfair competition do not impose costs on other nations (WTO, 2000b, p. 2, 2000j). US recommendations for domestic policies are designed to reduce trade distortion and 'release producers from restrictive government policies that prescribe what and how much to produce, freeing farmers to follow their judgement and the natural carrying capacity of their land' (WTO, 2000b, p. 2). Under the US proposal, farmers would enjoy expanded economic opportunities, supplies of food would be more secure, and consumers would also benefit.

Building on the disciplines in the URAA, the USA proposed to establish two categories for domestic support: exempt support, 'defined by criteria-based measures that have no or, at most, minimal trade distorting effects or effects on production' (similar to the green box), and non-exempt support, which must be reduced. The US proposal anticipated changes in the green-box requirements to ensure that exempt measures are 'targeted, transparent, and, at most, minimally trade-distorting' (WTO, 2000b, p. 4). Developing countries would have additional opportunities for exempt support.[5] In addition, the USA proposed further progressive, annual reductions of non-exempt support, based on the final bound aggregate measurement of support (AMS), by a fixed percentage over a fixed period. Though the proposal makes no specific reference to the blue box, the two-category scheme would clearly eliminate the blue box with its partially coupled support.

The accompanying Note on Domestic Support Reform (WTO, 2000c) explained the US proposal's formula-based approach for reducing non-exempt support under the AMS and identified the new policy directions for domestic support mentioned in the formal proposal (WTO, 2000b, p. 4). These include farm income safety-net and risk-management tools, targeted environmental and resource protection, rural development (investment in infrastructure, economic development, technical assistance), alternative technologies and bio-based products, and structural adjustment (decoupled income support and other adaptive measures) (WTO, 2000c, pp. 2–3). Yet another statement on reform of domestic support (WTO, 2000q) asserted that programmes that do not qualify for the green box must be included in AMS calculations and subject to reduction. The current green-box criteria 'may not fully reflect new directions in agricultural policy' or the needs of some countries, so the USA encouraged discussion of both old and new policy criteria for domestic support, to facilitate measures that will address Member objectives 'while having minimal trade and production effects' (WTO, 2000q, p. 1).

In July 2002 discussions in Geneva, the USA (WTO, 2002a) submitted its Proposal for Global Agricultural Trade Reform, with more specific plans for market access, export competition and domestic support. For domestic support, the USA reiterated its proposal for a simplified scheme with two categories. Exempt support is 'defined by criteria-based measures that have no, or at most minimal, trade-distorting effects or effects on production'. That is, the basic criteria of the green box would be maintained, with no maximum amounts. Non-exempt support would include both support that is part of the AMS and production-limiting (blue-box) support. The USA proposed specific reductions in non-exempt support. Over a 5-year period, non-exempt support should be reduced from the Member's AMS ceiling to 5% of the Member's average value of total agricultural production in the base period 1996–1998. Support eligible for the *de minimis* exemption (URAA, Art. 6, ¶ 4) would not be included in the 5% limit. By a date to be negotiated, Members would eventually eliminate all non-exempt domestic support (as well as all tariffs and export subsidies). Developing countries would receive special consideration, with additional exempt support necessary to meet domestic policy objectives. The USA recommended additional sector-specific reductions to address trade-distorting practices (WTO, 2002a).

Other documents

In these WTO documents, the USA has expressed its negotiating position with increasing detail, but without much focus on multifunctionality. In other contexts, the USA expressed reservations about multifunctionality as a rationale for agricultural support. For example, the US response to OECD's document, *Multifunctionality: a Framework for Analysis* (OECD, 1998b), reflects the US attitude and provides some background for the US negotiating position. In part, that position may stem from the fact that US policy-makers have paid more attention to 'negative environmental effects of agriculture than to agriculture's positive multifunctionality'. For WTO Members with more focus on the positive externalities of agriculture – e.g. the EU, Norway, Japan – multifunctionality is more central to agricultural policy (OECD, 1999a, p. 68).

In comments on OECD's *Framework*, the USA acknowledged that 'each nation may legitimately have unique goals in respect of a multi-functional agriculture' (OECD, 1999a, p. 63). For OECD, then, the main issue connected with multifunctionality should be whether current trade rules discipline domestic policies effectively. If the URAA green box provides appropriate discipline for domestic policies that encourage multifunctionality, then OECD has little reason for further study (OECD, 1999a, p. 63). The USA suggested a framework for analysis of multi-functionality that focuses on supply and demand for non-food services and appropriate (that is, efficient and trade-neutral) policies for encouraging those services (OECD, 1999a, pp. 65–67). Moreover, the USA wondered why multifunctionality in agriculture is a special case; that is, '[i]f interventions are to be made in the case of agriculture/rural economy, shouldn't they also be made in parallel cases in the rest of the economy?' (OECD, 1999a, p. 68). Other countries, including Australia and Canada, raised the same question (OECD, 1999a, pp. 5, 13). Indeed, the OECD (2001, pp. 133–142) recognized other areas characterized by management of resources for multiple uses. Though the term 'multifunctionality' is used mainly in agriculture, economic areas like forestry, fishing, banking and household production exhibit similar characteristics. Most involve marketable outputs, though forestry provides public goods.

The USA took issue with two common arguments that often justify government payments to support multifunctionality: that non-food services of agriculture are public goods and that those non-food services are necessarily joint products of agriculture. In the *Framework* document, the OECD had referred to non-food services of agriculture as public goods that are non-excludable (access to potential consumers cannot be denied) and non rival (consumption by one does not reduce availability or value to others) (OECD, 1998b, p. 9). The USA did not agree that food security and rural development, albeit valid and important, are pure public goods that arise from agricultural activities. In some cases, food security and rural development may be 'positive spillovers from agricultural activity', but neither spillovers nor the public-goods argument 'necessarily justifies a government role in maintaining production of the non-food goods' (OECD, 1999a, pp. 63–64). Some non-food products could be considered 'club goods', which fit in an intermediate category; they may be non-rival, but are excludable, and they could be provided by an organization that charges fees (see Bohman *et al.*, 1999, p. 11). Moreover, efficiency should play a role in a government decision to provide a non-market good (OECD, 1999a, p. 64).

On the issue of jointness, the USA noted that some of the non-food services produced in conjunction with agricultural activities are not exclusively joint products. Both rural development and food security can be achieved by means other than agricultural production (OECD, 1999a, p. 64). Thus, those activities 'are not in any way illuminated by the multifunctionality concept and analytic capacity' (OECD, 1999a, p. 67). Though food security and rural development might sometimes justify trade distortion, joint production is an inappropriate framework for analysis (OECD, 1999a, pp. 67–68). Moreover, if countries have unlimited freedom to define the joint products of agriculture (public goods), they may then provide payments to agriculture under various, perhaps invalid circumstances. Rational policy-making, the USA suggested, 'may require some form of transparent process by which member countries document the failure of its domestic market to provide the desired level of the joint product for which it seeks to foster production' (OECD, 1999a, p. 65). Perhaps policies for joint products need to be coordinated or standardized at the international level (OECD, 1999a, pp. 66–67). This process could ensure that products that are not necessarily joint with agriculture (e.g. rural employment) are excluded from the category of exempt support.

Despite the reluctance of the USA to embrace multifunctionality in the trade context, the USA recognizes the significance of the many functions of agriculture. A recent USDA publication articulates current US agricultural policies and priorities and encompasses a broad view of agriculture. *Food and Agricultural Policy: Taking Stock for the New Century* (USDA, 2001) highlights conservation and environmental quality, rural development and food quality and safety as important US agricultural goals. But the report never uses the term 'multifunctionality' and it assumes that these goals should be met by market-oriented policies, rather than through trade-distorting agricultural support (M. Bohman, Washington, DC, 2002, personal communication). In fact, the USA implements numerous market-based policies, as well as conservation and other policies not connected with agricultural production, to ensure that these goals are met in US rural areas (Normile and Bohman, 2002).

Cairns Group

Members of the Cairns Group, like the USA, express strong reservations about multifunctionality and NTCs as justifications for domestic support that distorts trade. An August 1999 communiqué from the Cairns Group ministerial meeting asserted that multifunctionality does not justify special treatment for trade related to manufacturing and services, and special treatment is not appropriate for agriculture either. Moreover, '[n]on-trade objectives should not be used as a smoke screen for protectionist policies which perpetuate poverty, hunger and environmental degradation', with particular harm to developing countries (WTO, 1999l, p. 2; Hill, 2000). Instead, multifunctional objectives should be supported only by targeted policies that address specific NTCs (Cairns Group, 1999).

Policy-makers and researchers in Australia, a Cairns leader, have been particularly active in analysing the effects of agricultural support to foster multifunctionality (e.g. Tielu and Roberts, 1998; ABARE, 1999). Australia indicated that multifunctionality presents a real dilemma to countries that offer only limited support to farmers. High-support countries operate from the 'misplaced notion' that production-linked support, rather than less trade-distorting instruments and measures, should be used to address NTCs (OECD, 1999a, p. 3). Australia identified two key dangers of multifunctionality: the 'clear risk of the issue being captured by protectionist interests using the environment, rural development and cultural heritage, etc., as a disguise' and 'the real scope for "government failure" in developing and applying indirect policies to address many of the unmeasurable externalities that fall under the ever enlarging list of "non-food services"' (OECD, 1999a, p. 5).

Recognizing that NTCs also exist in its own agricultural policy, Australia acknowledged that WTO Members can pursue policies to further NTCs, but insisted that WTO negotiations focus on how those NTCs are achieved – ideally through policies that are simple, transparent, targeted and decoupled (WTO, 2000v). Argentina, also a Cairns member, wrote a technical submission on Legitimate Non-Trade Concerns (WTO, 2000x), which identified three legitimate NTCs: rural poverty, unemployment and environmental protection. These NTCs are legitimate for Argentina and other WTO Members because they can be pursued along with 'a fair and market-oriented agricultural trading system' (WTO, 2000x, p. 1, quoting URAA Preamble). Even these legitimate concerns, however, must not be pursued at the expense of trading partners (WTO, 2000x, p. 3).

Thus, Cairns Group submissions to the WTO lack sympathy for multifunctional arguments, in so far as policies that promote multifunctionality distort trade. In its March 1999 Vision Statement (WTO, 1999a), the Cairns Group pleaded for further liberalization of agricultural trade and, in particular, for significant reductions in domestic support, including elimination of all trade-distorting support. Domestic support must be 'targeted, transparent, and fully decoupled' (WTO, 1999a, p. 2), and the 'uncapped' blue box, which permits production-linked support, should be eliminated (WTO, 1999b).

During the 2000 negotiations, the Cairns Group (WTO, 2000l) submitted a proposal on domestic support. Introduced by a statement from Australia (WTO, 2000n), the proposal noted that domestic support, including amber-box support and unlimited green- and blue-box support, occurs primarily in rich countries and continues to distort trade. Therefore it proposed a major reduction in support for all agricultural products that distorts trade and production, including support in the amber and blue boxes. Further, the Group proposed a review of green-box criteria to ensure that permitted policies have no, or only minimal, distorting effects on trade or effects on production, as well as special provisions for developing countries (WTO, 2000l; see WTO, 2000s). Canada (WTO, 2000aa) added the recommendation that Members be encouraged to move support to green-box programmes and to negotiate an overall limit on support for agriculture. The Cairns Group expected the peace clause to expire at the end of 2003, with WTO Members then permitted to take legal action against illegal domestic support.

In September 2002, the Cairns Group (WTO, 2002c) submitted another, more specific negotiating proposal on domestic support. Noting that high levels of domestic support in recent years reflected a failure of the URAA, the proposal called for 'real cuts' in all production- and trade-distorting domestic support during the Doha Round. The Group proposed reduction of AMS support to zero over 5 years (9 years, for developing countries); the *de minimis* exemption would be retained for developing countries and eliminated gradually for developed countries. Disciplines would ensure that product-specific support is not misclassified as non-product-specific support. Further, the Cairns Group would close the blue box entirely and amend the green box significantly, adding a limit on direct payments and adjusting language to ensure that green-box support does not distort trade or production.

European Communities

The EC have long emphasized the importance of the many functions of agriculture, including its positive effect on the environment, biodiversity and the stability of rural areas. Both Council and Commission have endorsed multifunctionality, which is central to the European model of agriculture. In September 1999, the Agriculture Council adopted a common position on the agricultural aspects of the WTO, which communicates the Member States' agreement to take an 'offensive approach' in the trade negotiations to defend the European model of agriculture, as reflected in the Agenda 2000 reform of the Common Agricultural Policy (CAP). The Council stressed the importance of safeguarding and developing the European model because of the 'multifunctional nature of Europe's agriculture' and the part it plays in the economy, environment, landscape and society (CEC, DG Agri, 1999b, p. 13). The EC demand much of their agriculture:

> European Agriculture as an economic sector must be versatile, sustainable, competitive and spread throughout Europe . . . It must be capable of maintaining the countryside, conserving nature and making a key contribution to the vitality of rural life. It must also be able to respond to consumer concerns and demands regarding food quality and safety, environmental protection and the safeguarding of animal welfare.
>
> (CEC, DG Agri, 1999b, p. 13)

These demands influence negotiating proposals from the EC, outlined in the Council's common position and later submitted to the WTO.

Agriculture Commissioner Fischler noted that the future model of agriculture in the EC will be based on 'the multiplicity of functions performed by the agricultural sector'. These include the market function (served by more competitive prices), the environmental function (served by rewarding farmers' efforts that benefit society) and the functions of rural areas (served by reformed policy instruments for rural development) (CEC, DG Agri, 1999b, p. 1). Significantly, the Agenda 2000 reform of the CAP identifies rural development as an important 'second pillar' of the CAP. The new rural-development policy (re)authorizes several types of support for farmers and rests on specific principles. First among these principles is the multifunctionality of agriculture, that is, agriculture's 'varied role over and above the production of foodstuffs', which carries with it 'the recognition and encouragement of the services provided by farmers' (CEC, DG Agri, 1999a, p. 1). If enacted, proposals revealed in the Mid-term Review of Agenda 2000 will consolidate and strengthen rural development (CEC, 2002).

WTO submissions

The EC's submissions to the WTO reflect this multifunctional emphasis. The EC submitted contributions to the Committee on Agriculture on the multifunctional character of agriculture (CEC, DG Agri, 1999c) and on instruments for safeguarding that multifunctional character (CEC, DG Agri, 1999d). In its contribution on the multifunctional character of agriculture, the EC outlined basic concepts, with emphasis on the various services performed by agriculture. These services, more than mere externalities, are interdependent with agricultural production and valued by society, sometimes as public goods. The EC argued that public intervention must ensure that the multiple functions of agriculture will continue. In its contribution on instruments to safeguard that multifunctional character, the EC expanded on the main functions of agriculture: production of food, the environment function and the rural function (reinforced by Agenda 2000). The paper explained some of the measures used to enhance the multifunctional role – e.g. agri-environmental payments and measures to support structural and rural development.

Accordingly, the EC's negotiating proposals are directed in part to measures that support the multifunctional character of agriculture and the related NTCs. Shortly before the Seattle negotiations, the EC identified NTCs – multifunctionality (not normally identified as an NTC), food safety and quality, environment and animal welfare – as important issues for negotiation (WTO, 1999f). The EC then made these NTCs the focus of three separate proposals submitted to the Committee on Agriculture during the 2000 negotiations. Another EC proposal focuses on export competition, including export subsidies, export credits and abuse of food aid (WTO, 2000k; accompanying statement, WTO, 2000o).

The EC (WTO, 2000d) addressed domestic support in its proposal on the blue box and other agricultural support measures. The EC proposed that the blue box, as well as the green box, be maintained in any new agricultural agreement. Blue-box measures, linked to production factors but not to price or output volume, limit production. These payments had been useful tools in domestic policy reform, especially reform of the CAP, and would continue to be useful for agricultural reform. The EC cited an OECD study to indicate that the shift from price supports to blue-box payments reduced the impact on trade of CAP agricultural support. Further, the EC (WTO, 2000g) asserted that objective OECD studies indicate that the blue-box has made a contribution to reform and that per-hectare blue-box payments do not distort trade more than per-hectare green-box payments. The EC chastised countries (presumably including the USA) that want to abolish the blue box 'simply because they do not use it', with the rationale that even countries that do not use the blue box gain from its availability. It would seem, however, that proposed changes in the CAP, as a result of the Mid-term Review, will make the blue box less important to the EC (CEC, 2002).

The EC proposal did not squarely address multifunctionality, noting only that the green box should be maintained (WTO, 2000d). In a statement introducing its proposals, the EC responded to the US proposal and the 'keen interest' of the USA in multifunctionality, adding 'for some reasons they do not like using the word' (WTO, 2000g, p. 2). This interest, however, suggested to the EC that negotiations could focus on the means of safeguarding the multifunctional functions of agriculture.[6]

Two further EC proposals deal with issues that are not normally linked with multifunctionality. A proposal on animal welfare (WTO, 2000f) was intended to ensure that trade liberalization does not weaken measures that regulate the conditions for animal production (WTO, 2000g). The EC hoped for discussion of animal welfare in the WTO negotiations, to 'promote high animal welfare standards, to provide clear information to consumers, [and maintain] the competitiveness of the EC farming sector and food industry' (WTO, 2000f, p. 2). To promote animal welfare, the EC recommended payments to farmers for additional costs imposed by high animal-welfare standards, as long as those payments have only minimal effects on trade and production. The other proposal concerns 'quality food specificity' to ensure that trade rules protect high-quality food products with specific names and regional or traditional origins (WTO, 2000e). The EC proposed that agricultural negotiations consider protection of product names, market access for named products and regulation of labelling to differentiate specific food products. In a sense, this proposal attempts to recognize intellectual property in the new agricultural agreement, though other aspects of intellectual property are governed by the Agreement on Trade-related Aspects of Intellectual Property Rights (TRIPS) (Agra-Europe Weekly, 2000b, p. A/2).

In its comprehensive negotiating proposal, submitted 14 December 2000, the EC reiterated the positions described above (WTO, 2000y). Strong public support for agricultural trade reform is possible, the EC asserted, only if that reform also helps to support the multifunctional role of agriculture. For domestic support, the EC proposed further reduction in the AMS, continuation of the peace clause and retention of both the blue box and the green box, with review of green-box criteria to ensure minimal distortion and effectiveness in meeting multifunctional goals. The EC proposed that measures to promote important NTCs – to protect environment, sustain vital rural areas and alleviate poverty – be included in the revised Agreement on Agriculture, but that those measures should be 'well targeted, transparent, and implemented in no more than minimally trade-distorting ways' (WTO, 2000y, p. 4). To the usual list of NTCs, the EC again added others: food safety (including measures that implement the precautionary principle), consumer concerns (food-labelling schemes) and animal welfare.

In September 2001, the EC submitted 'non-papers' on the green box and geographical indications. The non-paper on the green box emphasized its importance in achieving societal goals (NTCs) and recommended that current provisions should be maintained. The requirement that effects on trade be minimal led the EC to recommend that counter-cyclical payments or those determined by production, price or input levels not be included in the green box. The non-paper seemed to suggest that payments related to the additional costs of complying with animal-welfare rules should be included in the green box (WTO, 2001d; see CEC, DG Agri, 2001). The non-paper on geographical indications focused on food specificity (identified as an NTC) and recommended a system to protect geographical names and products (WTO, 2001e). The EC also raised the issue of labelling of food and agricultural products in connection with the Agreement on Technical Barriers to Trade (WTO, 2001f).

The EC published their Proposal for Modalities in December 2002 (WTO, 2002d). Their negotiating package continues an emphasis on multifunctionality and NTCs, especially in the context of domestic support. The EC favour a 55% reduction in AMS, starting from the URAA-bound commitment level (see Chapter 2 this volume), but with modulations for developing countries; the EC would eliminate *de minimis* exemption for developed countries. The blue box should be continued, to facilitate reductions in the most trade-distorting support; the green box should remain available for achievement of goals related to environment, rural development and animal welfare, as well as domestic-support needs of developing countries; and the peace clause should be continued. The 'EC proposals to liberalise trade and decrease trade-distorting domestic support are conditional upon key non-trade concerns being adequately addressed' (WTO, 2002d, pp. 6–7). Food safety is now listed as such an NTC, and the EC wants a strict definition of criteria for application of precaution. The EC also want an understanding of criteria and guidelines to implement mandatory labelling of food and agricultural products. Food security for developing countries, protection of the environment, rural development and animal welfare are again numbered among the NTCs that must be addressed. The EC continue to emphasize geographical indications for speciality agricultural products, but in the context of market access, rather than NTCs (WTO, 2002d).

Environment and rural development

The European Commission commented on important NTCs in two discussion papers presented at a conference on NTCs in Norway in July 2000. The EC participated in later conferences on NTCs in Mauritius (May 2001) and Doha (November 2001). The EC papers, along with four others from the 2000 conference and those from Mauritius, were submitted to WTO for consideration as part of negotiations (WTO, 2000m, 2001c).

In the paper entitled *Agriculture's contribution to environmentally and culturally related non-trade concerns*, the European Commission argued that farming and environmental conservation are 'inextricably linked' (CEC, 2000a, p. iv). The discussion paper considers and provides background on several NTCs connected with the environment: conservation of biological diversity, maintenance of farmed landscapes, preservation of cultural features and protection against natural or induced disasters. Different societies value these concerns differently, of course, and farming systems may or may not help to fulfil the various NTCs (CEC, 2000a, p. 1). But the Commission warned that a purely economic focus to agriculture that ignores important NTCs has resulted in several potentially harmful trends: concentration, specialization, intensification and marginalization (underuse or abandonment of farmland). The result is that 'high-value landscapes are lost; biological diversity suffers; pollution of water resources increases; and production methods become divorced from public expectations' (CEC, 2000a, p. 6).

Fulfilling NTCs imposes burdens on farmers. For example, to conserve biological diversity, farmers must often incur expenses or earn lower profits, and they may suffer lower production. Similarly, maintaining farmed landscapes and historical features of farmland may impose costs or other burdens on farmers (CEC, 2000a, pp. 4–5). The European Commission refers to the environmental and cultural results achieved through farming as joint products of agriculture. The Commission rejected the suggestion that farm products from these activities should be destroyed, rather than marketed (to avoid trade distortions), because of the waste involved. To avoid distortion, however, the Commission suggested that 'whenever society . . . demands that farmers undertake efforts to deliver the public good, then governments should only recompense farmers for their additional costs and income foregone, taking fully into account the farmer's income from selling commodities on the market' (CEC, 2000a, p. 5). When public goods (e.g. environmental benefits) are involved, policy measures are needed because the market will not provide an incentive for the non-trade objective (CEC, 2000a, p. 1).

The Commission concluded that government policies directed toward NTCs must focus on the public good. Policies must be 'targeted, have clear objectives, . . . [be] administered in a transparent manner, and . . . implemented in no more than minimally trade-distorting ways' (CEC, 2000a, p. 7). Policy instruments might include encouragement (e.g. extension services); compulsory regulation (to prohibit inappropriate farm activities); and voluntary programmes (to encourage production of public goods). Compensation for farmers' costs and income lost from practices more stringent than good agricultural practice should be allowed. Moreover, according to the Commission, WTO rules should encourage implementation of these policies (CEC, 2000a, p. 7).

In another discussion paper, *Agriculture's contribution to rural development*, the Commission asserted that farming contributes to rural development in many countries, helping to improve the economic situation of rural peoples (CEC, 2000b, p. 1). This occurs in four ways: providing employment, generating economic activity related to farming, maintaining economic viability in remote areas and providing environmental and cultural amenities. Both on-farm and ancillary (off-farm) activities help to provide these benefits. In many areas, it is important to maintain levels of farm employment to maintain social and political stability, even in instances where commodity production might be more economically efficient elsewhere (CEC, 2000b, pp. 1–2). Thus the Commission recommended policies to promote the farm sector, including 'encouraging investment, training, applied research and appropriate technology, and policies to manage structural adjustment such as land reform and generational change' (CEC, 2000b, p. 3). To make these policies possible within the constraints of agricultural trade, the Commission indicated that WTO rules must be flexible enough to promote the rural-development aspects of multifunctionality (CEC, 2000b, p. 4).

In a statement submitted to the WTO, the EC referred to these two papers and offered further comments. For the EC, NTCs are 'key elements' in the Article 20 negotiations, but key in the 'sense of an instrument for opening a door', rather than locking the door on other negotiating items. Agriculture, to the EC and other Members, is 'an engine of rural development', which 'shapes the environment'. Trade, the EC asserted, must not destroy these functions (WTO, 2000w, p. 1).

Norway

Norway's support for multifunctionality is explained in part by the nature of Norwegian agriculture, which is subject to difficult geographical and climatic conditions. Norway faces high production costs because of its small farms, small field sizes, steep slopes, short growing seasons and high labour costs. Only 3% of its area is cultivated land, and harsh natural conditions leave little room for structural change (WTO, 1999m, p. 5). As a result of these conditions, Norway's policies include high levels of border protection and domestic support, with government subsidies nearly equal to agricultural output. Norway's estimated support for 2001 was $35,000 per farmer ($2086/ha) (OECD, 2002, tables III.5, III.6).

Norway had emphasized the importance of NTCs as early as 1989, in a statement in the Mid-term Review of the Uruguay Round (Lindland, 1998, p. 3). In Norway's view, agriculture contributes important public goods, including food security, viability of rural areas, agricultural landscape, conservation of agro-biological diversity, good phytosanitary measures, zoo-sanitary measures and public health (Lindland, 1998, pp. 7–11). Norway characterized a number of these public goods as joint products of agriculture. In July 2000, Norway hosted an international conference on Non-trade Concerns in Agriculture. Norway's own contribution to that conference urged flexibility in WTO provisions, allowing Members to implement domestic agricultural policy that addresses NTCs connected with agricultural production (Norway, 2000).

Norway's blue-box payments, linked to factors of production, have helped to encourage farming in marginal agricultural areas. Though decoupled support can achieve income objectives, green-box measures cannot safeguard Norway's NTCs (WTO, 2000r, p. 1). Research on farms in Norway indicated that changing blue-box and AMS measures to green-box support would be ineffective in protecting NTCs. Farmers are unlikely to receive sufficient financial incentives for long-term production, and the required targeting measures would be infeasible and costly to administer (WTO, 1999m, pp. 6–12). Because Norway's support focuses on public goods:

> a combination of policy measures, including support coupled to the agricultural production, seems to be the most efficient way of ensuring a sufficient production level of public goods, to the extent that these public goods are joint products of the agricultural production.
>
> (Lindland, 1998, p. 23, italics omitted)

Agricultural policy in Norway uses two important principles. The 'polluter-pays principle' ensures provision of environmental goods up to a certain reference level (that of good agricultural practices). But where the public goods produced jointly with agriculture require activities that go beyond that reference level, the 'provider-gets principle' applies (Lindland, 1998, pp. 5, 22). Under this principle, the government may have to pay farmers who provide public goods with private production factors, to 'achieve the desired resources allocation' (WTO, 1999m, p. 3). Without support, these public goods would disappear.

Norway's WTO submissions reflect its support of multifunctionality. In a brief statement from July 1999, Norway insisted that negotiations be based on URAA Article 20 and, among other things, that they safeguard and promote 'the non-trade concerns of a multifunctional agriculture'. The public-good characteristics of these NTCs may justify government intervention and suggest a rationale for continued special treatment of agricultural trade in the WTO (WTO, 1999e, pp. 1–2). In the 2000 negotiations, Norway emphasized the varied agricultural conditions among Members and the importance of enacting policies that safeguard concerns of Members whose agriculture faces significant production constraints (WTO, 2000i, pp. 2–3). Norway (WTO, 2000r) strongly supported the EC proposal for retention of the blue box.

Norway's negotiating proposal, submitted in January 2001, rehearses the importance of NTCs, which 'include agriculture's multifunctional contributions to the viability of rural areas, food security, the cultural heritage and environmental benefits such as the agricultural landscape, agro-biological diversity, land conservation and high standards of plant, animal and public health' (WTO, 2001a, p. 2). The public-goods characteristics of these NTCs, their jointness with ongoing agricultural production and the biological, site-specific nature of agriculture justify special treatment for agricultural trade. Production conditions vary among WTO Members and demand consideration of countries with unfavourable production conditions and narrow product ranges. Thus, Norway called for maximum flexibility in national policy design. The importance of domestic production to safeguard NTCs led Norway to propose that the AMS be divided into two categories. Less stringent reduction commitments would apply to supporting measures for production for the domestic market, while export-oriented production would face reduction. Norway proposed retention of both the blue and green boxes (WTO, 2001a, pp. 3–4, 2001b, pp. 1–2).

Japan

Japan, which has the highest per-hectare producer support (OECD, 2002), also values multifunctionality, a key element in its agricultural policy. Indeed, Japan's basic law on agriculture, enacted in 1999, emphasized support for the multifunctional aspects of farming (Effland *et al.*, 2002, p. 36). Agricultural activities produce food, but they also help to maintain rural communities and to protect the countryside and the environment. Rice farmers, for example, contribute to flood mitigation and therefore receive significant support (ABARE, 1999, p. 1).

Japan's WTO submissions (WTO, 1999d, 2000h) called for 'due consideration' for the importance of multifunctionality and for differences in natural conditions (related to historical background). Like other Members, Japan (WTO, 1999d, p. 3) noted that a multifunctional agriculture involves externalities that cannot be realized by market mechanisms and argued that policy interventions are therefore necessary to promote these externalities. Multifunctionality is especially important when non-commodity outputs are closely related to (or inseparable from) agriculture, when they play an important role in agricultural production activities or when they are especially valued by a country's people. Japan favours WTO measures that allow domestic policies directed towards food security, as well as other multifunctional aspects of agriculture. Japan (WTO, 1999d, p. 4) argued that food security is threatened if all of agricultural production is subject to market mechanisms. Because food security comes through relationships among food imports, domestic production and stockholding, countries should have the right to enact policies to optimize these relationships (Japan and Republic of Korea, 2000, pp. iv, 6–7).

Japan's comprehensive negotiating proposal, submitted in December 2000, begins from a basic philosophy of 'the coexistence of various types of agriculture' and lists the multifunctionality of agriculture as the first of five major points (WTO, 2000z, p. 1). As a major agricultural policy issue worldwide, multifunctionality is characterized by joint production, public-good aspects and externalities that are difficult to value. Multifunctionality requires policy intervention, but there is no consensus about the concept of 'non-trade distortion'. Therefore the most difficult task in WTO negotiations will be 'to harmonize . . . non-trade distortion and the coexistence of various types of agriculture' (WTO, 2000z, p. 6). Japan would improve the green box to reflect the 'real situation of agriculture', by amending the requirement for decoupled income support. In language that seems paradoxical (Agra-Europe Weekly, 2001), Japan recommends that an amended green box allow decoupled support that would permit consideration of 'the actual situation concerning agricultural production . . . while at the same time decoupling it from the current level of production' (WTO, 2000z, pp. 12–13). In addition, Japan would ease the requirements for safety-net programmes under the green box and keep blue-box policies as a 'midpoint' for transforming amber policies to green policies (see WTO, 2000p). Japan indicated that multifunctionality and food security are implicated in many of its proposals, rather than only those concerning domestic support. This Japanese proposal on domestic support has been characterized as 'more entrenched and more reactionary than even that of the European Union' (Agra-Europe Weekly, 2001, p. A/1).

Issues in Multifunctionality

Though WTO Members generally agree that multifunctionality and the associated NTCs are valid concerns, much 'ambiguity and lack of precision' characterize the multifunctionality debate, in part due to the nature and breadth of the concept itself (Paarlberg *et al.*, 2000, pp. 3–4; see also OECD, 2001). Members do not agree about how trade rules should take account of multifunctionality, especially the role and legality of production-related payments to support multifunctional aspects of agriculture.

Much is at stake in the WTO resolution of the multifunctionality issue, because trade talks will determine what domestic support for agriculture will be free from challenge under the (existing or amended) green box, permitted under a renewed blue box or subject to reduction in the amber box. Proponents of multifunctionality are generally Members whose URAA amber boxes are nearly full; these countries have little opportunity to increase production-related support within the constraints of the amber box. Some Members reluctant to authorize domestic support for NTCs have room to increase production-related support within amber-box limits (Bohman *et al.*, 1999, pp. 6–7).

Two related issues seem central to the multifunctionality debate. One is whether the non-food services that are the objectives of multifunctionality and NTCs are inextricably connected with agricultural production – the joint-products issue. Joint production seems to be poorly understood, and yet an understanding is important for analysis of multifunctionality (OECD, 1998b, p. 12). Another issue is what kinds of domestic agricultural policies best ensure that multifunctional aspects of agriculture can be supported without distorting agricultural production or trade.

Joint products

A multifunctional agriculture produces food, fibre and other marketable products, but it also provides non-food services, including food security, environmental benefits and a viable rural area. As Norway indicated, these NTCs – in a sense, countries' interests in the continued existence of those non-food services – are often characterized as public goods that are joint products of agricultural production (WTO, 1999m, p. 2). Norway's negotiating proposal emphasized this jointness: most NTCs are unique or specific to agriculture and 'can only be safeguarded jointly with agricultural production' (WTO, 2001a, p. 7). This conclusion about jointness, however, is not uniformly accepted.

The related public-goods issue is important in a policy context because governments often justify support for the non-food outputs of agriculture with the rationale that these outputs are public goods. Public goods, which are often public services, are generally considered non-rival and non-excludable. Unlike private goods connected with agricultural production (e.g. marketable products, such as food and fibre), public goods may lack effective public markets (Bohman *et al.*, 1999, pp. 10–11). Without markets or corrective policy measures, 'there will be no signals that tell farmers how much of these outputs to produce' (OECD, 2001, p. 28).

Not all so-called public goods, however, are alike; the OECD, for example, identified six possible categories of public goods with different characteristics and management possibilities (OECD, 2001, p. 21). Not all public goods require public support. Some, like protection of environmental habitat, could perhaps be provided by non-profit or other organizations. Moreover, even true public goods may not require government intervention. Some have even argued that such intervention does harm:

> by subsidising the provision of a public good, producers are less inclined to produce it – production moves from being a virtuous action of the producer . . . to being a chore which can result in either compliance problems or at least the payment of considerably higher amounts than were initially considered . . . necessary.
>
> (OECD, 1999a, p. 33 (New Zealand))

The discussion of multifunctionality often focuses on the joint-products issue. That is, are these, often public, goods truly joint products with agricultural production (OECD, 1999b, p. 5)? Countries disagree about the nature of the relationship of various non-commodity outputs with agricultural production (Abler, 2001). Some WTO Members argue that most NTCs cannot be disassociated from agricultural production, but others suggest that some non-commodity outputs can be supplied independently by activities other than agriculture (OECD, 1999b, pp. 4–5). Most agree that the least-cost means of providing those goods should be preferred, whether the providers are farmers or others (OECD, 2001, p. 52).

Further, one may ask whether even non-food services that are produced jointly with agriculture are exclusively joint – that is, whether they can be, but need not always be, provided in conjunction with agriculture. Non-food outputs may be 'the result of particular aspects of the production process', including input use, farming practices, intensity or the level of food production (OECD, 1998a, p. 12). The correlation with agricultural production is not always clear, because higher agricultural production may result in an increase or a decrease in the jointly produced good or service (OECD, 1999a, p. 30 (New Zealand)). Even when products are joint, different types of jointness exist. OECD identified several: technical interdependencies (often causing negative externalities); multiple outputs from the same non-allocable input (wool and mutton, meat and manure); and allocable fixed factors (land, labour), which can be allocated in the production process to change the relationship of outputs. The first two are most relevant to multifunctionality. Some claimed multifunctional outputs of agriculture (rural employment and food security) do not fit easily into any of these categories (OECD, 2001, p. 16).

USDA researchers rejected the assertion that provision of a non-food objective must be linked to agricultural production (Bohman *et al.*, 1999, pp. 12–13). Some countries focus on joint products and argue that 'production-linked payments are necessary to obtain socially desired nonfood outputs'. But many of the non-food objectives that are the focus of multifunctionality do not require agricultural production, and links to agricultural production may change as technology develops (Normile, 2001). In some instances, the non-food output (e.g. a particular environmental amenity) could be achieved by paying the landowner to provide that output (Bohman *et al.*, 1999, p. 12).

From the US point of view, it is important to know how joint products are produced, both 'their relationship to agricultural production, and alternative (non-agricultural) avenues for the provision of each product' (OECD, 1999a, p. 66). Thus, research could profitably consider both positive and negative externalities of agriculture, as they relate to multifunctionality, as well as 'the process by which the various joint products of agriculture are produced, their degree of separability [from agricultural production], and the non-agricultural alternative sources of rural amenities' (OECD, 1999a, p. 66).

Canada, too, has doubts about the essential jointness of non food services. Canada identified two types of jointness. Technical jointness involves two products produced in fixed proportions (e.g. wool and mutton). Economic jointness occurs when two products share a common input. Interdependency between outputs is less rigid here, and appropriate public policy should depend on the nature of the externality. Canada suggested that externalities are not always generated by agricultural production itself, but sometimes by an associated activity. Thus, '[l]imiting the analysis of multifunctionality to the effects of joint products is probably too restrictive'. For example, food security can be satisfied from domestic production, but also from imports or stored food (OECD, 1999a, pp. 14–15).

The joint-products issue is far from resolution. OECD recommended continued work on 'the production relationships underlying the non-food outputs, and the externality and public goods aspects of these outputs' (OECD, 1999b, p. 8). Its own 2001 study, *Multifunctionality: Towards an Analytical Framework*, provides further analysis but no definitive resolutions of the issues connected with jointness, externalities and public goods connected with agricultural production. In part, OECD concluded that:

> the concept of joint production is relevant for the analysis of multifunctionality. The importance of the various causes of jointness, however, may differ between situations where only private goods are involved and others where some non-commodity outputs involve externalities or public goods.
>
> (OECD, 2001, p. 119)

To understand jointness in the agricultural context, it is important to know the 'exact nature of the production relationships underlying the multiple outputs of agriculture', whether provision of commodity and non-commodity outputs by farms is cheaper than separate production of those outputs, how farmers respond to incentives and whether 'alternative combinations of activities and practices . . . would produce . . . goods and services demanded by society more efficiently' (OECD, 2001, p. 119).

Appropriate policies

Even without complete understanding of the joint production process, WTO Members want to ensure continued production of the valued non-food services connected with agriculture. Stating the issue strongly, Norway asserted that all WTO Members 'must be given sufficient flexibility and room for manoeuvre in national agricultural policy design to ensure a viable domestic agricultural sector with domestic production required to properly address NTCs' (Norway, 2000, p. iv). Yet Norway and other friends of multifunctionality agree that this flexibility cannot be 'a *carte blanche*' (Norway, 2000, p. 6), and the European Commission (CEC, 2000a, p. 1) noted that '[f]or all non-trade concerns, care is needed to identify legitimate objectives which may be pursued and avoid abuse of the concept'.

Present policies

WTO negotiations under the Doha Declaration continue to address the issue of NTCs, but some see multifunctionality as a threat to real trade reform. Australian researchers assert that policies to promote multifunctionality (e.g. Japanese rice support connected with flood mitigation) are expensive for both consumers, who pay higher prices for food and other products, and taxpayers, who finance support measures. Support for agriculture deprives non-agricultural activities of financial resources; agricultural producers and exporters in other countries are hurt by lower world prices; and developing countries face lower employment when subsidies distort prices and production (ABARE, 1999, pp. 1–2).

US researchers, too, indicate that 'agricultural price support programs are an inferior means to multifunctional ends because they distort production and trade'. Specifically,

> trade distortions arise because the policies raise prices and domestic agricultural production. Increased production distorts trade and either lowers imports or raises exports. As a result, world prices fall and impact other countries' agricultural sectors. Thus, the requirement that domestic policies be minimally trade-distorting prevents one country's domestic policy from adversely affecting resource allocation in other countries.
>
> (Bohman *et al.*, 1999, p. 12)

One question raised by the USA is whether even decoupled payments (which, in theory, do not distort trade) affect farmers' decisions about production or lead to increased wealth and reduced risk, which may distort trade (United States, 1998, p. 4).

Recent USDA research focused on the connection between farm-programme benefits and farmers' planting decisions and agricultural markets. The research indicated that the effect of direct government payments on resource allocation and markets varies, depending on the programme. For example, production flexibility contracts (PFCs), green-box payments based on enrolled acreage, do not depend on prices or current production and should not influence production decisions. But, because 'PFC payments raise farmers' income and financial well-being, they can . . . enhance production' (Westcott and Young, 2000, p. 11). Farmers who receive payments have more access to loans, more income to invest and higher production in the long term, as well as more willingness to produce high-risk, high-return crops. Though the acreage impacts of PFC payments are rather small – at most, increased planting on less than 0.3% of total cropland – these payments may encourage farmers to keep land in agricultural use (Westcott and Young, 2000, p. 11). Direct decoupled payments (also in the green box) under new agriculture legislation (Farm Bill, 2002) are likely to have similar effects, and farmers' opportunity to update base acreage under that law may lead to trade distortion. Crop insurance, marketing loans and disaster assistance may also have effects on production decisions and agricultural markets. Canadian research, too, has suggested that direct payments to producers may lead to increased production by removing financial constraints that otherwise limit production (Rude, 2001, pp. 1020–1021).

In light of the cost of promoting NTCs and the effect of national measures on trade, numerous issues are involved in determining the best policies for achieving multifunctional objectives connected with agriculture. These include the appropriate level of policy (regional or local, perhaps, rather than national), role of property-rights structures and consistency with WTO guiding principles (e.g. green box). Effective policies should provide joint products associated with multifunctionality at the lowest cost and with least distortion of trade.

Looking beyond agriculture

Some NTCs can be satisfied by means other than support for agriculture. USDA researchers have indicated that a number of public and private instruments, other than production-linked payments, can achieve non-food objectives of agriculture, including environment, food security and rural development. Market-based incentives, government cost-share programmes and land-retirement schemes can help to ensure continued environmental amenities, and measures to protect farmland can ensure preservation of rural landscapes. Various instruments, not linked with production, can help to ensure rural viability and food security (Bohman *et al.*, 1999, pp. 13–21).

In its *Framework* discussion document, the OECD (1998b, p. 14) suggested that 'the more the non-food services can be dissociated from food production, the greater the likelihood that the multiple objectives can be achieved with minimal production and trade distortions'. Some desirable externalities can be provided by non-agricultural activities – for example, non-agricultural employment in rural areas and non–farm management of rural habitat and ecosystems. Food security can be assured by food from alternative sources (stocks, imports) rather than current production. But, as OECD noted in a later report, '[i]t is in society's interest that the demand for non-commodity outputs is satisfied with the least resource cost to the domestic and international economy' (OECD, 2001, p. 52).

In response to the OECD *Framework*, some countries agreed that the current approach to multifunctionality might indeed be too agricultural. That is, analysis should focus first on the non-commodity goods desired by society and thereafter on the best ways to provide them. 'Agriculture should be considered as one possible supplier of the goods but it should not be treated differently from other economic sectors of the economy' (OECD, 1999b, p. 4). Important questions are which non-commodity outputs can be provided by other economic sectors, how well outputs provided by those sectors substitute for outputs provided by agriculture, and where and how non-agricultural sectors can provide these outputs at a lower cost than agriculture (OECD, 2001, p. 52).

The nature of the output is critical, because non-food outputs vary considerably in their relationship to agricultural activity. Recent OECD analysis suggests that some non-food outputs not tied to agricultural land (e.g. viability of rural areas, rural employment) can perhaps be provided economically by non-agricultural enterprises. In contrast, for non-food

> outputs that are tied to agricultural land, including some ecological and amenity services, provision by non-agricultural groups or enterprises is only possible if these are granted access to the land and if the functions they perform do not conflict with the agricultural activity.
>
> (OECD, 2001, p. 18)

Access to land raises significant issues of property rights, especially in countries where the majority of agricultural land is privately owned.

Targeting objectives

In some instances, of course, NTCs may best be satisfied by support to agriculture and other endeavours in rural areas. Ideally, such instances should be characterized by both jointness and market failure (e.g. with public goods), as well as the absence of non-governmental incentives for farmers to provide the NTCs demanded by society (OECD, 2001, pp. 22–24). In cases that justify support, trade measures are rarely most efficient; direct instruments are more appropriate. Efficiency is relevant, because resources required to achieve NTC objectives would otherwise be available to achieve other goals (Anderson, 1998a, p. 4).

The OECD generally recommends market approaches or targeted, decoupled measures as a means of achieving agricultural policy objectives. Targeted policies usually involve lower costs and fewer international spillovers than price support. In designing targeted policies, 'the incentive or disincentive provided by the policy measure [must] be as closely as possible tied to the desired outcome . . . or to the element of the production process that creates the outcome' (OECD, 1998b, p. 18). Further, policies should be decoupled from the inputs or outputs associated with agricultural production. OECD guidelines that govern direct payments for environmental benefits are instructive. Under these guidelines, a country should pay farmers only if there is a well-established public demand for the benefit; relate the size of payment to the farmer's cost incurred or income foregone; target payments closely to specific outcomes; and decouple payments from production or input use (OECD, 1998a, pp. 18–19).

Similarly, economists agree that supporting agriculture in general to satisfy NTCs is usually indirect, costly and ineffective. Instead, direct payments can and should be targeted to the multifunctional objectives and even to the size of society's benefit from those objectives (ABARE, 1999, p. 3). Targeting specific policy objectives – that is, decoupling support – is often cheaper; for example, the farmer could be paid to maintain a hedgerow, rather than to farm on the adjacent land (Anderson, 1998a, p. 9). These recommendations are consistent with a general trade principle articulated by Australian economists: 'Trade policies should not be used for other goals – be they . . . environment, aid, regional employment, health and community welfare.' Instead, efficient policies should be targeted to achieve a specific goal (ABARE, 2001, p. 2).

Agreeing that agricultural price support, which distorts trade, is not the best means of achieving multifunctional objectives, US researchers also emphasize the efficiency of well-designed, targeted policies. Because multifunctional services from agriculture (for example, for the environment or rural areas) are often local, uniform national policies are not likely to be efficient. Instead, more local policies can target the desired aspects of environmental protection or rural development (Bohman *et al.*, 1999, p. 12). USDA researchers recommend principles for effective policy design:

> [P]olicies associated with minimal distortions target the specific objective associated with the nonfood output; the more distorting policy provides the nonfood objective indirectly through the market and thus creates other distortions; . . . [and] policies that have minimal trade distortions are also more efficient at meeting their objectives.
>
> (Bohman *et al.*, 1999, p. 22)

Not every WTO Member supports targeted policies to promote multifunctionality. For example, Norway (WTO, 2001a, p. 10) acknowledged that maximum targeting of policies is desirable, but asserted that targeted green-box measures designed to achieve NTCs related to the environment would carry high administrative and control costs. The trade-off between precision and high administrative costs means that extensive and detailed targeting would be infeasible and impractical, especially in countries with many small-scale farmers.

Building the green box and resolving the blue box

The green box often holds Member policies designed to achieve multifunctional objectives, and its structure is therefore central to WTO negotiations. Because green box measures are exempt from AMS reduction and challenge, countries have an incentive to put policies in the green box. Under the URAA, Members have shifted support from the amber and blue boxes to the green box, changing the form of support, instead of reducing its level. At the same time, blue-box support has been an important component of policy for the EU and a few other Members. A continued blue box is part of the EC negotiating proposal (WTO, 2002d), but the US proposal (WTO, 2002a) would allow only two categories of domestic support, effectively eliminating the blue box. Whether the blue box is continued under the Doha Declaration may be related to the extent of perceived distortion in blue-box policies and the level of flexibility in green-box criteria. In its Mid-term Review of Agenda 2002, the European Commission (CEC, 2002) proposed the introduction of a system of producer support intended to meet green-box requirements. This change would make the blue box less critical for the CAP, but other countries (e.g. Norway) still rely on the blue box.

Though amber-box (and, to a lesser extent, blue-box) policies distort trade, green-box policies, with at most a minimal effect on trade or production, are less distorting. Many non-commodity objectives of agriculture can be achieved through green-box policies, but disagreement exists about both the ability of the current green box to meet the objectives of multifunctionality and the real effect of green-box policies on trade. The level of production and trade impact that is considered minimal is not specified in the URAA and is subject to discussion (see Rude, 2001). Some WTO Members, of course, would prefer more stringent green-box criteria. Others, especially Members with full amber boxes, prefer more flexible criteria for the green box, even allowing production-related support to enhance multifunctionality and achieve the related NTCs. Increased flexibility, which could allow more than minimal trade distortion (e.g. as Japan seems to propose), would help those countries to meet national objectives (Bohman *et al.*, 1999, pp. 6, 23).

Some Members have asserted that green-box measures do not accommodate NTCs adequately, especially in developing countries and countries with difficult production conditions (WTO, 2000u, pp. 5–6). Developing countries criticized the present green box because it leads to higher overall domestic support levels, provides opportunities for misuse, lacks transparency and enjoys protection under the peace clause. But they also found the green box too narrow to protect NTCs in developing countries, especially the livelihood of small farmers and food security. Therefore they recommended that domestic support be combined into one category, called general subsidies, with criteria to establish legality and a common level (10%) of protected support. Developing countries should enjoy additional flexibility to address NTCs connected with rural employment and food security (WTO, 2000a). In 2002, a group of African Members also criticized the green box, because it disguises support from developed countries like the USA and the EU and leads to abuse. They proposed a 'development box' with more flexibility to provide targeted aid to meet NTCs and other concerns (WTO, 2002b).

Even without additional flexibility, some measures currently in the green box may not be production-neutral. USDA research (discussed above) confirms that some green-box programmes do affect production, as well as prices and trade patterns (Westcott and Young, 2000). Large land-retirement programmes reduce production of commodities; the Conservation Reserve Program, for example, idled 9.4% of cropland in 1995. Large-scale programmes with government cost-share payments for environmentally friendly technologies – such as the Environmental Quality Incentives Program, with increased funding under the 2002 Farm Bill – affect costs and therefore trade. But these programmes distort trade only minimally (Vasavada and Warmerdam, 1998, pp. 12–13).

Some analysts fear that opening the green box for renegotiation will result in expanded criteria, allowing more policies that distort production and trade. To avoid distortion, multifunctional policies could be required to conform strictly to existing green-box criteria (Josling, 1998, p. 14), or criteria could be made even stricter to eliminate policies, such as crop insurance, that encourage production (Anderson, 1998b, p. 21). Some analysts suggested rather stringent justifications for green-box policies, designed to avoid disguised protection and improve transparency.

> First, a nation would need to explicitly identify the externalities due to multifunctionality. Second, a nation would need to value those externalities using standard market and non-market valuation techniques. Third, the values of the externalities would have to be explicitly linked to commodity output levels.
>
> (Paarlberg *et al.*, 2000, p. 19)

Similarly, another analyst suggested that Members proposing green-box payments designed to correct a market failure should be required to explain the market failure, show why private markets do not resolve the failure and explain how the proposed programme will be effective. If a farmer could anticipate the payment and alter production decisions, further criteria (perhaps support limits or need-based targeting) should be added (Rude, 2001, p. 1028). These approaches are exacting, but they could ensure that green-box programmes have proved motivation and efficacy.

On the question of green-box criteria, a scholar (a 'long-time observer of trade negotiations') recently raised two significant questions. First, 'if payments are made whose primary aim or effect is to increase producer income, will such payments indeed have a minimal impact on production?' (Blandford, 2001, pp. 48, 54). USDA research mentioned above suggests that the answer to this question may be 'no'. Even decoupled payments increase financial resources available to farmers and thus influence the level of output (Blandford, 2001, p. 50). Second, 'if payments are made to producers in order to achieve other aims, e.g. environmental objectives, is it logical to require these to have a minimal impact on production?' (Blandford, 2001, p. 48). This question, of course, goes to the heart of the multifunctionality debate. Payments to producers may be necessary to ensure continued provision of a public good (e.g. agricultural landscapes) or a positive externality (e.g. wildlife habitat). Such payments, linked to continued farming or specific farming practices, may affect production even if targeted to achieve the desired multifunctional result. Indeed, if payments to ensure desired public goods and positive externalities are legitimate, the idea that such payments cause distortion may be wrong. That is, when 'levels of production, consumption, and trade are determined without taking into account unpriced but valuable outputs, then markets are already distorted', and the payments help to correct the distortion (Blandford, 2001, p. 52, bold omitted). Clearer 'structural and operational' rules for green-box payments may be necessary to limit distortion of production, especially for payments directed towards the supply of non-market goods. Such rules should allow payments for 'legitimate trade-affecting policies (those designed to ensure the supply of a public good or positive externality)', but ensure that those payments are not used to distort trade (Blandford, 2001, p. 55).

Discussions of trade policies to address multifunctional objectives and NTCs often pay little attention to underlying considerations of land ownership and property rights. In the context of the environmental and landscape benefits from agriculture, which are important multifunctional objectives, property-rights issues are relevant. In many countries, the majority of agricultural land is privately owned. Though the permissible scope of land-use regulation varies significantly under different legal systems, governments are often cautious about interfering with land-use decisions on privately owned agricultural land (Tobey and Smets, 1996, p. 72). The European Commission, for example, noted that it could be 'an infringement of private property rights' to require farmers to provide environmental benefits without compensation (CEC, DG Agri, 1999c, ¶ 19).

Governments can and do use regulation to prohibit activities that cause harmful externalities, such as pollution, but they are less likely to enact measures that require specific land uses or agricultural practices. Prohibition of harmful activities is consistent with the polluter-pays principle, which may require good farming practices.[7] In contrast, provision of land uses (non-commodity outputs) valued by society may require judicious application of the provider-gets principle, with payment for services. When landowners do not share governmental objectives for agricultural land use, payments to farmers may be necessary to generate the desired level of environmental amenities.

Researchers have suggested that multifunctional objectives might be achieved by a natural-resources policy that pays farmers and other land managers for positive non-commodity outputs (with payments targeted to output) and assesses penalties for negative outputs. Because of the 'spatial diversity' of non-commodity outputs, lower levels of government could most effectively design and manage the policy to reward or penalize land managers. Higher-level government could play a role in protecting public goods and ensuring compliance with international obligations. Such a programme would involve 'payment for services rendered' or payment for 'an unpriced output that has social value', rather than a subsidy. This type of natural-resources policy would comply with WTO requirements if Member programmes met criteria for payments established in a revised green box, perhaps with a required review of each Member programme (Blandford and Boisvert, 2002, pp. 110–116).

Concluding Observations

Recent WTO agricultural negotiations have focused, in part, on acceptable ways of ensuring that agriculture can continue to serve its multifunctional roles and satisfy related NTCs without causing distortions in production and trade. Perhaps too simply stated, a critical question is whether support for a multifunctional agriculture must meet the criteria of the (existing or amended) green box, or whether that support can instead be linked to production.

Using economic analysis to bridge the gap between countries that favour domestic support linked to agriculture and countries that prefer decoupled agricultural support, researchers concluded

> First, multifunctionality never justifies trade barriers . . . Second, multifunctionality may justify domestic output subsidies or taxes if the level of externality is tied to output levels . . . Third, the extent of support to domestic agriculture varies by nation . . . Fourth, nations have the incentive to inflate the importance of multifunctionality to disguise protection so strong disciplines must be negotiated.
>
> (Paarlberg *et al.*, 2000, pp. 18–19)

Broadly based domestic support for producers is a blunt instrument for satisfying NTCs associated with agriculture, and protectionist trade measures are generally inefficient in addressing NTCs (Anderson, 1998b, pp. 14, 25). Those who oppose indiscriminate support for multifunctionality assert that there are 'more effective and less costly ways of maintaining what people in society want [than broadly based agricultural protection]' and that subsidized agricultural production often increases negative environmental effects from agriculture (ABARE, 1999, p. 5). None the less, most WTO Members have NTCs connected with agriculture and seek legal ways to satisfy these important national concerns. For the USA in particular, a US Agency for International Development report suggested that solving world agricultural trade problems may require US 'acceptance of agriculture's "multifunctionality" as the basis for domestic policies that have clear social, environmental, or security rationales' (USAID, 2002, p. 15).

Reasonable WTO Members disagree about the means of ensuring that each country can meet its domestic objectives, as well as its obligations to the world trading community. In densely populated Europe, where people live and recreate in areas of agricultural production, the multiple functions of agriculture seem more significant, perhaps, than in nations where the population lives at a distance from the countryside. But, for most WTO Members, agriculture is a major use of land-based resources with amenities valued by society. National policies should therefore recognize agriculture's 'role in the supply of highly valued non-commodity outputs', as long as these policies do not harm trading partners (Blandford and Boisvert, 2002, p. 110). To achieve the often-conflicting goals of a multifunctional agriculture and liberalized trade, WTO Members, both friends of multifunctionality and other nations, must remain open-minded and exhibit a spirit of compromise in the Doha Round of agricultural trade negotiations.

Notes

1 The Cairns Groups consists of agricultural exporting countries: Argentina, Australia, Bolivia, Brazil, Canada, Chile, Colombia, Costa Rica, Fiji, Guatemala, Indonesia, Malaysia, New Zealand, Paraguay, the Philippines, Thailand, South Africa and Uruguay. It was founded in part to ensure that the Uruguay Round included agricultural trade.

2 The so-called 'last draft' of the Seattle negotiations was not released officially by the WTO. The draft referred to '[s]ubstantial progressive reductions in domestic support' (¶ 29(iii)). Negotiations were also to take into account NTCs, including 'the need to protect the environment, food security, the economic viability and development of rural areas, and food safety'. NTCs were to be addressed by measures consistent with the WTO, 'particularly targeted, transparent and non-trade distorting measures' (¶ 29*bis*) (WTO, 1999n).

3 In the UK, unlike other EU countries, population in the countryside is increasing. Despite pockets of poverty, many rural areas now enjoy significant prosperity (Countryside Agency, 2001, pp. 30–35).

4 The USA views its Proposal for Long-Term Comprehensive Agricultural Trade Reform as a food-security proposal, which would help to satisfy one important NTC (WTO, 2000b, p. 5). That is, liberalized agricultural trade and legitimate assistance programmes will help to strengthen food security, as will domestic measures that do not distort trade. Free trade will lead to greater food security through an expanded food supply, more efficient production, economic growth and innovation. Trade, alone, cannot solve all food-security problems, however, and the USA proposed continuation of food aid and related disciplines (WTO, 2000b, pp. 5–6).

5 Some countries that suffer from low agricultural production and lack of capital find standard green- or blue-box measures out of economic reach. These countries therefore propose additional exemptions (e.g. investment, input, interest subsidies), applicable while persistent economic problems plague agriculture. See, for example, the proposal by former centrally planned economies (WTO, 2000t).

6 An analysis published in *Agra Europe* was rather sceptical of the EC's blue-box arguments. Though the EC argues that the blue box was intended to allow domestic subsidies that were less damaging than other supports, *Agra Europe* noted that the real intention of the blue box was to serve as a transition between excess subsidies and no subsidies (Agra-Europe Weekly, 2000a, p. A/2). *Agra Europe* expressed doubt that EC farmers, who receive more than half of their income from blue-box payments, will respond to market signals, even under a reformed CAP. Though blue-box payments result in less trade distortion than earlier EC policies, 'direct payments are still more distorting than no support at all' (p. A/3). Further, *Agra Europe* insisted, subsidies are not neutral. Blue-box

payments under the CAP have raised costs for EU consumers and taxpayers, and the payments are 'production positive' (p. A/3).

7 In fact, it might be argued that the URAA (Annex 2, ¶ 12(b)) itself violates the polluter-pays principle because it allows environmental subsidies to pay 'up to the full cost of compliance with environmental measures' (Tobey and Smets, 1996, pp. 65, 82).

References

ABARE (1999) 'Multifunctionality' a pretext for protection? *ABARE Current Issues* 99(3), 1–6.

ABARE (2001) A vision for multilateral agricultural policy reform. *ABARE Current Issues* 2001(2), 1–4.

Abler, D. (2001) A synthesis of country reports on jointness between commodity and non-commodity outputs in OECD agriculture. In: *OECD Workshop on Multifunctionality*, Paris, 2–3 July 2001. From Website.

Agra-Europe Weekly (2000a) Blue box defence is key to EU's WTO position. *Agra Europe*, 30 June, A/1–A/3.

Agra-Europe Weekly (2000b) EU strengthens position on WTO issues. *Agra Europe*, 27 October, A/1–A/2.

Agra-Europe Weekly (2000c) 'Non-trade' farm issues take centre stage at WTO. *Agra Europe*, 4 November, EP/7.

Agra-Europe Weekly (2001) Domestic support key to WTO progress. *Agra Europe*, 16 February, A/1–A/2.

Anderson, K. (1998a) *Domestic Agricultural Policy Objectives and Trade Liberalisation: Synergies and Trade-offs*. OECD document: COM/AGR/CA/TD/TC/WS(98)101 (October).

Anderson, K. (1998b) *Agriculture, WTO, and the Next Round of Multilateral Trade Negotiations*. Prepared for World Bank/WTO Joint Trade and Development Centre website, http://www.itd.org.

Blandford, D. (2001) Are disciplines required on domestic support? *The Estey Centre Journal of International Law and Trade Policy* 2(1), 35–59.

Blandford, D. and Boisvert, R.N. (2002) Multifunctional agriculture and domestic/international policy choice. *The Estey Centre Journal of International Law and Trade Policy* 3(1), 106–118.

Bohman, M., Cooper, J., Mullarkey, D., Normile, M.A., Skully, D., Vogel, S. and Young, E. (1999) *The Use and Abuse of Multifunctionality*. Economic Research Service, USDA (November), at www.ers.usda.gov/briefing/wto/multifunc119.pdf.

Burrell, A. (2001) Synthesis of the evidence on the possible impact of commodity price decreases on land use and commodity production, and the incidence on the provision of non-commodity outputs. In: *OECD Workshop on Multifunctionality*, Paris, 2–3 July 2001. From Website.

Cairns Group (1999) *Multifunctionality*. Cairns Group Papers; Fact Sheet by Australia, at http://www.dfat.gov.au/trade/negotiations/cairns_group/paper_9901_multifunctionality.html (visited 22 September 2000).

CEC (2000a) Agriculture's contribution to environmentally and culturally related non-trade concerns. Discussion paper. International Conference on

Non-Trade Concerns in Agriculture, Ullensvang, Norway, 2–4 July 2000, at http://odin.dep.no/ld/engelsk/multifunctionality/.

CEC (2000b) Agriculture's contribution to rural development. Discussion Paper, International Conference on Non-trade Concerns in Agriculture, Ullensvang, Norway, 2–4 July 2000, at http://odin.dep.no/ld/ engelsk/multifunctionality/.

CEC (2002) *Mid-term Review of the Common Agricultural Policy*. COM(2002) 394 final (10 July). From EU Website.

CEC, Directorate-General for Agriculture (DG Agri) (1999a) *CAP Reform: Rural Development (Fact Sheet)*. From EU Website.

CEC, Directorate-General for Agriculture (DG Agri) (1999b) *EU Agriculture and the WTO: From the Uruguay Round to the Millenium Round (Fact Sheet)*. From EU Website.

CEC, Directorate-General for Agriculture (DG Agri) (1999c) *Contribution of the European Community on the Multifunctional Character of Agriculture*. Info-Paper, at europa.eu.int/comm/agriculture/external/wto/document/ip2_en.pdf.

CEC, Directorate-General for Agriculture (DG Agri) (1999d) *Safeguarding the Multifunctional Role of EU Agriculture: Which Instruments?* Info-Paper, at europa.eu.int/comm/agriculture/external/wto/document/ip3_en.pdf.

CEC, Directorate-General for Agriculture (DG Agri) (2001) *EU Agriculture and the WTO*. (September). From EU Website.

Countryside Agency (2001) *The State of the Countryside*. Cheltenham, Gloucestershire. 84 pp.

Effland, A., Normile, M.A., Young, E. and Dyck, J. (2002) Commodity policies of the US, EU, and Japan – how similar? *Agricultural Outlook* (December), 33–37.

Farm Bill (2002) *Farm Security and Rural Investment Act of 2002*. Pub. L. No. 107–171, 116 Stat. 134–540. US Government Printing Office, Washington, DC.

Gale, F. (2000) How important are farm payments to the rural economy? *Agricultural Outlook* (October), 15–18.

Green, R. (2000) Panel II C: Agreement on Agriculture: the Uruguay Round Agreement on Agriculture. *Law and Policy in International Business* 31, 819–835.

Hill, R. (2000) Cairns Group statement on multifunctionality. In: *8th Meeting of the Commission on Sustainable Development*, New York, 26 April, http://203.6.171.15/trade/negotiations.cainns_group/media/000426_statement.html (last visited 22 Sept 2000).

Japan and Republic of Korea (2000) Food security and the role of domestic agricultural production. Discussion paper. International Conference on Non-Trade Concerns in Agriculture, Ullensvang, Norway, 2–4 July 2000, at http://odin.dep.no/ld/engelsk/multifunctionality/.

Josling T. (1998) *The Uruguay Round Agreement on Agriculture: A Forward Looking Assessment*. OECD document COM/AGR/CA/TD/TC/WS(98)100 (October).

Lindland, J. (Minister of Agriculture, Norway) (1998) *Non-Trade Concerns in a Multifunctional Agriculture: Implications for Agricultural Policy and the Multilateral Trading System*. OECD document COM/AGR/CA/TD/TC/WS(98)124.

Mauritius (2000) Developing countries and non-trade concerns. Discussion paper. International Conference on Non-Trade Concerns in Agriculture,

Ullensvang, Norway, 2–4 July 2000, at http://odin.dep.no/ld/engelsk/multifunctionality/.

Normile, M.A. (2001) Multifunctionality: options for agricultural reform. In: Burfisher, M.E. (ed.) *Agricultural Policy Reform in the WTO – the Road Ahead.* Agricultural Economic Report No. 802, USDA, Washington, DC, pp. 80–81.

Normile, M.A. and Bohman, M. (2002) Non-trade concerns: international debate and US policy. *Agricultural Outlook* (June–July), 14–17.

Norway (2000) The need for flexibility in national policy design to address non-trade concerns. Discussion paper. International Conference on Non-trade Concerns in Agriculture, Ullensvang, Norway, 2–4 July 2000, at http://odin.dep.no/ld/engelsk/ multifunctionality/.

OECD (1998a) Communiqué: Agriculture in a changing world: which policies for tomorrow? Meeting of the Committee for Agriculture at Ministerial Level, 5–6 March, at http://www.oecd.org/agr/ministerial/commune.htm.

OECD (1998b) Secretariat. *Multifunctionality: a Framework for Policy Analysis.* OECD document AGR/CA(98)9 (November).

OECD (1999a) Secretariat. *Written Comments on the Document 'Multifunctionality: A Framework for Policy Analysis' [AGR/CA(98)9].* OECD document AGR/CA/RD(99)1 (March).

OECD (1999b) Secretariat. *Multifunctionality: Status Report and Proposal for Further Work.* OECD document AGR/CA(99)2 (April).

OECD (2000) Final Communiqué, OECD Council, Meeting at Ministerial Level, Paris (26–27 June), at http://www.oecd.org/agr/News/index.htm.

OECD (2001) *Multifunctionality: Towards an Analytical Framework.* OECD, Paris, 159 pp.

OECD (2002) *Agricultural Policies in OECD Countries: Monitoring and Evaluation 2002.* OECD, Paris, 238 pp.

Paarlberg, P.L., Bredahl, M. and Lee, J.G. (2000) Multifunctionality and agricultural trade negotiations. Manuscript.

Pruzin, D. (2000a) Ghosts of Seattle haunt WTO in form of draft ministerial text. *BNA International Trade Daily*, 13 June.

Pruzin, D. (2000b) Norway's trade policy marred by agricultural protectionism, WTO says. *BNA International Trade Daily*, 26 June.

Rude, J. (2001) Under the green box: the WTO and farm subsidies. *Journal of World Trade* 35(5), 1015–1033.

Runge, C.F. (1998) *Emerging Issues in Agricultural Trade and the Environment.* OECD document COM/AGR/CA/TD/TC/WS(98)103 (October).

Tielu, A. and Roberts, I. (1998) Farm income support: implications for gains from trade of changes in methods of support overseas. *ABARE Current Issues* 98(4), 1–8.

Tobey, J.A. and Smets, H. (1996) The polluter pays principle in the context of agriculture and the environment. *The World Economy* 19(1), 63–87.

United States (1998) *Domestic Support Policies.* OECD document COM/AGR/CA/TD/TC/WS(98)115 (October).

US Agency for International Development (USAID) (2002) *Foreign Aid in the National Interest: Promoting Freedom, Security, and Opportunity.* USAID, Washington, DC, 169 pp.

US Department of Agriculture (USDA) (2001) *Food and Agricultural Policy: Taking Stock for the New Century.* USDA, Washington, DC, 120 pp.

Vasavada, U. and Warmerdam, S. (1998) Environmental policy and the WTO: unresolved questions. *Agricultural Outlook* (November), 12–14.

Westcott, P.C. and Young, C.E. (2000) US farm program benefits: links to planting decisions and agricultural markets. *Agricultural Outlook* (October), 10–14.

WTO (1994) *Uruguay Round Agreement on Agriculture (URAA)*. From Website.

WTO (1996) *Trade Liberalization and the Environment*. WT/CTE/W/35 (23 July). From Website.

WTO (1998) *United States: General Council Discussion on Mandated Negotiations and the Built-In Agenda*. WT/GC/W/115 (19 November). From Website.

WTO (1999a) *Australia: Communication [Cairns Group Vision Statement]*. WT/GC/W/156 (24 March). From Website.

WTO (1999b) *Australia: Domestic Support*. WT/GC/W/177 (4 May). from Website.

WTO (1999c) *United States: Objective and Overall Framework for the Agriculture Negotiations*. WT/GC/W/186 (20 May). From Website.

WTO (1999d) *Japan: Negotiations on Agriculture*. WT/GC/W/220 (28 June). From Website.

WTO (1999e) *Norway: Agriculture*. WT/GC/W/238 (6 July). From Website.

WTO (1999f) *EC Approach on Agriculture*. WT/GC/W/273 (27 July). From Website.

WTO (1999g) *United States: Objectives for the Agriculture Negotiations: Export Competition*. WT/GC/W/286 (4 August). From Website.

WTO (1999h) *United States: Objectives for the Agriculture Negotiations: Market Access*. WT/GC/W/287 (4 August). From Website.

WTO (1999i) *United States: Measures Affecting Trade in Agricultural Biotechnology Products*. WT/GC/WW/288 (4 August). From Website.

WTO (1999j) *United States: Objectives for the Agriculture Negotiations: Domestic Support*. WT/GC/W/290 (4 August). From Website.

WTO (1999k) *United States: Trade and Sustainable Development*. WT/GC/W/304 (6 August). From Website.

WTO (1999l) *Cairns Group: Communiqué, Nineteenth Ministerial Meeting*. WT/L/312 (3 September). From Website.

WTO (1999m) Norway: Appropriate policy measure combinations to safeguard non-trade concerns of a multifunctional agriculture. Informal Process of Analysis and Information Exchange of the WTO Committee on Agriculture (28 September), at http://odin.dep.no/ld/engelsk/multifunctionality.

WTO (1999n) *Text on Agriculture*. Last draft of WTO negotiations, Seattle, at http://www.agriculturelaw.com/specials/agtext.htm (visited 19 April 2001).

WTO (2000a) *Cuba* et al.: *Agreement on Agriculture: Green Box/Annex 2 Subsidies*. G/AG/NG/W/14 (23 June). From Website.

WTO (2000b) *United States: Proposal for Comprehensive Long-Term Agricultural Trade Reform*. G/AG/NG/W/15 (23 June). From Website.

WTO (2000c) *United States: Note on Domestic Support Reform*. G/AG/NG/W/16 (23 June). From Website.

WTO (2000d) *EC: The Blue Box and Other Support Measures to Agriculture*. G/AG/NG/W/17 (28 June). From Website.

WTO (2000e) *EC: Food Quality – Improvement of Market Access Opportunities*. G/AG/NG/W/18 (28 June). From Website.

WTO (2000f) *EC: Animal Welfare and Trade in Agriculture*. G/AG/NG/W/19 (28 June). From Website.

WTO (2000g) *EC: Introduction of EU Papers*. G/AG/NG/W/24 (11 July). From Website.

WTO (2000h) Japan: statement. Second Special Session of the Committee on Agriculture, G/AG/NG/W/27 (11 July). From Website.

WTO (2000i) *Norway: Statement*. G/AG/NG/W/30 (11 July). From Website.

WTO (2000j) *United States: Statement Introducing a Proposal for Comprehensive Long Term Agricultural Trade Reform*. Greg Frazier; August Schumacher. G/AG/NG/W/32 (12 July). From Website.

WTO (2000k) *EC: Export Competition*. G/AG/NG/W/34 (18 September). From Website.

WTO (2000l) *Cairns Group: Negotiating Proposal: Domestic Support*. G/AG/NG/W/35 (22 September). From Website.

WTO (2000m) Barbados *et al*.: Note on non-trade concerns, G/AG/NG/W/36 (22 September) and Rev.1 (9 November) (six discussion papers from International Conference on Non-Trade Concerns in Agriculture, Ullensvang, Norway, 2–4 July 2000). From Website.

WTO (2000n) *Australia: Statement Introducing the Cairns Group Proposal on Domestic Support*. G/AG/NG/W/40 (5 October). From Website.

WTO (2000o) *Introduction of the EC paper on Export Competition*. G/AG/NG/W/45 (5 October). From Website.

WTO (2000p) *Comments by Japan on the Negotiation Proposals*. G/AG/NG/W/46 (5 October). From Website.

WTO (2000q) *United States: Domestic Support Reform*. G/AG/NG/W/49 (5 October). From Website.

WTO (2000r) *Norway: Comments on Negotiating Proposals Submitted by Members*. G/AG/NG/W/50 (6 October). From Website.

WTO (2000s) *Bulgaria* et al.: *Statement and Cairns Group proposal on Domestic Support (NG/W/35)*. G/AG/NG/W/53 (16 October). From Website.

WTO (2000t) *Albania* et al.: *WTO Negotiations on Agriculture: Domestic Support – Additional Flexibility for Transition Economies*. G/AG/NG/W/56 (14 November). From Website.

WTO (2000u) *Statement by Norway*. G/AG/NG/W/77 (28 November). From Website.

WTO (2000v) *Statement by Australia. Non-trade Concerns*. G/AG/NG/W/59 (29 November). From Website.

WTO (2000w) *Statement by the European Communities*. G/AG/NG/W/66 (29 November). From Website.

WTO (2000x) *Argentina: Legitimate Non-trade Concerns*. G/AG/NG/W/88 (30 November). From Website.

WTO (2000y) *EC Comprehensive Negotiating Proposal*. G/AG/NG/W/90 (14 December). From Website.

WTO (2000z) *Negotiating Proposal by Japan on WTO Agricultural Negotiations*. G/AG/NG/W/91 (21 December). From Website.

WTO (2000aa) *Canada: WTO Negotiations on Agriculture – Domestic Support*. G/AG/NG/W/92 (21 December). From Website.

WTO (2001a) *Proposal by Norway*. G/AG/NG/W/101 (16 January). From Website.

WTO (2001b) *Statement by Norway*. G/AG/NG/W/120 (15 February). From Website.

WTO (2001c) *Communication from Mauritius*. G/AG/NG/W/186 (20 July). From Website.

WTO (2001d) *Green Box: Non-paper from the European Communities, 6865* (24 September). From Website.

WTO (2001e) *Geographical Indications: Non-paper from the European Communities, 6866* (24 September). From Website.

WTO (2001f) *European Communities: Mandatory Labelling for Agricultural Products* (3–5 December). From Website.

WTO (2001g) *Ministerial Declaration*. Adopted on 14 November 2001, Doha, Qatar. WT/MIN(01)/DEC/1. From Website.

WTO (2002a) *United States: Proposal for Global Agricultural Trade Reform* (July), available at http://www.fas.usda.gov/itp/wto/actual.htm.

WTO (2002b) *Senegal: Common Position of the Member States of the West African Economic and Monetary Union (WAEMU) in the Multilateral Trade Negotiations on Agriculture*. G/AG/NG/W/188 (26 September). From Website.

WTO (2002c) *Domestic Support, Specific Input: Cairns Group Negotiating Proposal* (27 September), at http://www.cairnsgroup.org.

WTO (2002d) *The EC's Proposal for Modalities in the WTO Agriculture Negotiations*. Ref. 625/02 (16 December), available at europa.eu.int/comm/trade/goods/agri/modalities.pdf.

Multifunctionality of Agriculture: a European Community Perspective

6

Michael Cardwell

Introduction

The objectives of the Common Agricultural Policy (CAP) have never been limited to food production. However, there has been general recognition that, in historical terms, the weight of legislation and of the budget has focused on the various common organizations of the market (in particular, cereals, beef and veal and milk and milk products); and that the mechanisms of those common organizations of the market have combined to provide substantial price support. More recently, there has again been general recognition that the focus of agricultural policy has broadened, with the development of a 'European Model of Agriculture', multifunctional in nature. None the less, it would seem that agriculture itself remains of central importance; and it is notable that the Community has retained a 'Common Agricultural Policy', as opposed to, for example, adopting a 'Common Agricultural and Rural Policy for Europe', as advocated in the Buckwell Report (Buckwell Report, 1997).

At a theoretical level, some support for this approach may be derived from *Multifunctionality: a Framework for Policy Analysis*, issued by the Organisation for Economic Co-operation and Development (OECD). Not least, there was the assertion that: '[a]t the heart of multifunctionality is the agricultural activity' (OECD, 1998b, para. 5). Further, emphasis was laid upon the joint production process, which 'determines the way in which the multiple outputs of agriculture are linked to the farming activity, including the complementarities and trade-offs among the various non-food outputs, and their relationship with food production'

(eds M.N. Cardwell, M.R. Grossman and C.P. Rodgers)

(OECD, 1998b, para. 42). That said, differing views have been forcefully expressed. For example, New Zealand has stated that these non-food outputs may be generated without any agricultural production (OECD, 1999, para. 130); and it is perhaps significant that, even within the Community, the English Ministry of Agriculture, Fisheries and Food (MAFF) was replaced in 2001 by the Department for Environment, Food and Rural Affairs.

In this context, five aspects may be considered: first, the development of multifunctional agriculture in the European Community prior to the Agenda 2000 reforms; secondly, the commencement of the Agenda 2000 reforms, including the pressures that have promoted changes in policy; thirdly, the multifunctional elements of the measures agreed both at the Special Berlin European Council of 24 and 25 March 1999 (Berlin Summit) and in the period prior to the *Mid-term Review of the Common Agricultural Policy*, issued in July 2002 (Commission of the European Communities (CEC), 2002a); fourthly, the impact of the Mid-term Review; and, fifthly, specific World Trade Organization (WTO) implications.

The Development of Multifunctionality of Agriculture in the European Community Prior to the Agenda 2000 Reforms

As indicated, the CAP has always embraced more than food production. At the inception of the Community, Article 39 of the EEC Treaty (now Article 33 EC) provided that:

> 1. The objectives of the common agricultural policy shall be:
> (a) to increase agricultural productivity by promoting technical progress and by ensuring the rational development of agricultural production and the optimum utilisation of the factors of production, in particular labour;
> (b) thus to ensure a fair standard of living for the agricultural community, in particular by increasing the individual earnings of persons engaged in agriculture;
> (c) to stabilise markets;
> (d) to assure the availability of supplies;
> (e) to ensure that supplies reach consumers at reasonable prices.
> 2. In working out the common agricultural policy and the special methods for its application, account shall be taken of:
> (a) the particular nature of agricultural activity, which results from the social structure of agriculture and from the structural and natural disparities between the various agricultural regions;
> (b) the need to effect the appropriate adjustments by degrees;
> (c) the fact that in the Member States agriculture constitutes a sector closely linked with the economy as a whole.[1]

Accordingly, the objectives to be taken into account have always included social concerns and structural policy; and it has long been established by such cases as *Balkan* v. *Hauptzollamt Berlin-Packhof* (1973) that there is no strict hierarchy to be observed. Thus, should there be conflict between objectives, the Community institutions enjoy considerable discretion in according priority. Likewise, these provisions have throughout shown full appreciation of the diversity of farming practices in different regions of the Community. Moreover, as the Community has expanded, so has this diversity. With the accession of Austria, Finland and Sweden on 1 January 1995, the 15 Member States extended from the Arctic to the Mediterranean; and eastward enlargement will bring in a further ten Member States in 2004.

While price and market support proved central to the CAP prior to the MacSharry reforms of 1992, more multifunctional aspects did effect periodic resurgences. In particular, the importance of such aspects was reiterated in proposals and other documentation emanating from the Commission (Fennell, 1997). Further, although admittedly a pale reflection of the radical changes advocated by the Mansholt Plan (CEC, 1968), three structural directives were implemented in 1972: Council Directive 72/159/EEC, on the modernization of farms (OJ 1972, L96/1); Council Directive 72/160/EEC, concerning measures to encourage the cessation of farming and the reallocation of utilized agricultural area for the purposes of structural improvement (OJ 1972, L96/9); and Council Directive 72/161/EEC, concerning the provision of socio-economic guidance for and the acquisition of occupational skills by persons engaged in agriculture (OJ 1972, L96/15). Significantly, by 1985 the Commission could state that '[i]n the development of the common agricultural policy, attention has to be paid not only to the stabilisation of agricultural markets but also to the demands of consumers in terms of quality of food'; and that '[t]he challenge for the Community now is to reconcile the success of the CAP in achieving its economic objectives with the need to continue to fulfil the social objective of assuring a fair standard of living for the agricultural population' (CEC, 1985, pp. 1 and 2). Moreover, in the same year, Council Regulation (EEC) 797/85 (OJ 1985, L93/1) was enacted, which not only provided a framework for improving the efficiency of agricultural structures, but also rendered it possible for Member States to take special measures in environmentally sensitive areas with a view to initiating or maintaining agricultural practices compatible with the protection of the countryside. For these purposes 'environmentally sensitive areas' meant, in particular, areas of recognized importance from an ecological and landscape point of view.

The MacSharry reforms of 1992 unequivocally accorded the wider role of agriculture greater emphasis. Thus, the Reflections Paper drawn up by the Commission declared, in language that was to become ever more prevalent, that '[s]ufficient numbers of farmers must be kept on the land. There is no other way to preserve the natural environment, traditional landscapes and a model of agriculture based on the family farm as favoured by the society generally' (CEC, 1991, pp. 9–10). The environment was singled out for particular attention. Thus, '[c]oncern for the environment means that we should support the farmer also as an environment manager through use of less-intensive techniques and the implementation of environment-friendly measures' (CEC, 1991, p. 10). This concern was addressed by legislation introducing 'accompanying measures', alongside the various common organizations of the market. Council Regulation (EEC) 2078/92 (Agri-environmental Regulation) (OJ 1992, L215/85) marked a new departure in terms of the importance attached to conservation initiatives; and structural considerations were promoted by Council Regulation (EEC) 2079/92 (OJ 1992, L215/91), instituting a Community aid scheme for early retirement from farming, and Council Regulation (EEC) 2080/92 (OJ 1992, L215/96), instituting a Community aid scheme for forestry. These matters are considered further by Rodgers in Chapter 11.

Indeed, aspects of multifunctional agriculture were incorporated by a series of amendments into the EC Treaty itself. The 1986 Single European Act first created a title on the environment (Article 130r-130t, now Articles 174–176 EC); and, following the 1997 Treaty of Amsterdam, one of the general purposes of the Community as set out in Article 2 EC is now the promotion of 'a high level of protection and improvement of the quality of the environment'. Likewise, Article 3(l) EC stipulates that the activities of the Community shall include 'a policy in the sphere of the environment'. The 1997 Treaty of Amsterdam also annexed a Protocol on Protection and Welfare of Animals, with the Community and Member States being required, when implementing, *inter alia*, the agricultural and transport policies of the Community, to 'pay full regard to the welfare requirements of animals, while respecting the legislative or administrative provisions and customs of the Member States relating in particular to religious rites, cultural traditions and regional heritage'.

That said, budgetary expenditure was still concentrated on payments under the common organizations of the market. This may be illustrated by the figures for 1997, the year that the Agenda 2000 reforms commenced. Of the 40,423.4 million Ecu spent under the European Agricultural Guidance and Guarantee Fund (EAGGF) Guarantee Section, only 2064.8 million Ecu was spent on the 'accompanying measures' (CEC, 1999a, pp. 143–151 and T/103).[2] This allocation of resources was reflected in the figures specific to the UK. In 1997/98, of the £3320.5 million expenditure funded via the Intervention Board on market regulation and other agricultural support measures under the CAP, only £41.3 million was spent on agri-environmental and other measures (MAFF *et al.*, 1999, Table 9.1).

A major development following the MacSharry reforms, however, was that the weight of expenditure under the common organizations of the market switched from payments to support prices (and, therefore, production) to direct payments to producers. The key cereals sector saw the intervention price fall, while farmers in return received compensatory payments, based on area and, in the case of most farmers, dependent upon placing a proportion of eligible land into set-aside (Council Regulation (EEC) 1765/92 (OJ 1992, L181/12) and Council Regulation (EEC) 1766/92 (OJ 1992, L181/21)) (Neville and Mordaunt 1993, pp. 15–51).

The Commencement of the Agenda 2000 Reforms

In any event, there can be little doubt that as from the commencement of the Agenda 2000 reforms the concept of multifunctional agriculture has found ever more concrete expression (McMahon, 2002). A clear example was provided by the *Explanatory Memorandum*, which accompanied proposals for eight regulations, issued on 18 March 1998 (1998 *Explanatory Memorandum*). The proposals were to put in place a European Model of Agriculture; and:

> the main aims of this model should be:
>
> - a competitive agriculture sector which can gradually face up to the world market without being over-subsidised, since this is becoming less and less acceptable internationally;
> - production methods which are sound and environmentally friendly, able to supply quality products of the kind the public wants;
> - diverse forms of agriculture, rich in tradition, which are not just output-oriented but seek to maintain the visual amenity of our countrysides as well as vibrant and active rural communities, generating and maintaining employment;
> - a simpler, more understandable agricultural policy which establishes a clear dividing line between the decisions that have to be taken jointly and those which should stay in the hands of the Member States;
> - an agricultural policy which makes clear that the expenditure it involves is justified by the services which society at large expects farmers to provide.
>
> (CEC 1998a, para. 3)

Such aims have also been expressly reiterated in the context of the Millennium Round agriculture negotiations. For example, on the submission of the *EC Comprehensive Negotiating Proposal*, the contemporaneous Memorandum could state that:

> the specific role of agriculture as a provider of public goods should be recognised. In this context, the multifunctional role of agriculture, which includes its contribution to sustainable development, the protection of the environment, the sustained vitality of rural areas and poverty alleviation, both in developing and developed countries should be taken into account.
>
> (European Community, 2001a).

As has been seen, taken individually, many of these policy objectives had already been articulated. What would seem new was their bundling into a coherent 'package' and subsequent translation from policy objectives into implemented legislation. As again stated in the 1998 *Explanatory Memorandum*, the Agenda 2000 proposals were to give 'concrete form' to a European Model for Agriculture (CEC, 1998a, para. 3).

Before considering the Agenda 2000 reforms in greater detail, five pressures promoting these changes may be highlighted, three internal and two external. First, within the Community itself there was widespread acceptance that many aspects of the CAP were politically unpopular. In particular, farmers were perceived to be 'farming' subsidies, with the bulk of payments being made to those operating on a large scale (Gardner, 1996, pp. 40 and 44; European Court of Auditors, 1998, Diagram 3). This pattern of support may have been replicated elsewhere (ABARE, 2000), but within the Community a heightened sensitivity might be detected. Thus, as stated in the Agenda 2000 document itself, while the MacSharry reforms of 1992 had rendered financial support to agriculture more transparent, there was an increasing need for such support to be socially acceptable (CEC, 1997a). Further, the 1998 *Explanatory Memorandum* went so far as to declare that '[m]aking the CAP more acceptable to the citizen in the street, to the consumer, is one of our primary objectives in the years ahead' (CEC, 1998a, para. 1). This undoubtedly remains a primary objective; and considerable exasperation has been expressed at the persistent criticism of the CAP, notwithstanding the reform process. By way of illustration, in January 2002 Commissioner Fischler forcefully argued that 'the Common Agricultural Policy is better than the press it gets' (European Community, 2002).

Secondly, there was a genuine anxiety that European agriculture was becoming uncompetitive. Not least, production controls, such as milk quotas, and the high level of internal prices had precluded full participation by the Community in expanding world markets. For example, in 1989 the European Community enjoyed a net 11.9% share in world trade in cereals (excluding rice) (CEC, 1993, p. T/144); but by 1995 that net share had fallen to 7.6% (CEC, 1998b, p. T/178).

Thirdly, reform would meet growing demands for decentralization and the return of decision-making to Member States. This would be especially appropriate in the context of rural development, with individual Member States enjoying the ability to tailor schemes to the specific requirements of their own regions (Scott, 1999). That said, the Community institutions remained conscious of the need to avoid wholescale renationalization of the CAP, a development viewed with increasing suspicion since the MacSharry reforms of 1992 (Kjeldahl and Tracy, 1994; Grant, 1995).

Turning to external factors, the prospect of eastward enlargement was already proving of considerable influence (McMahon, 2000). The agricultural sector in many of the acceding countries is substantially greater than in the current 15 Member States, but also substantially less developed. For example, data available at the commencemnt of the Agenda 2000 reforms showed that during 1995 agriculture in Poland accounted for 8% of gross domestic product (GDP), while employing 26.9% of the workforce (CEC, 1997c). This general picture should not, however, mask considerable variation. Thus, in contrast, during 1995 agriculture in Hungary accounted for 7.2% of GDP, while employing only 8% of the workforce (CEC, 1997b). In any event, according to the Agenda 2000 document, the application of existing CAP instruments would create difficulties. It affirmed that '[i]nordinate cash injections through direct payments would risk creating income disparities and social distortions in the rural areas' of the acceding Member States; and that '[i]n addition, surpluses would increase, in particular for sugar, milk and meat, reinforcing the growing market imbalances predicted after 2000' (CEC, 1997a, Part One, III, 2). At the same time, a multifunctional model of agriculture has been perceived as being of great importance to the acceding Member States (Fischler, 2000). Indeed, the diversity that they offer may reinforce that model.

A second, and very important, external factor has been the Millennium Round; and the Community institutions have openly acknowledged that the Agenda 2000 reforms have been directed to creating a defensible position in these negotiations. An example may again be provided by the Agenda 2000 document itself, which took the view that '[c]utting border protection, reducing export subsidies and reshaping support towards more 'decoupled' instruments will enhance the Union's negotiating stance in the new Round' (CEC, 1997a, Part One, III, 2). Such WTO implications will be addressed more fully below.

Multifunctionality and the Berlin Summit

General

As indicated, proposals for eight regulations were issued on 18 March 1998 and, following intense discussion, agreement was finally reached at the Berlin Summit of 24 and 25 March 1999 (CEC, 1999b, I.1–I.44). That said, it at once became clear that the Community institutions did not regard the measures agreed at the Berlin Summit as in any sense completing the required changes. Further aspects of multifunctional agriculture demanded attention, and Commissioner Fischler could announce in his opening address for International Green Week 2001 that '[r]eforming farm policy is an ongoing process which does not stop at Agenda 2000' (Fischler, 2001a).

The measures agreed at the Berlin Summit may be regarded as addressing primarily the common organizations of the market (with considerable emphasis on competitiveness) and rural development (including agri-environmental initiatives). Less attention was devoted to matters of food quality, food safety and animal welfare, now moving towards centre stage. It may be observed, however, that the objectives of the CAP as originally set out in Article 39 of the EEC Treaty (now Article 33 EC) have remained unchanged.

The common organizations of the market

The ability to compete on world markets remains for the Community a feature of a multifunctional agriculture that is not confined to non-trade concerns; and, with this aim, the measures agreed at the Berlin Summit saw a further reduction in many of the intervention prices (Ackrill, 2000, Tables 4.1–4.3). Thus, in the cereals sector, the intervention price was cut by 15% in two steps, commencing in the 2000/01 marketing year (Council Regulation (EC) 1253/1999 (OJ 1999, L160/18), amending Council Regulation (EEC) 1766/92 (OJ 1992, L181/21)). As a result, by the 2001/02 marketing year, this stood at only 101.31 Euro/tonne (although it may be noted that the Agenda 2000 document had proposed a deeper 20% cut in a single step). In the beef sector, market support was again cut, to the extent that, as from 1 July 2002, only a 'safety-net' intervention system operated.[3] In the dairy sector, it was agreed that the reduction in intervention prices for butter and skimmed milk be 15%, 5% greater than proposed in the Agenda 2000 document. However, implementation was materially delayed, with this reduction to be carried into effect in three equal steps commencing on 1 July 2005 (Council Regulation (EC) 1255/1999, Art.4 (OJ 1999, L160/48)). As a result, the intervention price for butter would fall from 328.20 to 278.97 Euro/100 kg over the period from 30 June 2005 to 1 July 2007, while, over the same period, that for skimmed milk powder would fall from 205.52 to 174.69 Euro/100 kg. Moreover, with a view to expanding Community production to fill demand on world markets, milk quotas would be increased as a general rule by 1.5% in three equal instalments commencing on 1 April 2005 (Council Regulation (EC) 1256/1999 (OJ 1999/L160/73), amending Council Regulation (EEC) 3950/92 (OJ 1992, L405/1)).[4]

While the price structure may have become more competitive, farmers saw their loss of income to a considerable degree compensated by higher direct payments. Thus, in the cereals sector, area payments were increased and simplified (Council Regulation (EC) 1251/1999, Art. 4 (OJ 1999, L160/1)). Further, despite initial proposals to fix the set-aside rate at 0% (CEC, 1997a, Part One, III, 4), an element of supply control was maintained by fixing this at 10% from the 2000/01 marketing year to the 2006/07 marketing year (Council Regulation (EC) 1251/1999, Art. 6(1) (OJ 1999, L160/1)). In the beef sector, the reduction in price security was addressed by an increase in the amount and in the range of direct payments to producers (Council Regulation (EC) 1254/1999, Arts. 3–25 (OJ 1999, L160/21)). Likewise, in the dairy sector, reductions in the intervention prices for butter and skimmed milk powder would be compensated for by direct payments to producers introduced in the 2005 calendar year (Council Regulation (EC) 1255/1999, Arts. 16–25 (OJ 1999, L160/48)).

Accordingly, at a general level, the common organizations of the market survived largely intact. Their structure remained for the most part that implemented by the MacSharry reforms; and the changes agreed at the Berlin Summit may be more accurately regarded as an expansion of earlier initiatives than as a root-and-branch rewriting of the CAP. Indeed, the Agenda 2000 document saw its proposals as 'deepening and extending the 1992 reform' (CEC, 1997a, Part One, III, 4). None the less, while the overall framework of the common organizations of the market did survive, there were changes of substance; and four such changes may be highlighted. Of some significance is that, in particular, the third and fourth of these provide clear illustration of multifunctionality gathering strength at the heart of the CAP.

First, the measures of broad application already discussed and, not least, the scaling down of intervention, reinforced the move away from price and market support in favour of direct payments to producers.

Secondly, while basic payments under such key schemes as the beef special premium scheme and the suckler-cow premium scheme remained payable on a headage basis, the more detailed reforms increased the scope to make direct payments on an area basis. For example, under the common organization of the market in beef and veal, Member States were allocated a 'national envelope' to distribute in the form of additional direct payments, and the legislation permitted distribution on a headage and/or area basis. Indeed, this switch was found not just in the context of the common organizations of the market but also in the context of rural development. For example, in England there was a major overhaul of targeted payments for hill farmers. Previously such farmers might qualify for Hill Livestock Compensatory Allowances paid on a headage basis. As from 2001, any entitlement was to Hill Farm Allowances paid on an area basis, the legislation to first implement this change being the Hill Farm Allowance Regulations 2001 (SI 2001, No. 476). The implications of these changes are discussed further by Rodgers in Chapter 11.

Thirdly, Council Regulation (EC) 1259/1999 (1999 Horizontal Regulation) (OJ 1999, L160/113) introduced compulsory environmental-protection requirements. These applied to a wide range of direct support schemes and took a significant stage further the 'greening' of the CAP. For example, when the MacSharry reforms introduced compensatory payments in the arable sector, the obligation to apply appropriate environmental measures had extended only to land set aside, and the efficacy of so limited an obligation had been doubted (Rodgers, 1996; European Court of Auditors, 2000, p. 14). In contrast, arable-area payments under the 1999 Horizontal Regulation were rendered subject to environmental-protection requirements in the case of both land in production and land set aside. Further, the multifunctional credentials of the 1999 Horizontal Regulation were trumpeted in its Preamble, these broader environmental obligations being expressly enacted 'with a view to better integrating the environment into the common market organisations'.

Fourthly, it has been noted that immediately prior to the commencement of the Agenda 2000 reforms the bulk of EAGGF expenditure was still allocated to the common organizations of the market. Concerns to this effect were in part addressed by the introduction of 'modulation', again under the 1999 Horizontal Regulation. Member States were granted authority to reduce the amounts of specified direct payments to farmers by up to 20%, with any sums realized to be made available to the Member States concerned for specified rural-development programmes. Governing criteria were the size of the labour force on the holding and/or the overall prosperity of the holding and/or the total amount of payments received. Accordingly, these provisions opened a financial conduit from the common organizations of the market to rural development, but limited implementation by Member States curbed its practical effect. Thus, modulation was introduced by only a minority of Member States and, even where it was introduced, the sums realized were less than had been hoped. For example, implementation was delayed until 2003 in Germany, while the percentage reduction was to reach only 4.5% for 2005 and 2006 in the UK.[5]

Rural development

Since the Berlin Summit rural development has constituted the 'second pillar' of the CAP. Council Regulation (EC) 1257/1999 (Rural Development Regulation) (OJ 1999, L160/80) integrates and expands existing schemes, and its importance in promoting multifunctional agriculture is very evident from the Preamble. For example, it is stated that, 'in the coming years, a prominent role should be given to agri-environmental instruments to support the sustainable development of rural areas and to respond to society's increasing demand for environmental services'. Moreover, in this context, the initiatives implemented extend well beyond an agri-environmental or, indeed, an agricultural focus. As again recited:

> over the coming years, agriculture will have to adapt to new realities and further changes in terms of market evolution, market policy and trade rules, consumer demand and preferences and the Community's next enlargement; whereas these changes will affect not only agricultural markets but also local economies in rural area in general.

The recitals reflect working documents of the Directorate-General for Agriculture (CEC Directorate-General for Agriculture, 1997, 1998). They also reflect in the Agenda 2000 document itself, which declared that not just agriculture but the countryside more generally should be considered multifunctional, with farmers encouraged 'to exploit all opportunities for rural entrepreneurs' (CEC, 1997a, Part One, III, 3).

Of the programmes to be implemented following the Berlin Summit under the Rural Development Regulation, agri-environmental initiatives were alone compulsory. These replace and build upon initiatives under the Agri-environmental Regulation. Key attributes are that they are long-term (the minimum length of commitment being 5 years) and, significantly, that they require the generation of 'added value' in terms of their environmental effects. Thus, as enacted following the Berlin Summit, Article 23 of the Rural Development Regulation expressly stipulated that '[a]gri-environmental commitments shall involve more than the application of usual good farming practice', and that '[t]hey shall provide for services which are not provided for by other support measures, such as market support or compensatory allowances'. In like manner to the MacSharry reforms, these initiatives are expressed to be complementary to other instruments of the CAP.

Other programmes implemented following the Berlin Summit under the Rural Development Regulation were those providing for: investment in agricultural holdings; setting up of young farmers; training; early retirement; support for less-favoured areas and areas with environmental restrictions; improving processing and marketing of agricultural products; forestry; and measures for promoting the adaptation and development of rural areas. Many of these were familiar from earlier legislation. For example, investment in holdings, early retirement and training echoed the schemes introduced by the three structural directives of 1972. None the less, there would appear to be a difference in terms of legislative approach. Not least, in comparison with earlier measures, the Rural Development Regulation had the clear objective of being more comprehensive in terms of coverage and yet less prescriptive in terms of implementation, with Member States enjoying considerable autonomy in the drawing up of rural development plans. Thus, under its authority the UK soon established a range of schemes extending to, by way of example in England, assistance for energy crops, under the Energy Crops Regulations 2000 (SI 2000, No. 3042), and, assistance for dealing with farm waste in nitrate-vulnerable zones, under the Farm Waste Grant (Nitrate Vulnerable Zones) (England) (No. 2) Scheme 2000 (SI 2000, No. 2911).

That said, the difference in legislative approach was only in part matched by an increase in the budgetary allocation for rural-development measures. As indicated, in 1997 only 2064.8 million Ecu out of a total of 40,423.4 million Ecu EAGGF Guarantee Section expenditure was allocated to 'accompanying measures' (CEC, 1999a, pp. 143–151 and T/103). The financial perspective for 2000 drawn up at the Berlin Summit did see a higher ceiling of 4300 million Euro applied to expenditure on rural development and ancillary measures, the ceiling on total CAP costs being 40,920 million Euro (CEC, 1999b, I.12, 21–23).[6] Further, in England it was projected that £1.6 billion would be spent on rural development over the 7-year period to 2006/07, amounting to a 60% increase (MAFF, 1999). Despite such increases, however, sensitivity as to the small proportion of the CAP budget spent on the 'second pillar' could soon be detected. In the words of Commissioner Fischler just 2 years after the Berlin Summit, '[w]hy do only 10% of budget resources go into rural development?' (Fischler, 2001b).

Food quality, food safety and animal welfare

Food quality, food safety and animal welfare did not feature large in the package of measures agreed at the Berlin Summit, notwithstanding that the Agenda 2000 document saw them as an integral part of multifunctional agriculture. It regarded prices as 'only one aspect of competitiveness', with food safety and food quality being 'at least as important'. Indeed, a fundamental obligation of Community policy was 'to guarantee the safety of food to consumers both within and outside the Union, and this must therefore be a top priority for the CAP'. Emphasis was also placed on the environmental friendliness of production methods and animal-welfare considerations; and, significantly, it was concluded that, '[i]n all these respects, European farmers are able to offer quality products deserving to be known worldwide' (CEC, 1997a, Part One, III, 3). Such issues, therefore, were perceived as providing a competitive edge in world trade. Moreover, there was the opportunity to build upon earlier legislation promoting quality products, often linked to specific geographical origins or specific production methods identifiable by consumers, such as Council Regulation (EEC) 2092/91 (OJ 1991, L198/1), on organic production of agricultural products, including labelling; and Council Regulation (EEC) 2081/92 (OJ 1992, L208/1), on the protection of geographical indications and designations of origin for agricultural products and foodstuffs.

Although the measures agreed at the Berlin Summit only touched upon such matters, they have been the object of very considerable policy development and legislation since 1999. Added momentum has undoubtedly been supplied by mounting public concerns over the biotechnology revolution and the bovine spongiform encephalopathy (BSE) and foot-and-mouth disease (FMD) catastrophes. As enunciated by Commissioner Fischler, '[c]itizens do expect quality and this goes much further than quantity and safety alone. They expect quality in production methods, in tastiness and wholesomeness and in the respect of nature' (Fischler, 2001d). Accordingly, in this context, certain multifunctional aspects of food quality, food safety and animal welfare may be highlighted. The impact of the biotechnology revolution will be considered by Hilson and French in Chapter 9.

A legislative advance in the field of food quality was amendment of Council Regulation (EEC) 2092/91 (OJ 1991, L198/1), on organic production of agricultural products. This was extended by Council Regulation (EC) 1804/1999 (OJ 1999, L222/1) so as to cover organic livestock production. Importantly, much emphasis was accorded to animal-welfare considerations, including, in principle, a ban on tethering.

Food safety saw even greater developments and was characterized as a 'non-negotiable' aspect of Community policy (Fischler, 2001c). An umbrella regulatory framework was advocated in the 1999 *White Paper on Food Safety*, which, echoing the Agenda 2000 document, stated that '[a]ssuring that the EU has the highest standards of food safety is a key policy priority for the Commission' (CEC, 1999c, p. 1). Moreover, the importance of agriculture could not be overestimated as the first link in the food-chain from 'farm to table' (Lauterburg, 2001, pp. 38–39). Proposals for legislation were issued in 2000 (CEC, 2000b); and in 2002 Regulation (EC) 178/2002 of the European Parliament and of the Council (Food Law Regulation) (OJ 2002, L31/1) was enacted to lay down the general principles and requirements of food law, establish the European Food Safety Authority and lay down procedures in matters of food safety. Again, the integral role of agriculture was accorded prominence, the Preamble reciting as follows: '[i]n order to ensure the safety of food, it is necessary to consider all aspects of the food production chain as a continuum from and including primary production and the production of animal feed.' With more specific reference to world trade, it was also recited that the Community had 'chosen a high level of health protection as appropriate in the development of food law, which it applies in a non-discriminatory manner whether food or feed is traded on the internal market or internationally'. This approach would seem consistent with ever more frequent recourse to the precautionary principle in Community legislation. In this context, the *Communication from the Commission on the Precautionary Principle*, issued in 2000, was most illuminating (Salmon, 2002). Cast very much against the background of international law and, in particular, the law of international trade, it asserted that:

> each Member of the WTO has the independent right to determine the level of environmental or health protection they consider appropriate. Consequently a member may apply measures, including measures based on the precautionary principle, which lead to a higher level of protection than that provided for in the relevant international standards or recommendations.
>
> (CEC, 2000a, p. 11)

The heightened profile of food safety may, in part at least, be attributed to the growing legislative importance of the European Parliament.

Following the 1992 Treaty on European Union and the 1997 Treaty of Amsterdam, the 'co-decision' procedure set out in Article 251 EC now governs, *inter alia*, matters of public health, consumer protection and the environment, effectively conferring on the European Parliament the power of veto. Accordingly, as agriculture has become more multifunctional, the European Parliament has enjoyed greater opportunity to shape its regulation. This opportunity it has not been slow to take. Not least, it was very active during the passage of the Food Law Regulation (*Agra Europe*, 2001, pp.A/1–A/2); and has been responsible for numerous amendments in the genesis of legislation governing genetically modified organisms (GMOs) (European Community, 2001b; CEC, 2002b; Scott, 2003).

Likewise, even prior to the Mid-term Review, momentum had been generated in the regulation of animal welfare. Thus, the umbrella Council Directive 98/58/EC (OJ 1998, L221/23) laid down minimum standards for the protection of animals bred or kept for farming purposes; Council Directive 1999/74/EC (OJ 1999, L203/53) provided for the improvement of the condition of battery hens; and Council Directive 2001/88/EC (OJ 2001, L316/1) provided for the improvement of the condition of pigs. Moreover, there was a concerted effort to improve the welfare of animals during long-distance transportation, including the adoption of a Commission proposal in April 2001 (European Community, 2001c).

The Impact of the Mid-term Review

The *Mid-term Review of the Common Agricultural Policy* proposed numerous measures that reflected the Community vision of a multifunctional agriculture; and two such proposals may be highlighted: first, the attachment of a wide range of multifunctional conditions to direct payments; and, secondly, the creation of new multifunctional programmes under the Rural Development Regulation. However, before considering these aspects, it may also be noted that the policy document affirmed Community commitment to a competitive agricultural sector. This commitment was bolstered by encouraging statistics as presented in *Prospects for Agricultural Markets 2002–2009* (CEC, 2002c). In particular, while it was conceded that full account could not yet be taken of the effect of the US Farm Security and Rural Investment Act of 2002 (FSRI Act), it was none the less projected that in the medium term the outlook for cereals was generally favourable, with rye being the only crop giving material cause for concern. Moreover, the policy document took the view that such competitiveness in the cereals sector should be further promoted by a general 5% reduction in the intervention price as from the 2004/05 marketing year (CEC, 2002a, p. 13). That said, when the reform package was agreed in Luxembourg on 26 June 2003, it saw no reduction in intervention price

(European Community, 2003). Likewise, although there was agreement to cut the intervention price for butter by 25% over the period 2004–2007, this marked a considerable retreat from the 35% cut which had been envisaged at the stage of the proposed legislation (CEC, 2003, p.9).

Turning first to the attachment of a wider range of multifunctional conditions to direct payments, the *Mid-term Review of the Common Agricultural Policy* consciously built upon the 1999 Horizontal Regulation, which, as indicated, had imposed compulsory environmental-protection requirements. It advocated that direct payments be subject to 'respect of statutory environmental, food safety and animal health and welfare standards, as well as occupational safety requirements for farmers'. In addition to these statutory management requirements, farmers would be obliged to maintain land in good agricultural condition. Although cross-compliance was to reflect regional differences, with Member States defining and enforcing standards, a common Community framework was to be introduced so as to avoid distortion of competition.

At the stage of the proposed legislation, some 38 statutory management requirements were listed. In the event, however, the agreed reform package saw the number reduced to 18. Indeed, all those relating to occupational safety were omitted. Further, while under the proposed legislation the statutory management requirements would all apply as from 1 January 2004, they will now be implemented in 3 stages, the first commencing on 1 January 2005. By contrast, however, the agreed reform package did see an enhanced role for multifunctional agriculture in the obligation that farmers maintain their land in not only good agricultural but also good environmental condition (European Community, 2003).

It was unequivocally accepted that a key factor driving the introduction of such extended cross-compliance was the need to increase the public acceptability of the CAP. In the words of Commissioner Fischler, the Mid-term Review served:

> not least to provide more solid justifications for public spending on the farm sector. As well as supporting farm incomes, such payments must yield something in return – whether it is safer food, an intact environment, compliance with animal welfare rules, upkeep of the countryside, preservation of the cultural heritage, or greater social fairness and balance
>
> (Fischler, 2002b)

Accordingly, these measures will bring multifunctionality into the heart of the 'first pillar' of the CAP; and a recurrent theme of the Mid-term Review has been greater integration of non-trade concerns into a legislative framework now associated with producer (as opposed to market) support. Indeed, an express aim was that '[f]ood safety must be fully integrated into the CAP through cross-compliance' (CEC, 2002a, p. 11).

It may be emphasized, however, that extended cross-compliance is not the only major change to direct payments to be effected under the

Mid-term Review. A key reform is the introduction of the single farm payment (SFP), in principle as from 1 January 2005. Based on historical entitlements over a 2000–2002 reference period, this will eventually encompass most agricultural sectors. It is expressly contended that the SFP will be green-box compatible. A matter of some significance is that extended cross-compliance will apply to all direct payments, not just the SFP.

Secondly, multifunctionality is to be promoted by the addition of two new chapters to the Rural Development Regulation, addressing, respectively, food quality and 'meeting standards'. The former will specifically encourage farmers to participate in quality-assurance schemes recognized by the Member States or the Community, including those which govern geographical indications and designations of origin. The latter will assist farmers to adapt to demanding standards based on Community legislation in such fields as the environment, food safety and animal welfare. Implementation will be at the discretion of Member States (although there had been proposals for the chapter on food quality to be compulsory).

Nevertheless, in the *Mid-term Review of the Common Agricultural Policy* there was also recognition that the balance of funding still greatly favoured the 'first pillar' of the CAP, it being estimated that only 16% of total EAGGF expenditure was devoted to rural development. Measures were proposed to redress this imbalance, not least the compulsory imposition of 'dynamic modulation' as from 2004. This would not apply to smaller farms by reason of a 'franchise' based on employment levels. Indeed, it was estimated that approximately three-quarters of the farms in the Community would be exempt. However, where dynamic modulation was applicable, it would encompass all direct payments, which would be reduced progressively at the rate of 3% per annum up to the 20% maximum agreed at the Berlin Summit. Sums realized would be made available to address specific rural needs. Notwithstanding the radical nature of this initiative, the sums involved would not be so vast as to see anything like parity of expenditure on the 'second pillar'. Thus, it was estimated that only in the region of 500–600 million Euro would be realized in 2005.

Such dynamic modulation proved particularly controversial and, as a result, the proposed legislation opted instead for 'degression'. This would again be targeted to larger farms, with the full rate only being payable by those which received over 50,000 Euro per annum. Moreover, implementation would be delayed until 2006. Where the full rate was applicable, the percentage removed would rise from 1% in that year to 19% in 2012. However, very importantly, only a proportion of the sums realized would be modulated to the 'second pillar': 1% in 2006, rising to 6% in 2011 (CEC, 2003, pp. 11–12).

In the event, the agreed reform package saw a return in focus to modulation. Degression is to replaced by a mechanism to impose financial discipline, only triggered where forecasts indicate that expenditure on market measures and direct payments (before the application of modulation) will come within 300 million Euro of their budgetary ceiling. By contrast, modulation will commence at the higher rate of 3% in 2005, reaching 5% in 2007 and remaining at that rate until 2012. Only farms receiving 5000 Euro or less per annum will escape its effect. That said, it has been conceded that the sums realized remain relatively small. Even when the 5% rate applies, only 1.2 billion Euro per annum will be generated for rural development funding (European Community, 2003).

The WTO

General

As has been seen, the Community institutions have openly acknowledged that the development of multifunctional agriculture is driven, to a considerable degree, by the current WTO negotiations. This was graphically set out in the 1998 *Explanatory Memorandum*, which declared that:

> [t]he Union has to prepare its agriculture sector for these negotiations. This has two vital consequences: First, with this reform the Union has to lay down the agricultural policy that it intends carrying out in the years ahead in a way that satisfies its own interests and takes a realistic view of developments in the international context. This needs to be done before the opening of the WTO negotiations so that the Union can negotiate on a solid basis and knows where it wants to go. Secondly, it must be made quite clear to all that the reform to be adopted will outline the limits of what the Union is able to agree in the forthcoming international negotiations.
>
> (CEC, 1998a, para. 1)

During the implementation of the Agenda 2000 reforms, these imperatives were recognized by, for example, the European Court of Auditors (1998, para. 4) and the European Committee of the Regions (1999, para. 7). They were also highlighted on the issue of the proposed regulations to implement the Mid-term Review (CEC, 2003, p. 5).

In this context may be considered, first, the manner in which the Community has presented the multifunctional role of agriculture before the WTO and, secondly, the extent to which Community multifunctionality is compatible with existing WTO commitments and likely to be compatible with the developing demands of the current negotiations.

Community proposals

The establishment of a multifunctional European Model of Agriculture has permitted the Community to advance negotiating proposals in the Millennium Round based upon a far more coherent vision than in the Uruguay Round. Indeed, even before the commencement of the

Millennium Round this European Model was being characterized as distinctive and expressly contrasted with the ethos and regulatory framework of agriculture in the USA and the Cairns Group. For example, in the 1998 *Explanatory Memorandum* it was affirmed that:

> [t]he fundamental difference between the European model and that of our major competitors lies in the multifunctional nature of Europe's agriculture and the part it plays in the economy and the environment, in society and in preserving the landscape, whence the need to maintain farming throughout Europe and to safeguard farmers' incomes.
>
> (CEC, 1998a, para. 3)

Of the proposals submitted by the Community during the first phase of the negotiations, five may be highlighted: first, *The Blue Box and Other Support Measures to Agriculture* (WTO, 2000b); secondly, *Food Quality – Improvement of Market Access Opportunities* (WTO, 2000c); thirdly, *Animal Welfare and Trade in Agriculture* (WTO, 2000d); fourthly, *Export Competition* (WTO, 2000e); and, finally, the *EC Comprehensive Negotiating Proposal* (WTO, 2000g). In addition, multifunctionality featured large in three early contributions by the Community during the second phase: *Food Safety* (WTO, 2001a); *Green Box* (WTO, 2001b); and *Geographical Indications* (WTO, 2001c). Further, it formed an integral part of *The EC's Proposal for Modalities in the WTO Agriculture Negotiations*, issued in December 2002 (CEC Directorate-General for Trade/Agriculture, 2002).

While the debate as to the role of multifunctionality in world trade has largely been conducted in the context of such non-trade concerns as the environment, food safety, food quality and animal welfare, it may again be reiterated that competitive production is regarded as a facet of the European Model of Agriculture. Accordingly, the Community has asserted compliance with reduction commitments imposed on export subsidies by the Uruguay Round Agreement on Agriculture (URAA), namely a 21% reduction in subsidized volume and a 36% reduction in budgetary expenditure, as from the 1986–1990 base period (WTO, 2000e). Likewise, in the November 2001 Doha Declaration, the Community committed itself to comprehensive negotiations aimed at, *inter alia*, 'reductions of, with a view to phasing out, all forms of export subsidies' (WTO, 2001d, para. 13). However, this commitment was, at the insistence of the Community, subject to the proviso that there would be no 'prejudging the outcome of the negotiations' (European Community, 2001d). Subsequently, in *The EC's Proposal for Modalities in the WTO Agriculture Negotiations*, it was felt possible to foresee an average substantial cut in the volume of export subsidies and an average 45% cut in the level of budgetary outlays (CEC Directorate-General for Trade/Agriculture, 2002, p. 5). Moreover, by the time of Cancún, the community affirmed willingness to eliminate export subsidies for certain products of importance to developing countries (Fischler, 2003a). At the same time, there has been a consistent endeavour to widen the range of measures subject to export-competition commitments. In particular, concern has been expressed at the level of export credits employed by other Members of the WTO and at the lack of progress in bringing such credits within WTO disciplines (WTO, 2000g; CEC Directorate-General for Trade/Agriculture, 2002, p. 5).

Competitiveness was also a feature of the proposal on food quality (WTO, 2000c). This was phrased in terms of market access and emphasized the need for consumers to be able to choose products on the basis of their specific characteristics and, concomitantly, for producers to be able to protect denominations linked to food quality or food specificity. Similarly, the *EC Comprehensive Negotiating Proposal* stated that '[i]mproved market access also demands, as a counterpart, fair competition opportunities for those products whose quality and reputation are linked to their geographical origin and traditional know-how', and the regulation of labelling was advocated as an effective instrument to achieve consumer protection and fair competition (WTO, 2000g, para. 3). Further, in *The EC's Proposal for Modalities in the WTO Agriculture Negotiations*, it was proposed that amendment to the URAA should see the establishment of a list of protected geographical indications (CEC Directorate-General for Trade/Agriculture, 2002, p. 4).

Notwithstanding these proposals, the promotion of multifunctionality by the Community has largely been conducted in the context of domestic support, and this feature is likely to be accentuated following the Mid-term Review.

However, even before the commencement of the Mid-term Review, the *EC Comprehensive Negotiating Proposal* had argued that direct aids, as opposed to price and production support, could 'contribute to some of the objectives of multifunctional agriculture, namely protecting the environment and contributing to the sustained vitality of rural areas and poverty alleviation' (WTO, 2000g, para. 10). This was a reason why blue-box and green-box payments were to be preferred to those falling in the amber box, which, being variable in relation to market prices, could be particularly trade-distorting (on the URAA amber, blue and green boxes, see Grossman in Chapter 2). As a result, while there was a clear commitment to negotiating further reductions in domestic support, the retention of the blue and green boxes was to be a precondition. At the same time, this emphasis on the blue and green boxes met the switch of Community support from production to producer. As affirmed by Commissioner Fischler:

> [i]n 1991 almost 70% of our agriculture budget went on export refunds and intervention. Now that we have opted out of this production race, in future only 20% of the farming budget will be spent on market support. Instead, when the Agenda 2000 reform takes full effect, almost 70% will be used for direct payments to farmers.
>
> (Fischler, 2001a)

Indeed, if it can be substantiated that the SFP is green-box compatible, that category of support will assume greatly heightened importance. However, a matter of some interest is that the Community has intimated that it does not want to scrap the blue box (CEC Directorate-General for Trade/Agriculture, 2002, p. 6; Fischler, 2002c); and such defence of the blue box is likely to be the more vigorous following acceptance of partial implementation of SFP.

Multifunctionality appeared even more prominently in those aspects of the proposals which addressed food safety and animal welfare, graphically illustrating the degree to which the agriculture negotiations have already moved beyond agricultural boundaries as traditionally understood. Perhaps in consequence, some defensiveness on this account may be detected. For example, as has been seen, the proposal on food quality was couched more in terms of market access than multifunctionality as such (WTO, 2000c). Likewise, while that on animal welfare asserted the relevance of Article 20 of the URAA, it also acknowledged that other WTO agreements might be equally applicable. Further, it was openly accepted that raising the issue of animal welfare might be regarded as hidden protectionism. Any allegations to this effect were countered, in particular, by declaring that such a policy would be unwise when the Community is the second largest exporter of agriculture and food products in the world; and the overriding Community concern was stated to be to 'ensure that the process of liberalising world trade supports what we are building in the EC about the protection of animals' (WTO, 2000d).

Accordingly, a recurrent theme in the proposals has been concern lest world-trade considerations should create a 'race to the bottom', and again the proposal on animal welfare was illustrative. It highlighted fears that, in the absence of a WTO framework within which to address such issues, animal-welfare standards 'could be undermined if there is no way of ensuring that agricultural and food products produced to domestic animal welfare standards are not simply replaced by imports produced to lower standards' (WTO, 2000d). On this basis, there has been a vigorous defence of the right of the Community to impose a high level of animal welfare; and, flowing from this, three initiatives were advocated: first, the development of multilateral agreements addressing animal welfare; secondly, appropriate labelling, to facilitate consumer choice; and, thirdly, the consideration of compensation for producers to contribute to their additional costs stemming directly from higher animal-welfare standards, provided that such compensation would have no, or at most minimal, effects on trade and production. A similar approach was taken in the *EC Comprehensive Negotiating Proposal* with regard to food safety. It noted that '[t]here is public concern that WTO could be used to force onto the markets products about whose safety there are legitimate concerns' (WTO, 2000g, para. 18). These themes were taken up in *The*

EC's Proposal for Modalities in the WTO Agriculture Negotiations, which enunciated that:

> [i]n order to maintain support for trade reform, society needs to be reassured that certain societal goals such as specific domestic support needs of developing countries, the protection of the environment, rural development and animal welfare may be achieved without obstacles created by the WTO. Support granted for the achievement of such goals should therefore primarily be provided through the *green box* on the condition that such measures are well targeted, transparent, and implemented in no more than minimally trade-distorting ways.
>
> (CEC Directorate-General for Trade/Agriculture, 2002, p. 6).

The multifunctional European Model of Agriculture: compatibility with WTO commitments

Introduction

The development of a multifunctional European Model of Agriculture and, not least, the growing emphasis on non-trade concerns have not escaped criticism. Multifunctionality has been regarded as a bulwark for 'Farm Fortress Europe', changing the parameters of the debate and replacing price and market support with more subtle and less tangible subventions. Exasperation has been expressed that, so soon after price and market support had been subjected to tariffication in the Uruguay Round, new forms of support have been put in play, considerably more difficult to subject to tariffs and, accordingly, capping and reduction (ABARE, 1999; Bohman *et al.*, 1999).

There has, however, been recognition that in this context it is a formidable task to create a truly 'level playing-field'. Not least, geographical differences preclude even similar solutions among all Members of the WTO, and, as acknowledged by the OECD, a key reason for preserving and strengthening the multifunctional role of agriculture is to combat territorial imbalances (OECD, 1998a). Even within the Community itself there is a wide divergence of agricultural structures (Lowe *et al.*, 2000); and this has been openly accepted since the inception of the CAP. It may be reiterated that Article 39(2)(a) of the EEC Treaty (now Article 33(2)(a) EC) required that account be taken of 'the particular nature of agricultural activity, which results from the social structure of agriculture and from the structural and natural disparities between the various agricultural regions'. Accordingly, great weight has been placed on the need to recognize the different natural conditions and historical backgrounds pertaining among the Members of the WTO; and support for the Community viewpoint has been supplied by several other 'Friends of Multifunctionality'. For example, Japan has strongly advocated that 'the diversity and coexistence of agriculture among various countries need to be preserved' (WTO, 2000h).

A similar, and arguably more intangible, feature has been growing awareness that in different countries there are different degrees of popular demand for the non-trade aspects of agriculture. Again, this has been expressly articulated by the OECD, which countenances the fact that the increasing importance of multifunctionality is a consequence of increasing demand for certain non-food services, in particular among more affluent societies (OECD, 1998b, paras. 32–41). A good example of such a society would be Norway, which has submitted that '[t]he importance of multifunctionality may be related to the value and preciousness of the different functions that agriculture is providing, as judged by the country's population' (OECD, 1999, para. 157). In the case of the Community, it has been seen that the reform process has been in part a response to the changing public perceptions of the role of agriculture; and the *EC Comprehensive Negotiating Proposal* expressly articulated that, to meet the goals of further liberalization and expansion of trade for agricultural products, 'it is vital to muster strong public support, which can only be achieved if other concerns are met, in particular the multifunctional role of agriculture' (WTO, 2000g). The USA too has accepted 'that each nation may legitimately have unique goals with respect to a multifunctional agriculture' (OECD, 1999a, paras. 240–241), but, significantly, has also taken the view that the only relevant question is whether or not domestic policies in this arena have no, or at most minimal, trade-distorting effects or effects on production (United States, 2002).

With the Community perspective on the agri-environmental element of multifunctionality being considered by Rodgers in Chapter 11, the focus in this chapter will be largely directed to issues of food quality, food safety and animal welfare. In particular, consideration will be given to two aspects: first, the effect of attaching a wider range of multifunctional conditions to direct payments; and, secondly, the world-trade impact of promoting 'higher standards', with special reference to the fields of food quality, food safety and animal welfare.

The effect of attaching a wider range of multifunctional conditions to direct payments

As indicated, the 1999 Horizontal Regulation attached compulsory environmental-protection requirements to direct payments and the Mid-term Review, once implemented will extend cross-compliance to matters of public, animal and plant health and to matters of animal welfare. Accordingly, there has been continuation of the 'greening' of the CAP, as advocated by the European Court of Auditors (2000); and multifunctionality is being more fully integrated into the common organizations of the market, which still account for a major proportion of agricultural expenditure. Thus, in 1999, the year of the Berlin Summit, arable crops alone accounted for 17,866 million Euro out of total agricultural expenditure of 40,940 million Euro (CEC Directorate-General for Agriculture, 2000, p. 24). Further, according to figures issued with the proposed legislation during the course of the Mid-term Review, total direct payments to farmers in the existing 15 Member States will account for 32,371.9 million Euro by 2013 (CEC, 2003, Annex A-2).

It has been seen that the Commission considers the SFP to be green-box compatible. However, in defence of this claim, far great reliance has been placed on decoupling of support than on the wider range of multifunctional conditions attached to the payment. Indeed, the detailed requirements would seem calculated to track the criteria for decoupled income support as set out in paragraph 6 of Annex 2 to the URAA, with particular reference to the criterion that '[t]he amount of such payments in any given year shall not be related to, or based on the type or volume of production (including livestock units) undertaken by the producer in any year after the base period' (para. 6(b)). It has also been expressly understood that the SFP will meet the criterion that '[n]o production shall be required in order to receive such payments' (para. 6(e)). Thus, Commissioner Fischler declared that 'farmers will not be obliged any more to grow the crops or keep the animals that they got the direct payments for' (Fischler, 2002a).

Some doubts may, however, be expressed as to the extent that green-box compatibility will be achieved, and three such doubts may be noted. First, although the SFP will embrace a broad swathe of pre-existing support schemes, there will remain some differentiation in accordance with the type of production. For example, protein crops will enjoy a crop-specific area payment of 55.57 Euro/ha (European Community, 2003). Secondly, as demonstrated by Rodgers in Chapter 11, it is not easy to conceive of a regime that requires no production at all. Not least, the obligation to maintain land in good agricultural and environmental condition may of itself involve a degree of production. Thirdly, over and above the detailed criteria set out in paragraph 6 of Annex 2 to the URAA, green-box compatibility will only be achieved if the SFP meets the fundamental requirement set out in paragraph 1, that it has 'no, or at most minimal, trade-distorting effects or effects on production'. As a preliminary point it may be observed that, on a theoretical level, there has been considerable debate as to the extent that decoupling is a genuine possibility (OECD, 2001a,b).[7] In this regard, however, the Community approach has been robust. Even blue-box payments have been contended to have a minor trade impact, as clearly asserted in *European Communities Proposal: the Blue Box and Other Support Measures to Agriculture* (WTO, 2000b); and similar views have been expressed by the OECD. Moreover, the SFP would seem directed to meeting two factors considered by the OECD to have a material effect in neutralizing trade distortions: linkage with fixed historical periods; and elimination of the requirement to plant. None the less, the OECD has also warned against 'second-level' distortions. These are associated, for example, with farmers' perceptions of future payments and/or payment conditions (OECD, 2001c, p. 33). More broadly, on a practical level, both logic and evidence suggest that the sheer scale of direct payments could not fail to have a trade impact (Rude, 2001). In addition, the agreed reform package saw Member States granted authority, subject to prescribed conditions, to effect partial implementation of the SFP. Accordingly, this already constitutes a significant retreat from the principle of full decoupling.

There remains the question of the multifunctional conditions attached to direct payments. As agreed under the Mid-term Review, these will definitely result in further 'greening' of support, a policy also advocated by the UK House of Commons Agriculture Committee in 2000 (House of Commons Agriculture Committee, 2000). However, argument that direct payments will be rendered green-box compatible is hampered by the fact that the URAA is relatively reticent on multifunctionality. Cross-compliance does not emerge as a criterion for decoupled income support under paragraph 6 of Annex 2; and, while payments under environmental programmes are addressed under paragraph 12, this provision would seem more apt for targeted measures under the Rural Development Regulation. Paragraph 2(f) does cover 'marketing and promotion services'; but food safety and animal welfare as such receive no specific mention. In this light, it is not perhaps surprising that the Community has sought to reconfigure the URAA so as to accommodate the multifunctional role of agriculture (WTO, 2000g). Thus, it has argued for adaptation of the green-box criteria so as to address issues of animal welfare (WTO, 2001b); and met with some success in the draft modalities prepared by Chairman Harbinson (WTO, 2003). Not for the first time, a certain scepticism towards this approach may be detected on the part of the Cairns Group (WTO, 2000f).

Before focusing on the world-trade impact of promoting 'higher standards', it may be observed that even the Mid-term Review has not sought to take advantage of certain green-box measures employed by other Members of the WTO. In this regard may be highlighted '[p]ayments (made either directly or by way of government financial participation in crop insurance schemes) for relief from natural disasters', as set out in paragraph 8 of Annex 2 to the URAA. Although the Community has issued a major working document on *Risk Management Tools for EU Agriculture with a Special Focus on Insurance* (CEC Directorate-General for Agriculture, 2001), such developments are at an early stage, notwithstanding that the working document itself specifically affirmed the importance of risk-management tools in the WTO arena (CEC Directorate-General for Agriculture, 2001, pp. 64–65), and that sophisticated crop insurance has long been a feature of agriculture in the USA (Harwood *et al.*, 1999).

The world-trade impact of promoting 'higher standards'

There is a growing body of evidence to suggest that the future direction of agricultural trade negotiations will shift discernibly from tariffs as traditionally understood, financial in nature, to the agricultural products themselves, the method of their production and, importantly, the extent to which such matters may legitimately constitute impediments to free trade (Blandford and Fulponi, 1999). It may also be reiterated that the work programme for the second phase of the current agriculture negotiations included food safety as one of the priority issues to be addressed (WTO, 2001e). While this has long been the position of the Community, it is also in line with recognition by the USA that trade measures may address legitimate health and safety concerns, so obviating the need to reopen the Agreement on the Application of Sanitary and Phytosanitary Measures (SPS Agreement) (WTO, 2000a).

None the less, while it may not prove necessary to reopen the SPS Agreement, it would seem clear that the implementation of higher standards in the fields of food quality, food safety and animal welfare will require exploration of the linkage between the various WTO agreements. Indeed, Article 14 of the URAA itself acknowledges linkage, stipulating that 'Members agree to give effect to the Agreement on the Application of Sanitary and Phytosanitary Measures'. Likewise, the negotiating proposal of the Community on animal welfare maintained that animal-welfare issues have the capacity to bring into consideration not only the URAA and the SPS Agreement, but also the Agreement on Technical Barriers to Trade and Article XX of the General Agreement on Tariffs and Trade (GATT). The same proposal also argued forcefully for a global approach (WTO, 2000d); and this would reflect the more integrated treatment of animal welfare as subsequently agreed under the Mid-term Review.

What would also seem clear is that the SPS Agreement permits Members of the WTO, in principle, to impose higher standards of sanitary and phytosanitary protection (Scott, 2000). Article 3.3 provides that 'Members may introduce or maintain sanitary or phytosanitary measures which result in a higher level of sanitary or phytosanitary protection than would be achieved by measures based on the relevant international standards, guidelines or recommendations.' According to the Appellate Body in *EC Measures Concerning Meat and Meat Products (Hormones) (Beef Hormones Dispute)* (1998), this is an autonomous right, not merely an exception from the general obligation in Article 3.1 to base sanitary and phytosanitary measures on international standards, guidelines or recommendations, where they exist. However, for such measures to be valid, there must either be 'scientific justification' or, in the alternative, the Member must have determined a higher level of protection to be appropriate in accordance with the relevant provisions of Article 5. These include a risk assessment. Indeed, the Appellate Body in the same dispute decided that a risk assessment is also required should a Member seek to rely on scientific justification. This would seem consistent with its conclusion that '[t]he ultimate goal of the harmonization of SPS measures is to prevent the use of such measures for arbitrary or unjustifiable discrimination between Members or as a disguised restriction on international trade' (para. 177). In the current agriculture negotiations, such sentiments have already been echoed by the US administration (Barshefsky, 2000).

Allied to the ability to impose higher standards, the Community has argued strongly in the world-trade context for a precautionary approach to food safety (WTO, 2000g, 2001a). Reliance has been placed on both the Cartagena Protocol on Biosafety and Article 5.7 of the SPS Agreement. Under Article 5.7:

> [i]n cases where relevant scientific evidence is insufficient, a Member may provisionally adopt sanitary and phytosanitary measures on the basis of available pertinent information, including that from the relevant international organizations as well as from sanitary or phytosanitary measures applied by other Members.

In the *Beef Hormones Dispute* (1998) the Appellate Body held that, notwithstanding the absence of express mention, the precautionary principle is reflected in this provision. However, it also held that the precautionary principle does not override the general provisions governing risk assessment as set out in Articles 5.1 and 5.2, including an obligation to take into account available scientific evidence. Against this background, there may be merit in Community emphasis on the need to clarify the application of Article 5.7, with particular reference to achieving proportionality and consistency (WTO, 2001a; CEC Directorate-General for Trade/Agriculture, 2002, p. 7).

Support for the promotion of higher standards may also be derived from more recent jurisprudence under Article XX of the GATT 1994 (and its predecessor, the GATT 1947), which permit certain exceptions to GATT/WTO disciplines. Of these exceptions, the most relevant in this context would be Article XX(b), which permits Members to adopt and enforce measures 'necessary to protect human, animal or plant life or health', and Article XX(g), which permits Members to adopt and enforce measures 'relating to the conservation of exhaustible natural resources if such measures are made effective in conjunction with restrictions on domestic production or consumption'. Perhaps most significant have been the rulings of the Appellate Body in *United States – Import Prohibition of Certain Shrimp and Shrimp Products (Shrimp/Turtles I)* (1998) and *United States – Import Prohibition of Certain Shrimp and Shrimp Products: Recourse to Article 21.5 of the DSU by Malaysia (Shrimp/Turtles II)* (2001). It has been persuasively argued that these rulings have opened the door for trade restrictions based on processes and production methods, with particular reference to those directed to preserving the environment (Charnovitz,

2002). None the less, it would seem uncontroversial that for such restrictions to be effective, appropriate disciplines must be applied. Thus, in *Shrimp/Turtles I* (1998) emphasis was placed on the *chapeau* of Article XX, subjecting the exceptions to the proviso that 'such measures are not applied in a manner that would constitute a means of arbitrary or unjustifiable discrimination between countries where the same conditions prevail, or a disguised restriction on international trade'. Further, in *Shrimp/Turtles II* (2001) the Appellate Body found the restriction in question valid so long as, in particular, the USA continued to make serious, good-faith efforts to reach a multilateral agreement. Accordingly, a cooperative approach was to be preferred.

In this context perhaps the most distinctive feature of the Community negotiating position has been its proposal that, subject to conditions, compensation for additional costs incurred in meeting animal-welfare standards be exempt from the reduction commitments applied to domestic support (WTO, 2000d,g; CEC Directorate-General for Trade Agriculture, 2002, p. 7). This would appear to be a novel approach to animal-welfare issues which, as noted, found a favourable response in the draft modalities prepared by Chairman Harbinson (WTO, 2003). In the past it has been generally understood that a Member that sought to attain higher standards might endeavour to protect its producers by trade restrictions, conditioning imports. Instead, under this new approach, such a Member would have the ability to preserve the competitiveness of its producers on world markets by means of domestic support. In consequence, higher standards of animal welfare would, to a considerable extent, be relocated within the WTO legislative framework. It may be observed, however, that the Community has throughout accepted that the additional costs incurred must stem directly from the adoption of higher standards and thus have no, or at most minimal, effects on trade. Accordingly, there may again be found tracking of the green-box criteria.

The Community policy development which would seem most capable of benefiting from this approach is the new voluntary chapter on 'meeting standards', to be implemented under the Rural Development Regulation. As indicated, this will specifically help farmers to adapt to demanding standards based on Community legislation in the field of, *inter alia*, animal welfare. In comparison with the cross-compliance conditions attached to direct payments, there would be greater ease in demonstrating a quantifiable link between the support payment and achievement of the goal of promoting animal welfare. Nevertheless, real difficulties would seem to remain as to whether or not such support would have no, or at most minimal, effects on trade. For example, producers might receive not only domestic support but premium prices (although it must be noted that the Community legislative framework is directed to preventing overcompensation).

Conclusion

Multifunctionality is central to the stance adopted by the Community in the current agriculture negotiations. The URAA expressly recognizes that non-trade concerns are to be taken into account. Further, as affirmed by Commissioner Fischler, European agriculture 'is more than food production. It responds to demands from our society. Therefore we need different trade rules than industry' (European Community, 2000b). There is also a clear determination to defend this multifunctional model. As again stated by Commissioner Fischler, 'while the EU will play a constructive role' in the WTO negotiations, 'this does by no way mean that the EU would be prepared to sacrifice the European model of agriculture on the altar of liberalisation' (European Community, 2000a). Similar sentiments were, expressed following Cancúun (Fischler, 2003b).

This may legitimately be regarded as a response to popular demand. The combination of such disasters as BSE and FMD and widespread apprehension over biotechnology has left the European consumer ill-prepared to accept a new Agreement on Agriculture driven by trade concerns alone. Indeed, shortly after the first phase was completed at Geneva, Commissioner Fischler could state that '[w]e have to begin by asking what people want on their plates. We must examine the impact of the ever tighter concentration on trade and what standards should be met by the processing sector and agricultural producers and suppliers' (Fischler, 2001b). However, the *Mid-term Review of the Common Agricultural Policy* itself had to admit that 'there remains a gap between the preference for quality that consumers express and the way they behave in the marketplace' (CEC, 2002a, p. 6).

Notwithstanding some of the rhetoric to the contrary, this redirection of agricultural policy may be regarded as a journey to be shared with the USA. Thus, the '[m]aintenance of the family farm organization as a dominant part of the production system' was one of the farm policy goals set by the Commission on 21st Century Production Agriculture in their *Report to the President and Congress*, and, likewise, the promotion and enhancement of food safety, a clean environment and animal and plant health and safety were considered an appropriate role for government (Commission on 21st Century Production Agriculture, 2001, pp. xv, xvi). This theme was taken up by the policy document, *Food and Agricultural Policy: Taking Stock for the New Century*, which included recognition that 'Americans consider environmental quality as a kind of 'non-market' good that is extremely important in consumer choices' (United States Department of Agriculture (USDA), 2001, p. 2). Moreover, it cannot be ignored that the FSRI Act has seen the environmental aspects of US farm policy significantly expanded. For example, the Conservation Reserve Program may now cover 39.2 million acres, not far short of the total agricultural area of the UK. In effect, the Mid-term Review and FSRI Act have seen the two greatest agricultural exporters entrench a system of support for their farmers which *de facto* extends well beyond traditional market mechanisms. On the part of the Community at least, multifunctionality forms a major plank in the justification of such support, not only for the purposes of world-trade negotiations, but also to consumers.

Notes

1 Articles of the EC Treaty were renumbered by the 1997 Treaty of Amsterdam, with effect from 1 May 1999. Reference is made throughout to the new articles, except where otherwise indicated.

2 The EAGGF Guarantee Section has traditionally been employed for price and market support (such as intervention). The Guidance Section, in contrast, has traditionally been employed to finance structural adjustment (Evans, 1999, pp. 115–144). The Guarantee Section constitutes by far the greater proportion of the Common Agricultural Policy budget: as indicated, expenditure in 1997 reached 40,423.4 million Ecu, whereas for the same year Guidance Section expenditure was estimated at 4239.6 million Ecu. Indeed, in 1997 the Guarantee Section accounted for almost exactly half of the total Community budget of 80,880 million Ecu (CEC, 1999a, p. T/103). Under the Agenda 2000 reforms considerable changes have been effected to the financing of the Common Agricultural Policy. In particular, the Guarantee Section has financed most rural development measures outside the least developed regions (Council Regulation (EC) 1258/1999 (OJ 1999, L160/103)).

3 The 'safety net' has applied where, for a period of 2 consecutive weeks, the average market price in a Member State or in a region of a Member State fell short of 1560 Euro/tonne (Council Regulation (EC) 1254/1999, Art. 27 (OJ 1999, L160/21)).

4 By way of exception, Greece, Ireland, Italy, Northern Ireland and Spain received specific, accelerated increases, commencing in two steps as from 1 April 2000.

5 The implementing provisions in England were the Common Agricultural Policy Support Schemes (Modulation) Regulations 2000 (SI 2000 No. 3127), and similar provisions applied in Northern Ireland, Scotland and Wales.

6 It may be reiterated, however, that in 1997 there was also 4239.6 million Ecu estimated expenditure under the Guidance Section; and, following the Berlin Summit, responsibility for financing some measures previously covered by the Guidance Section was transferred to the Guarantee Section.

7 The OECD has identified difficulties in reaching an operational definition of 'decoupling' (OECD, 2001b, pp. 8–13). A distinction is, however, made between 'full decoupling' and 'effective full decoupling'. The former requires not only that the equilibrium level of production (or trade) remains the same as without the measure, but also that the quantity adjustment due to any outside shock should not be altered. The latter requires that production (or trade) does not differ from the level that would have occurred without the measure.

List of Cases

Case 5/73, *Balkan* v. *Hauptzollamt Berlin-Packhof*, [1973] ECR 1091.
EC Measures Concerning Meat and Meat Products (Hormones) (1998), WT/DS26/AB/R and WT/DS48/AB/R.
United States – Import Prohibition of Certain Shrimp and Shrimp Products (1998), WT/DS58/AB/R.
United States – Import Prohibition of Certain Shrimp and Shrimp Products: Recourse to Article 21.5 of the DSU by Malaysia (2001), WT/DS58/AB/RW.

References

ABARE (1999) 'Multifunctionality': a pretext for protection? *ABARE Current Issues* 99(3) (August), 1–6.

ABARE (2000) US and EU agricultural support: who does it benefit? *ABARE Current Issues* 2000(2) (October), 1–5.

Ackrill, R. (2000) *The Common Agricultural Policy*. Sheffield Academic Press, Sheffield, 240 pp.

Agra Europe (2001) Parliament and Commission clash on role of food authority. *Agra Europe* 1956, A/1–A/2.

Barshefsky, C. (2000) *US Agricultural Agenda at the WTO*. Testimony, Trade Subcommittee of the Senate Committee on Finance, at Washington, DC, 7 March 2000 (http://usinfo.state.gov/topical/econ/wto99/ag0307.htm, visited on 23 January 2001).

Blandford, D. and Fulponi, L. (1999) Emerging public concerns in agriculture: domestic policies and international trade commitments. *European Review of Agricultural Economics* 26, 409–424.

Bohman, M., Cooper, J., Mullarkey, D., Normile, M.A., Skully, D., Vogel, S. and Young, E. (1999) *The Use and Abuse of Multifunctionality*. Economic Research Service, USDA, Washington, DC. (http://www.ers.usda.gov/briefing/wto/PDF/multifunc1119.pdf, visited on 23 January 2001).

Buckwell Report (1997) *Towards a Common Agricultural and Rural Policy for Europe* (http://europa.eu.int/comm/dg06/publi/buck_en/index.htm, visited 20 December 2000).

CEC (1968) *Memorandum sur la réforme de l'agriculture dans la Communauté Économique Européenne*. COM(68)1000.

CEC (1985) *Perspectives for the Common Agricultural Policy*. COM(85)333.

CEC (1991) *The Development and Future of the CAP*. COM(91)100.

CEC (1993) *The Agricultural Situation in the Community: 1992 Report*. Brussels, Luxembourg.

CEC (1997a) *Agenda 2000: for a Stronger and Wider Union*. COM(97)2000 (*Bulletin of the European Union, Supplement* 5/97, Brussels).

CEC (1997b) *Commission Opinion on Hungary's Application for Membership of the European Union*. COM(97)2001 (*Bulletin of the European Union, Supplement* 6/97, Brussels).

CEC (1997c) *Commission Opinion on Poland's Application for Membership of the European Union*. COM(97)2002 (*Bulletin of the European Union, Supplement* 7/97, Brussels).

CEC (1998a) *Explanatory Memorandum: the Future for European Agriculture*. (http://europa.eu.int/en/comm/dg06/ag2000/agprop/mot_en.htm, visited on 3 June 1998).

CEC (1998b) *The Agricultural Situation in the European Union: 1997 Report*. Brussels, Luxembourg.

CEC (1999a) *The Agricultural Situation in the European Union: 1998 Report*. Brussels, Luxembourg.

CEC (1999b) *Bulletin of the European Union* 3/99. Brussels.

CEC (1999c) *White Paper on Food Safety*. COM(99)719.

CEC (2000a) *Communication from the Commission on the Precautionary Principle*. COM(2000)1.

CEC (2000b) *Proposal for a Regulation of the European Parliament and of the Council Laying Down the General Principles and Requirements of Food Law, Establishing the European Food Authority and Laying Down Procedures in Matters of Food*. COM(2000)716.

CEC (2002a) *Mid-term Review of the Common Agricultural Policy*. COM(2002)394.

CEC (2002b) *Amended Proposal for a Regulation of the European Parliament and of the Council on Genetically Modified Food and Feed*. COM(2002)559.

CEC (2002c) *Prospects for Agricultural Markets 2002–2009*. Brussels.

CEC (2003) *A Long-term Policy Perspective for Sustainable Agriculture*. COM(2003)23.

CEC Directorate-General for Agriculture (1997) *CAP 2000 Working Document: Rural Developments*. July (http://europa.eu.int/comm/agriculture/publi/pac2000/rd/rd_en.pdf, visited on 29 January 2001).

CEC Directorate-General for Agriculture (1998) *Working Document VI/7655/98, State of Application of Regulation (EEC) No. 2078/92: Evaluation of Agri-environment Programmes* (a summary of which may be found at http://europa.eu.int/comm/dg06/envir/programs/evalrep/sum_en.htm, visited on 12 December 2000).

CEC Directorate-General for Agriculture (2000) *The Common Agricultural Policy: 1999 Review*. Luxembourg.

CEC Directorate-General for Agriculture (2001) *Working Document, Risk Management Tools for EU Agriculture with a Special Focus on Insurance*. Brussels.

CEC Directorate General for Trade/Agriculture (2002) *The EC's Proposal for Modalities in the WTO Agriculture Negotiations*. Ref.625/02, Brussels, December.

Charnovitz, S. (2002) The law of environmental 'PPMs' in the WTO: debunking the myth of illegality. *Yale Journal of International Law* 27, 59–110.

Commission on 21st Century Production Agriculture (2001) *Directions for Future Farm Policy: the Role of Government in Support of Production Agriculture – Report to the President and Congress*. Washington, DC.

European Committee of the Regions (1999) *Opinion of the Committee of the Regions on the 'Proposals for Council Regulations (EC) concerning the Reform of the Common Agricultural Policy'* (OJ 1999, C93/1).

European Community (2000a) WTO farm negotiations: 'EU constructive but firm' Franz Fischler says. IP/00/295, Brussels, 24 March.

European Community (2000b) Fischler calls for dialogue in farm trade talks. IP/00/1160, Brussels, 13 October.
European Community (2001a) *EU Makes Proposal for 'Win–win' Solution to WTO Agricultural Negotiations.* MEMO/01/28, Brussels, 5 February.
European Community (2001b) *Commission Welcomes the Adoption of New Rules for GMOs.* MEMO/01/42, Brussels, 16 February.
European Community (2001c) Animal transportation: commission adopts proposal to improve conditions for the long distance transportation of animals. IP/01/527, Brussels, 9 April.
European Community (2001d) 'New WTO Round slap in the face for isolationism', says EU Farm Commissioner Fischler. IP/01/1584, Doha, 14 November.
European Community (2002) Commissioner Fischler: the Common Agricultural Policy is better than the press it gets. IP/02/32, Berlin/Brussels, 10 January.
European Community (2003) EU fundamentally reforms its farm policy to accomplish sustainable farming in Europe. IP/03/898, Luxembourg, 26 June.
European Court of Auditors (1998) *Opinion No. 10/98 of the European Court of Auditors on Certain Proposals for Regulations within the Agenda 2000 Framework* (OJ 1998, C401/1).
European Court of Auditors (2000) *Special Report No. 14/2000 on Greening the CAP.* Luxembourg.
Evans, A. (1999) *The EU Structural Funds.* Oxford University Press, Oxford, 335 pp.
Fennell, R. (1997) *The Common Agricultural Policy: Continuity and Change.* Clarendon Press, Oxford, 440 pp.
Fischler, F. (2000) New prospects for an enlarged Europe, Speech/00/335, at Krakow, 25 September.
Fischler, F. (2001a) The CAP after Agenda 2000: the achievements and challenges, Speech/01/10, at Berlin, 18 January.
Fischler, F. (2001b) Food quality, Speech/01/165, at Östersund, Sweden, 10 April.
Fischler, F. (2001c) Quality production: the new challenge of the Common Agricultural Policy, Speech/01/301, at Brussels, 21 June.
Fischler, F. (2001d) Round Table on Agriculture and Food, Speech/01/359, at Brussels, 26 July.
Fischler, F. (2002a) The Mid-term Review: towards a policy that pleases everybody, Speech/02/339, at Wageningen, 12 July.
Fischler, F. (2002b) Presentation of the CAP Mid-term Review at the Agricultural Council, Speech/02/342, at Brussels, 15 July.
Fischler, F. (2002c) Adjusting the CAP to better meet its objectives, Speech/02/412, at Brussels, 20 September.
Fischler, F. (2003a) EU position on agriculture before kick-off of Cancún Ministerial, Speech/03/395, at Cancún, 9 September.
Fischler, F. (2003b) The breakdown of the Cancún Ministerial Meeting, Speech/03/408, at Cancún, 14 September.
Gardner, B. (1996) *European Agriculture: Policies, Production and Trade.* Routledge, London, 230 pp.
Grant, W. (1995) The limits of Common Agricultural Policy reform and the option of denationalization. *Journal of European Public Policy* 2, 1–18.

Harwood, J., Heifner, R., Coble, K., Perry, J. and Somwaru, A. (1999) *Managing Risk in Farming: Concepts, Research and Analysis*. Agricultural Economic Report No. 774, March, Economic Research Service, USDA, Washington, DC.

House of Commons Agriculture Committee (2000) *Sixth Report of the Agriculture Committee: the Implications for UK Agriculture and EU Agricultural Policy of Trade Liberalisation and the WTO Round*. Session 1999–2000, HC 246-I.

Kjeldahl, R. and Tracy, M. (eds) (1994) *Renationalisation of the Common Agricultural Policy?* Institute of Agricultural Economics, Valby, Denmark, and Agricultural Policy Studies, Genappe, Belgium, 100 pp.

Lauterburg, D. (2001) *Food Law: Policy and Ethics*. Cavendish, London, 275 pp.

Lowe, P. Flynn, B., Just, F., Valadas de Lima, A., Patrício, T. and Povellato, A. (2000) National cultural and institutional factors in CAP and environment. In: Brouwer, F. and Lowe, P. (eds) *CAP Regimes and the European Countryside*. CAB International, Wallingford, UK, pp. 257–280.

McMahon, J.A. (2000) The eastern enlargement of the European Union and the Common Agricultural Policy. In: Usher, J.A. (ed.) *The State of the European Union*. Longman, Harlow, pp. 58–77.

McMahon, J.A. (2002) The Common Agricultural Policy: from quantity to quality? *Northern Ireland Legal Quarterly* 53, 9–27.

MAFF (1999) *Nick Brown Announces a New Direction for Agriculture*. News Release 435/99, 7 December.

MAFF, Scottish Executive Rural Affairs Department, Department of Agriculture and Rural Development (Northern Ireland) and National Assembly for Wales Agriculture Department (1999) *Agriculture in the United Kingdom 1998*. The Stationery Office, London.

Neville, W. and Mordaunt, F. (1993) *A Guide to the Reformed Common Agricultural Policy*. Estates Gazette, London, 150 pp.

OECD (1998a) *Communiqué of the Meeting of the Committee for Agriculture at Ministerial Level*, 5–6 March (http://www.oecd.org/agr/ministerial/commune/htm, visited on 3 March 2001).

OECD (1998b) *Multifunctionality: a Framework for Policy Analysis*. OECD Document AGR/CA(98)9, November.

OECD (1999) *Written Comments on the Document 'Multifunctionality: a Framework for Policy Analysis' [AGR/CA(98)9]*. OECD Document AGR/CA/RD(99)1, March.

OECD (2001a) *Multifunctionality: Towards an Analytical Framework*. OECD, Paris. 159 pp.

OECD (2001b) *Decoupling: a Conceptual Overview*. OECD, Paris. 29 pp.

OECD (2001c) *Market Effects of Crop Support Measures*. OECD, Paris. 119 pp.

Rodgers, C.P. (1996) Environmental gain, set-aside and the implementation of EU agricultural reform in the United Kingdom. In: Rodgers, C.P. (ed) *Nature Conservation and Countryside Law*. University of Wales Press, Cardiff, pp. 111–137.

Rude, J. (2001) Under the green box: the WTO and farm subsidies. *Journal of World Trade* 35, 1015–1033.

Salmon, N. (2002) A European perspective on the precautionary principle, food safety and the free trade imperative of the WTO. *European Law Review* 27, 138–155.

Scott, J. (1999) Regional policy: an evolutionary perspective. In: Craig, P. and de Búrca, G. (eds) *The Evolution of EU Law*. Oxford University Press, Oxford, pp. 625–652.

Scott, J. (2000) On kith and kine (and crustaceans): trade and environment in the EU and WTO. In: Weiler, J.H.H. (ed) *The EU, the WTO, and the NAFTA: Towards a Common Law of International Trade*? Oxford University Press, Oxford, pp. 125–167.

Scott, J. (2003) European regulation of GMOs and the WTO. *Columbia Journal of European Law* 9, 213–239.

United States (2002) *US Proposal for Global Agricultural Trade Reform*. (http://www.fas.usda.gov/itp/wto/actual.htm, visited on 3 October 2002).

USDA (2001) *Food and Agricultural Policy: Taking Stock for the New Century*. Washington, DC.

WTO (2000a) *Proposal for Comprehensive Long-term Agricultural Trade Reform: Submission from the United States*. G/AG/NG/W/15, 23 June.

WTO (2000b) *European Communities Proposal: the Blue Box and Other Support Measures to Agriculture*. G/AG/ NG/17, 28 June.

WTO (2000c) *European Communities Proposal: Food Quality – Improvement of Market Access Opportunities*. G/AG/ NG/18, 28 June.

WTO (2000d) *European Communities Proposal: Animal Welfare and Trade in Agriculture*. G/AG/NG/19, 28 June.

WTO (2000e) *European Communities Proposal: Export Competition*. G/AG/NG/34, 18 September.

WTO (2000f) *Cairns Group Negotiating Proposal: Domestic Support*. G/AG/NG/W/35, 22 September.

WTO (2000g) *EC Comprehensive Negotiating Proposal*. G/AG/NG/90, 14 December.

WTO (2000h) *Japan: Negotiating Proposal*. G/AG/NG/W/91, 21 December.

WTO (2001a) *Food Safety: Note by the European Communities*. D(2001)(DIVERS/500186 ur), 20 July.

WTO (2001b) *Green Box: Non-paper from the European Communities*. 6865, 24 September.

WTO (2001c) *Geographical Indications: Non-paper from the European Communities*. 6866, 24 September.

WTO (2001d) *Ministerial Declaration*. WT/MIN(01)/DEC/W/1, 14 November.

WTO (2001e) *Agriculture: Negotiations 2000–March 2001: Work Programme for the Second Phase*. 27 March (http://www.wto.org/english/tratop_e/agneg7_e.htm, visited on 2 April 2001).

WTO (2003) *Negotiations on Agriculture: First Draft of Modalities for the Further Commitments*. TN/AG/W/1, 17 February.

TRIPS and the Protection of Intellectual Property in Biotechnology in the United States

7

Theodore A. Feitshans

Introduction

The Agreement on Trade-related Aspects of Intellectual Property Rights (TRIPS) has required changes in domestic patent law in the United States. TRIPS is Annex 1C of the Agreement Establishing the World Trade Organization (WTO), concluded at Marrakesh on 15 April 1994. The Agreement Establishing the WTO, also known as the Marrakesh Agreement, was negotiated in the Uruguay Round of multilateral trade talks under the auspices of the General Agreement on Tariffs and Trade (GATT) and entered into force on 1 January 1995.

TRIPS addresses virtually the entire array of intellectual property, including copyrights, trademarks, indications of geographical origin, industrial designs, patents, integrated-circuit topographies and trade secrets. The portions of TRIPS relevant to the protection of intellectual property in biotechnology include the section on patents and, to a lesser degree, the section on trade secrets. This chapter addresses how TRIPS has altered US patent law, with specific reference to the protection of intellectual property in biotechnology, and discusses each type of US intellectual-property protection available to protect biotechnology. This chapter also discusses the Convention on Biological Diversity (United Nations, 1992), an instrument that has entered into force in 161 countries. The USA has not ratified the Convention on Biological Diversity, and ratification in the foreseeable future is unlikely. This chapter therefore discusses the impact of US non-participation in the Convention on coordination of intellectual-property protection under TRIPS.

Basic Principles of TRIPS

TRIPS sets minimum standards for the enforcement of intellectual-property rights. Countries may, but need not, provide higher levels of protection so long as the protection is not inconsistent with TRIPS. TRIPS offers considerable flexibility as to both the form and the content of intellectual-property systems. This flexibility is particularly broad in biotechnology.

Fundamental principles of TRIPS include national treatment (TRIPS, Art. 3) and most-favoured-nation treatment (TRIPS, Art. 4). National treatment requires that a country treat citizens of other countries at least as favourably as it treats its own nationals. Most-favoured-nation treatment requires that benefits accorded to one signatory nation be extended to all signatory nations. Thus, domestic intellectual-property law must be even-handed in its treatment of all citizens of all signatory countries. The Paris Convention (1967) (patents) and the Berne Convention (1971) (copyrights) provide limited exceptions to national treatment related to judicial and administrative procedure and jurisdiction. Laws such as those that require a domestic address for or appointment of an agent for service of process are permitted. TRIPS, however, limits these exceptions by prohibiting practices of judicial and administrative procedure and jurisdiction that are in reality disguised restrictions on trade (TRIPS, Art. 3(2)). Limited experience under TRIPS makes it difficult to provide concrete examples of practices that might be prohibited. The most-favoured-nation treatment provision of TRIPS is consistent with the principles expressed in the Agreement Establishing the WTO, though TRIPS sets forth limited exceptions. Two of these exceptions are important to intellectual property in biotechnology. The first is that most-favoured-nation treatment need not be offered in the area of judicial assistance or law enforcement of a general nature provided under international agreements so long as those agreements do not relate specifically to intellectual property (TRIPS, Art. 4(a)). This clause limits the effect of TRIPS on law enforcement and the criminal justice system to those issues that are specific to intellectual property (TRIPS, Art. 4(a)). More important is the second exception; TRIPS excepts from most-favoured-nation treatment benefits

> deriving from international agreements related to the protection of intellectual property which entered into force prior to the entry into force of the WTO Agreement, provided that such agreements are notified to the Council for TRIPS and do not constitute an arbitrary or unjustified discrimination against nationals of other Members.
>
> (TRIPS, Art. 4(d))

Article 8 of TRIPS includes general provisions that provide exceptions for health and safety and for antitrust enforcement. Article 8 is short and provides minimal guidance for applying these provisions in practice.

Given the widespread disagreement over the health and environmental impacts of bioengineered organisms, this exception has been, and is likely to continue to be, a rich source of controversy.

What constitutes an abuse of an intellectual-property right is also a rich source of controversy, in light of the pressing needs of many countries for patented pharmaceuticals to fight acquired immuno deficiency syndrome (AIDS) and other diseases. If the recently suspended litigation by the pharmaceutical industry in South Africa is any indication, however, some of these issues may be decided in the court of public opinion long before they are decided by judicial tribunals (Myers, 1999). Article 40 expands on the restrictions that Members may apply to intellectual property rights to prevent adverse effects on trade, technology transfer and information dissemination. Article 40 provides for consultations between Members if a Member believes that its nationals have been treated unfairly by another Member's laws designed to prevent abuse of intellectual property.

Article 62 addresses general principles of acquisition and maintenance of intellectual-property rights. In essence, this article requires fairness and reasonableness in the application of procedures and formalities. Article 63, governing transparency, requires that countries publish or make available all law, rules, regulations and decisions related to intellectual property, but it does not require translation from the national language. Article 65 provides for transitional arrangements, including a delay for developing countries and those making the transition to market economies. More controversial is the 10-year delay for least-developed countries provided in Article 66. Articles 67–71 provide for technical and international cooperation, institutional arrangements, transitional arrangements related to rights created prior to the entry into force of the TRIPS Agreement and means for review and amendment of the TRIPS provisions. Article 72 prohibits Members from making reservations to the TRIPS Agreement without the consent of the other Members. Article 73 allows Members to take actions necessary to protect essential security interests and to honour obligations to the United Nations for the purpose of maintaining international peace and security.

Article 64 requires the application of Articles XXII and XXIII of GATT 1994 as elaborated in the Dispute Settlement Understanding (DSU), except during a 5-year transition period. These dispute-settlement procedures are a key component of TRIPS and, together with harmonization of key provisions of national laws, represent an important departure from the approach used in previous intellectual-property agreements administered through the World Intellectual Property Organization (WIPO), a United Nations agency (Felgueroso, 2002). Article 64 adopts the judicial bodies and procedures of the WTO, with the result that decisions involving intellectual property are enforceable through trade sanctions (Felgueroso, 2002). The DSU offers no forum for private parties; use of the

DSU is reserved to Member States. Since it began functioning in 1995, the DSU has resolved six disputes (Felgueroso, 2002). Because no forum for private parties exists, the impact of TRIPS on the domestic law of Member States will depend upon the willingness of such states to use the DSU mechanism.

Utility Patent Protection[1]

Basic requirements

Title 35 USC §§ 1–376 (2002) (the Patent Act) authorizes the US Patent and Trademark Office (the Patent Office) to issue a patent to any person who invents a product or process that is novel, non-obvious and useful (Patent Act, §§ 101–103). An invention must meet the statutory definition of patentable subject-matter to receive patent protection (*In re Frank R. Bonczyk* (2001)). Although the inventor may assign patent rights to an employer or others, the application for patent protection must be made in the name of the actual inventor(s). Where large teams of researchers collaborate, a common situation with research in biotechnology, determining inventorship may be an issue (*Burroughs Welcome* v. *Barr Laboratories, Inc.* (1994)). Patents awarded under section 101 of the Patent Act are often called utility patents to distinguish them from special types of patents discussed below. Numerically and economically, utility patents are by far the most important type of patent.

For a product or process to be novel it must be new, meaning that no other person has made, sold or published a description of the product or process prior to the application (Patent Act, § 102). The Supreme Court has determined that a living organism or a part of a living organism may be patented[2] (*Diamond* v. *Chakrabarty* (1980)). Indeed, many patents have been granted for genes of particular organisms. Some plant varieties, such as Roundup Ready® soybeans and Bacillus thuringiensis (*Bt*) cotton, now contain patented genes. In addition to patenting genes and entire organisms, biotechnology companies may also obtain patent protection on the equipment and processes developed to create novel genes and organisms.

Patent protection is available only if the inventor files a patent application with the Patent Office within 1 year of the first commercial use (the on-sale bar) or publication of the invention (Patent Act, § 102(b)). Because a patent is entitled to a presumption of validity, one who seeks to prove the invention was either anticipated by another or is subject to the on-sale bar must demonstrate this by 'substantial evidence that is clear and convincing' (*Finnigan Corp.* v. *International Trade Commission* (1999), p. 1370). 'Generally, oral testimony of prior public use must be corroborated in order to invalidate a patent' (*Juicy Whip, Inc.* v. *Orange Bang, Inc.* (2002), pp. 737–738). The Supreme Court articulated the standard to be applied in determining whether the grant of a patent is invalid based on the on-sale bar (*Pfaff* v. *Wells Electronics, Inc.* (1999)).

> The Court established two conditions that must be satisfied to begin the one-year statutory period for the on-sale bar: 1) the invention must be the subject of a commercial offer for sale; and 2) the invention must be ready to be patented. [footnote omitted] The [C]ourt then stated that the second condition 'may be satisfied in at least two ways': 1) by proof of a reduction to practice; or 2) by proof that the inventor developed drawings or other materials sufficient to permit one skilled in the art to practice the invention.
>
> (Gonzalez, 2000, p. 88)

Since the relationships between customers and developers of technology in the biotechnology business are often complex, with sharing of information common, the *Pfaff* requirements may be violated inadvertently. To avoid this harsh result, biotechnology companies should have a comprehensive intellectual-property policy in place.

Jurisdictional reach of US patent law

A patent issued by the Patent Office is effective only within the territory of the USA. US patent law provides some protection against the import of non-patented products produced abroad by a process patented in the USA (Patent Act, § 271(g)), but without foreign patent protection the manufacture and sale of such products outside the USA cannot be prohibited. To obtain patent protection in foreign countries, an application must be filed in each country where protection is desired. The USA is party to international agreements that facilitate this process (Patent Act, §§ 351–376). Unlike the USA, most foreign countries offer no grace period for prior use or publication of the invention. Foreign rights may be lost as a result of any prior commercial use or publication of the invention prior to filing of foreign patent applications.

For the first time, TRIPS provides means for the US government to ensure that US citizens receive similar treatment of their intellectual property in Member States and at home. TRIPS has added the tool of trade sanctions to the arsenal that the US government has to compel foreign countries to provide intellectual-property protection. Trade sanctions usually consist of punitive tariffs applied by the US government to selected imports from the offending country. None the less, TRIPS excludes significant areas of biotechnology, including diagnostic, therapeutic and surgical methods for the treatment of humans or animals, plants and animals other than microorganisms, and essentially biological processes for the production of plants or animals (TRIPS, Art. 27(3)).

Simultaneous invention by competitors

In a field as competitive as biotechnology, more than one applicant may claim the same invention. When applications by multiple applicants for the same invention are pending simultaneously or a pending application interferes with an unexpired patent, the Commissioner of the Patent and Trademark Office must declare an interference (Patent Act, § 135). In *Singh* v. *Brake* (2000), the Federal Circuit overturned a Patent and Trademark Office Board of Patent Appeals and Interferences decision awarding priority of invention in a DNA construct. The Federal Circuit determined that the decision of the Board was not supported by substantial evidence and remanded so that the Board could reweigh the sufficiency of the evidence and reach factual conclusions (*Singh* v. *Brake* (2000)). At issue was the requirement that an inventor's testimony be corroborated (*Singh* v. *Brake* (2000)). The Federal Circuit concluded that the inventor's laboratory notebook, not witnessed[3] until several years after the fact, could provide corroboration (albeit weak) of the inventor's testimony regarding conception but not reduction to practise (*Singh* v. *Brake* (2000)). The case illustrates the importance of keeping good records, promptly witnessed, of all aspects of research in biotechnology to support subsequent applications for patent protection.

Barton v. *Adang* (1998) involved a three-way interference over priority of invention in a method of designing a synthetic *Bt* gene to be more highly expressed in plants. The interference was declared between two pending applications, assigned to Agracetus and Monsanto, respectively, and an issued patent assigned to Mycogen Plant Science. Shortly after the interference was declared, Monsanto purchased Agracetus, notified the Patent Office of common ownership and in the notification declared that good cause existed to continue the three-party interference. The Patent Office determined that good cause to continue the three-party interference did not exist and required Monsanto to elect between the two applications. While finding that the Patent Office has discretion as to whether to declare an interference or continue one, once begun, the Federal Circuit found that, in the absence of discovery, Monsanto could not determine which application would be the best evidence to establish priority and that this was 'good cause' to continue the interference. The implication of this decision is that the Patent Office could force Monsanto to elect once discovery was complete and it had obtained the information needed to make the election.

Genentech, Inc. v. *Chiron Corp.* (1997) arose from an interference involving claims to technology related to the production of human insulin in yeast. The Federal Circuit addressed the complex issue of interpretation of a count in an interference. The count in an interference is the matter for which the Patent Office has determined that priority is in issue. As with determination of the scope of claims in an issued patent, the proper construction of the count is a question of law for the court.

In *Kridl* v. *McCormick* (1997) the Federal Circuit reviewed and upheld the award of priority, by the Board of Patent Appeals and Interferences, to McCormick and fellow inventors, Barton and Swain. The Federal Circuit noted that priority is a question of law subject to review *de novo* on appeal. At issue in the interference was priority to an antisense recombinant DNA technology useful for giving plants resistance to certain viruses. The case contains a good review of the law applicable to the corroboration required for inventors' testimony. The complexity of priority claims in biotechnological inventions is illustrated in *Fiers* v. *Revel* (1993), an appeal from a three-way interference in which British, Israeli and Japanese teams of inventors contested priority of invention in DNA that codes for human fibroblast β-interferon.

Duration of patent protection and rights conferred

Patent protection is generally available for a term of 20 years from the date of filing the patent application (Patent Act, § 154(2)). This term was modified from 18 years from issue to comply with TRIPS. During the 20-year period, the owner of the patent has the right to exclude all others from making, using or selling any product or process that contains or uses the patented technology (Patent Act, § 271). A patent does not confer a right to use; for example, use of a patented organism may be banned as too hazardous to public health or the environment. Any other person who makes, uses or sells any part of that patented technology is an infringer, who is liable to the patent owner for damages, even if the infringer was unaware of the patent or the infringement. A court may treble damages and award attorney fees against one who knowingly infringed a patent (Patent Act, §§ 271–273). Patent rights are not self-executing; the owner of the patent must enforce these rights through an action in federal district court.

Issues arising from patenting living organisms

Patenting of living organisms poses special problems for the patent system. The so-called enablement requirement of section 112 of the Patent Act reads:

> The specification shall contain a written description of the invention, and of the manner and process of making and using it, in such full, clear, concise, and exact terms as to enable any person skilled in the art to which it pertains, or with which it is most nearly connected, to make and use the same, and shall set forth the best mode contemplated by the inventor of carrying out his invention.

To enable the public to practise an invention embodied in a self-replicating organism, a deposit of that organism must be made in an acceptable depository (MPEP, 1998, § 2404). Acceptable depositories are any International Depository Authority (IDA), as established under the Budapest Treaty on the International Recognition of the Deposit of Microorganisms for the Purposes of Patent Procedure (1977), or any depository deemed suitable by the US Commissioner of Patents and Trademarks (MPEP, 1998, § 2405). Anyone who intends to seek protection in countries in addition to the USA would be well advised to use an IDA, which triggers the provisions of the Budapest Treaty.

Diamond v. *Chakrabarty* (1980) established that the Patent Office must grant patent protection to living organisms. By interpretative rule the Patent Office determined that section 101 of the Patent Act also requires that it grant patent protection to inventions embodied in multicellular organisms, including animals. The Federal Circuit rejected a challenge to the Patent Office's interpretation on grounds of standing (*Animal Legal Defense Fund* v. *Quigg* (1991)).

Conflict with other law

The situation regarding the patenting of plants is complicated by the existence of the Plant Patent Act of 1930 and the Plant Variety Protection Act of 1970. The Supreme Court held that utility patent protection is available for plants despite partial overlap with the Plant Patent Act and the Plant Variety Protection Act (*J.E.M. Ag Supply Inc.* v. *Pioneer Hi-Bred International Inc.* (2001)). The defendants objected that Pioneer had obtained utility patent protection under the Patent Act for seed-produced varieties of maize capable of protection under certificates of protection under the Plant Variety Protection Act. They argued that the enactment of the Plant Patent Act and the Plant Variety Protection Act had removed plants from the realm of patentable subject-matter under section 101 of the Patent Act. The Supreme Court rejected this argument, noting that it had held that 'when two statutes are capable of coexistence, it is the duty of the courts, absent a clearly expressed congressional intention to the contrary, to regard each as effective' (*J.E.M. Ag Supply Inc.* v. *Pioneer Hi-Bred International Inc.* (2001), p. 155).

Standardization of symbols and format in applications for patents on genes

In regard to gene patents, the Patent Office requires 'the use of standard symbols and a standard format for sequence data in most sequence-type patent applications' (MPEP, 2001, § 2420). This is a departure from general Patent Office practice, which allows the inventor to be his own lexicographer (*Elekta Instrument* v. *O.U.R. Scientific Int.* (2000)). The standard symbols and format that the Patent Office requires for gene patents are likely both to simplify the role of courts in claim interpretation and to enhance the ability to search for gene patents.

Interpreting claims in a patent

Outside the realm of gene patents, the Federal Circuit held that the rules of claim construction require courts to look first to the plain meaning of the claim language and then to any different meaning that should be ascribed to the claim language based on a definition clearly set forth by the patentee (*Rexnord Corp.* v. *The Laitram Corp.* (2001)). Intrinsic evidence within the patent application may be used to resolve ambiguities in claim language (*Pickholz* v. *Rainbow Technologies, Inc.* (2002)).

Clarification of gene-patent issues by the patent office

The Patent Office has also recently clarified the utility requirements for gene patents under the Patent Act (§§ 101, 112; Utility Examination Guidelines, 2001), and the written description requirement under section 112 (para. 1) (Guidelines, 2001). As both clarifications govern internal practices, the Patent Office decided that they were exempt from notice and comment rule-making. None the less, these changes may have significant implications for some applicants. In clarifying the utility requirement, the Patent Office decided against developing a standard specifically for gene patents and stated that the utility must be 'specific and substantial' (*In re Ziegler* (1993); Revised utility, 1999).

> A rejection based on lack of utility should not be maintained if an asserted utility for the claimed invention would be considered specific, substantial, and credible by a person of ordinary skill in the art in view of all evidence of record.
>
> (Utility Examination Guidelines, 2001)

Statements of fact made by the applicant are treated as true unless one skilled in the art would doubt them. A lack of utility may also be the basis for a rejection based on a failure to disclose how to use the invention under the enablement requirement (Patent Act, § 112).

The Utility Examination Guidelines are consistent with Supreme Court precedent. The Supreme Court's rationale for requiring specific utility is the fear that an inventor's patent claims might occupy the entire field (*Brenner* v. *Manson* (1966)). The Federal Circuit has clarified that '[t]he threshold of utility is not high: An invention is "useful" under section 101 if it is capable of providing some identifiable benefit' (*Juicy Whip, Inc.* v. *Orange Bang, Inc.* (1999), p. 1366).

The disclosure in the patent application must be sufficient to enable one skilled in the art to practise the invention (*Genentech, Inc.* v *Novo Nordisk A/S* (1997)). '[W]hether a patent specification adequately describes the subject matter claimed is a question of fact' (*In re Alton* (1996), pp. 1171–1172). 'Patent protection is granted in return for an enabling disclosure of an invention, not for vague intimations of general ideas that may or may not be workable' (*Genentech, Inc.* v. *Novo Nordisk A/S* (1997), p. 1366). 'Tossing out the mere germ of an idea does not constitute enabling disclosure' (*Genentech, Inc.* v. *Novo Nordisk A/S* (1997), p. 1366). While every aspect of a generic claim certainly need not have been carried out by an inventor or exemplified in the specification, reasonable detail must be provided to enable members of the public to understand and carry out the invention. The specification of the patent need not contain sufficient detail to allow the public to practise the invention, but it must contain information about the novel steps that are essential to allow one skilled in the art to practise the invention. Omission of minor detail will not cause the specification to be inadequate, but it must not be so inadequate that one skilled in the art is required to engage in undue experimentation to practise the invention. The disclosure in the patent application may substantially limit the scope of the claims since the claims may be no broader than the disclosure (*The Gentry Gallery, Inc.* v. *The Berkline Corp.* (1998)). Claims are typically further narrowed during the process of prosecution of the patent application as the applicant gives up material in the process of negotiation with the patent examiner (*Festo Corp.* v. *Shoketsu Kinzoku Kogyo Kabushiki Company, Ltd.* (2002)).

The unpredictability of the art is a key issue in determining the scope of claims allowable (*Enzo Biochem, Inc.* v. *Calgene, Inc.* (1999)). The Patent Office as well as the courts have generally classified gene technology in the same category as chemistry, an inherently unpredictable art.

For the written-description requirement, as for the utility requirement, the Patent Office decided to develop neutral standards that apply across all arts (Guidelines, 2001). The written description must be sufficient for one skilled in the art to be able to practise the invention (Guidelines, 2001); to avoid confusion, the Patent Office elected not to attempt to define the word 'gene' (Revised Interim, 1999). These requirements, taken together, will prevent applicants from obtaining patent protection on nucleotide sequences with no known applications other than as the subject of further research (Grubb, 1999; Feitshans, 2001).

Non-obviousness

The Patent Act (§ 103) imposes the further requirement that the subject-matter of the invention be non-obvious at the time the application for the patent was filed. Subsection (b) is directed specifically to biotechnological-process inventions. The history of the non-obviousness standard has been discussed in detail elsewhere (Wigley, 2000). Whether a claim in a patent is obvious is a question of fact for a jury, and the jury's decision may be set aside only if there is no substantial evidence to support it (*Sibia Neurosciences* v. *Cadus Pharmaceutical* (2000)). In *Sibia Neurosciences*, a divided Federal Circuit found that claims to a cell-based screening method were obvious as a matter of law; the dissent protested that the court substituted its judgement for that of the jury. *In re Hiniker Co.* (1998) affirmed a finding of obviousness by the Patent and Trademark Office Board of Patent Appeals and Interferences. This case, decided by the Federal Circuit, illustrates the fact-intensive nature of analyses into the obviousness of claimed inventions. Unexpected results are one argument for non-obviousness of the claimed invention (*In re Mayne* (1997)). Whether the results of the claimed invention are unexpected is a question of fact. The mere fact that a claimed invention is simple in nature will not make that invention obvious if it was not obvious to one skilled in the art at the time the invention was made (*The Gentry Gallery, Inc.* v. *The Berkline Corp.* (1998)). Those reviewing claims for obviousness must avoid after-the-fact analysis.

The standard for Federal Circuit review of factual findings of obviousness depends on the route by which the issue came to the Federal Circuit – that is, whether the decision was made by the Patent Office, an agency or a federal district court. In *Dickinson* v. *Zurko* (1999), the Court held that findings of fact made by the Patent Office are subject to review under the arbitrary, capricious or abuse of discretion, or unsupported by the substantial-evidence standard, while factual findings of district courts are subject to review under the higher, clearly erroneous standard. The Supreme Court stated that it would be possible for a decision to be clearly erroneous while supported by substantial evidence, although such cases would be rare.

Inequitable conduct

Applicants for patent protection owe the Patent Office a duty of candour, good faith and honesty (*Life Technologies* v. *Clontech Laboratories, Inc.* (2000); *Perseptive Biosystems* v. *Pharmacia Biotech* (2000)). When this duty is breached, inequitable conduct has occurred. Inequitable conduct during prosecution of the patent application renders the patent unenforceable. 'Inequitable conduct can consist of affirmative misrepresentations of material fact, submission of false material information, or the failure to disclose known material information during the prosecution of a patent, coupled with intent to deceive the PTO [Patent and Trademark Office]' (*Life Technologies* v. *Clontech Laboratories, Inc.* (2000), p. 1324). Whether inequitable conduct has occurred is a question of fact that must be proved by clear and convincing evidence.

Plant Patents

A special type of patent is available for new varieties of plants. Section 161 of the Plant Patent Act of 1930 provides that:

> [w]hoever invents or discovers and asexually reproduces any distinct and new variety of plant, including cultivated sports, mutants, hybrids, and newly found seedlings, other than a tuber propagated plant or a plant found in an uncultivated state, may obtain a patent.

Plant patents are available for asexually reproduced plants, as well as plants capable of reproducing by seed, if they can also be reproduced asexually (MPEP, 2000, § 1601). Plant patents cannot be obtained on tuber crops, such as Irish potatoes and Jerusalem artichokes. The new plant must be a distinct variety (*Imazio Nursery, Inc.* v. *Dania Greenhouses* (1995)). No deposit is required for plants that are the subject of plant patents (MPEP, 1998, § 2403.2), but the applicant may be required to provide a specimen of the plant (MPEP, 2000, § 1607). Only a single claim is allowed in a plant patent.

Certificates of Protection under the Plant Variety Protection Act of 1970

Certificates of protection are available through the Plant Variety Protection Office of the US Department of Agriculture. This patent-like form of protection is available for true-breeding plants and tuber crops, but not for fungi and bacteria (Plant Variety Protection Act, § 2402). A sample of the seed or tuber of the variety for which a certificate of plant-variety protection is sought must be deposited with the Plant-Variety Protection Office of the US Department of Agriculture (§ 2422(4)). The term of a certificate of protection is 20 years for most crops and 25 years for trees, shrubs and vines (§ 2483(b)).

Trade-secret Protection

A trade secret is information that has value to a business and is not generally known to the public. The law of trade secrets is a matter of state law and varies from state to state. Trade secrets are of potentially infinite duration because they last as long as secrecy can be maintained. Most inventions will be held as trade secrets prior to obtaining patent protection. To preserve trade-secret status the owner of the trade secret must take affirmative steps to preserve secrecy. If litigation ensues, confidentiality agreements with employees, collaborators and sources of capital are key components of proof that affirmative efforts to preserve trade secrets have been made.

Trade-secret protection may also be a permanent alternative to patent or other formal protection for biotechnology inventions. Trade-secret protection is particularly appropriate for process inventions where the process remains under the control of the owner. The pre-grant publication practices of some foreign patent offices may also indicate that trade-secret protection is the better means for protecting certain biotechnology inventions; the pre-grant publication destroys the trade secret and there is no guarantee that a patent will ever be granted.

Copyright Protection

> Copyright protection subsists . . . in original works of authorship fixed in any tangible medium of expression, now known or later developed, from which they can be perceived, reproduced, or otherwise communicated, either directly or with the aid of a machine or device.
>
> (Copyright Act, § 102)

Copyright protection exists from the time that the original work is fixed in a tangible medium of expression.

Genetically modified organisms (GMOs) have not been protected to date using copyright because the sequences incorporated into most GMOs were found in other organisms and are therefore not original. As technology becomes more sophisticated, however, there is no reason why artificial (and original) sequences of DNA might not be protected through copyright.

The duration of copyright protection is much longer than that of patent protection. In general, a copyright in a 'work created on or after January 1, 1978, subsists from its creation and . . . endures for a term consisting of the life of the author and 70 years after the author's death' (Copyright Act, § 302). Where available, copyright exists in addition, rather than as an alternative, to patent protection. Protection is weak, however, because actual copying must be proved to prevail in an infringement action.

Enforcement of Intellectual-property Rights

Infringement of patents

The same law governs infringement for both utility and plant patents. Infringement includes acts of making, using, offering to sell or selling any patented invention within the territory of the USA. Limited exceptions apply to inventions involving biotechnology (Patent Act, § 271). The effect of these exemptions is to permit potential manufacturers of generic products to begin the process of regulatory review prior to the expiration of patents covering the product.

Infringement actions, except those against the US government, are brought in federal district court under federal-question jurisdiction (Judiciary and Judicial Procedure, § 1338(a)). 'Any civil action for patent infringement may be brought in the judicial district where the defendant resides, or where the defendant has committed acts of infringement and has a regular and established place of business' (§ 1400(b)). Actions seeking compensation for infringement by the US government are within the exclusive jurisdiction of the US Court of Federal Claims (§ 1498(a)). To the extent that states may be sued for infringement, such suits may be brought in federal district court. Congress attempted (Patent Act, § 296) to abrogate the states' Eleventh Amendment immunity to patent-infringement suits, but the Supreme Court held that attempt invalid, (*Florida Prepaid Postsecondary Education Expense Board* v. *College Savings Bank* (1999)). Appeals of patent-infringement actions, as well as appeals from adverse actions of the Patent Office, are exclusively within the jurisdiction of the US Court of Appeals for the Federal Circuit (Judiciary and Judicial Procedure, 1295(a)(1),(3),(4)(A)). Where the issue of patent infringement is raised for the first time in a compulsory counterclaim, appeal is to the US Court of Appeals in the circuit from which the case arose, not the Federal Circuit (*The Holmes Group, Inc.* v. *Vornado Air Circulation Systems, Inc.* (2002)).

Before finding infringement, the court must first determine the proper scope of the claims to be applied. *Markman* v. *Westview Instruments, Inc.* (1996) is the leading Supreme Court opinion on the subject of claim interpretation. *Markman* established that interpretation of claims is an issue of law 'exclusively within the province of the court, with no right to a jury determination' (p. 372). Once the court determines the scope of the claims, the second question, whether infringement has occurred, is for the jury.

Infringement may be either literal, if the accused device includes every limitation of the claim, or an equivalent of each limitation under the doctrine of equivalents. (*Regents of the University of California* v. *Eli Lilly & Company* (1997)). The doctrine of equivalents is an equitable doctrine that may be used to find infringement where the accused device does not literally infringe the claims but is none the less so similar to the claimed invention that fairness requires a finding of infringement.

> Each element contained in a patent claim is deemed material to defining the scope of the patented invention, and thus the doctrine of equivalents must be applied to individual elements of the claim, not to the invention as a whole.
>
> (*Warner-Jenkinson Company* v. *Hilton Davis Chemical* (1997), p. 29)

Courts have struggled with the proper application of the doctrine of equivalents because it 'conflicts with the definitional and public-notice functions of the statutory claiming requirement' (*Warner-Jenkinson Company* v. *Hilton Davis Chemical* (1997)). The Supreme Court discussed these limitations in *Warner-Jenkinson Company* v. *Hilton Davis*. The applicant is stopped from using the doctrine of equivalents to reclaim matter that was given up during the prosecution of the patent application.[4] Intent of the alleged infringer is irrelevant to the analysis under the doctrine of equivalents.

> An analysis of the role played by each element in the context of the specific patent claim will thus inform the inquiry as to whether a substitute element matches the function, way, and result of the claimed element, or whether the substitute element plays a role substantially different from the claimed element.
>
> (*Warner-Jenkinson Company* v. *Hilton Davis Chemical* (1997), p. 40)

This test is particularly difficult to apply to inventions in genes and organisms and may limit the application of the doctrine of equivalents in infringement actions involving patents on such inventions.

The effect of a finding of infringement is draconian and potentially disastrous for the defendant in an infringement suit. Attorney fees may be awarded to the prevailing party (Patent Act, § 285); typical attorney fees in an infringement suit run into six figures for each side. The court in an infringement action may calculate damages on the basis of a reasonable royalty, rather than the profits made by the infringer. Treble damages may be awarded if the infringement was wilful (§ 284).

Enforcement of plant patents

Plant patents are governed by the same law as utility patents except where the statute indicates otherwise, and the remedies for infringement of plant patents are the same as for infringement of utility patents (Plant Patent Act, § 161). The analysis required to find infringement is different, however, because plant-patent protection is limited to a single 'variety' (*Imazio Nursery, Inc.* v. *Dania Greenhouses* (1995)). The asexual-reproduction requirement restricts protection to a single plant – all protected specimens must have been asexually reproduced from the original plant. For that reason, it is insufficient to prove that an alleged infringing cultivar is similar to the patented variety. The scope of the single claim in a plant patent is always limited to asexual progeny of the original patented variety. Infringement is established by proving that the alleged infringing plant is an asexual progeny of the patented variety. Independent creation is a defence to an allegation of infringement in a plant-patent case. Plant patents therefore provide weaker protection than utility-patent protection.

Enforcement of certificates of protection

Separate law governs infringement of a certificate of protection under the Plant Variety Protection Act of 1970. Despite Congress's unfortunate use of the same term, 'variety', in both the Plant Patent Act and the Plant Variety Protection Act, the Federal Circuit has concluded that analysis of infringement under the two laws is quite different. (*Imazio Nursery, Inc.* v. *Dania Greenhouses* (1995)):

> It is true that both the Plant Patent Act and the PVPA [Plant Variety Protection Act] use the term 'variety' and grant some form of intellectual property protection. However, the two statutes differ significantly in their purposes. The Plant Patent Act grants a plant patent to one who 'invents or discovers and asexually reproduces any distinct and new variety of plant.' . . . Conversely, one is entitled to plant variety protection under the PVPA if he has sexually reproduced the variety and has otherwise met the requirements of [that law]. The term 'variety' in both statutes cannot be read divorced from the very different circumstances in which that term is used.
>
> (*Imazio Nursery, Inc.* v. *Dania Greenhouses* (1995), p. 1568)

Asexually reproduced plants are genetically identical to their parent, whereas sexually reproduced plants are not. For that reason the analysis of infringement under the two laws cannot be the same.

Acts of infringement under the Plant Variety Protection Act include selling or marketing the protected variety, or offering it or exposing it for sale, delivering it, shipping it, consigning it, exchanging it or soliciting an offer to buy it, or any other transferring of title or possession of it (§ 2541).

Exceptions exist for contractors who have seed as the result of a breach of contract by the owner of the protected variety, private non-commercial uses and state governments. There is also a fairly broad saved-seed exemption for farmers who save their own seed for use on their own farms (Plant Variety Protection Act, § 2543). For varieties registered after the effective date of the 1994 amendments to the Plant Variety Protection Act, farmers may not sell seed for reproductive purposes to other farmers (§ 2541). The Supreme Court's decision in *Asgrow Seed Company* v. *Winterboer* (1995) set the standard for pre-1994 amendment varieties – farmers are allowed to sell seed saved for the purpose of planting on their own acreage.

Research and intermediary exemptions also apply (Plant Variety Protection Act, §§ 2544, 2545). These limitations and exceptions make the practical definition of infringement under the Plant Variety Protection Act much more limited than the definition under the Patent Act. The definition of damages, including treble damages, is the same as under the Patent Act (§ 2564), but the availability of attorney fees is limited to 'exceptional cases' (§ 2565). The Supreme Court decision in *Asgrow Seed Company* v. *Winterboer* is the leading case analysing infringement under the Plant Variety Protection Act.

Enforcement of trade secrets

Due to constitutional divisions of responsibility between federal and state governments, trade secrets are generally enforced in state courts and, even if in federal court, almost always with the application of state law. Because state laws vary widely, the process of enforcement is complicated. Nothing in TRIPS would require uniformity of state laws regarding trade secrets.

Enforcement of copyrights

Copyrights are enforced in federal district courts (or the US Court of Federal Claims, for claims against the US government) (Judiciary and Judicial Procedure, §§ 1338, 1498). In addition to civil actions, criminal actions play a significant role in copyright enforcement (Copyright Act, § 506).

Enforcement of rights in landraces, stock and other genetic source materials and related traditional knowledge

The Convention on Biological Diversity (United Nations, 1992) was agreed at Rio de Janeiro on 5 June 1992. It was signed by the USA on 4 June 1993 and subsequently submitted to the Senate for ratification. The Convention entered into force on 29 December 1993, without US ratification. Although 161 nations, including all other industrialized nations, have ratified the Convention, the USA has not done so to date, nor is it likely to do so in the foreseeable future.

The Convention recognizes the importance of biological diversity both to the environment and to human well-being. The Convention also recognizes the adverse impact that human activity has had and continues to have on biological diversity. The Convention further recognizes the substantial investment required for preserving biological diversity and the substantial environmental, economic and social benefits that may be obtained by doing so.

The Convention is not the only international agreement that addresses issues of biological diversity. The USA has a long history of supporting these international efforts. In 1940, for example, the USA became a party to the Convention on Nature Protection and Wildlife Preservation in the Western Hemisphere (the Western Hemisphere Convention). Under the Western Hemisphere Convention, the USA agreed to establish protected wilderness areas and wildlife habitat. The USA has also ratified the Convention on Wetlands of International Importance especially as Waterfowl Habitat (Ramsar Convention), concluded on

2 February 1971 at Ramsar and entered into force in the USA on 18 April 1987. Under the terms of the Ramsar Convention, the USA agreed to enter certain wetlands in a List of Wetlands of International Importance. The primary criteria for selecting these wetlands are that they are used at some part of the year by waterfowl. The USA has also ratified the Convention for the Protection of the World Cultural and Natural Heritage, concluded in Paris on 21 November 1972 and entered into force in the USA on 17 December 1975. Under the terms of this Convention, the USA agreed to submit sites having either cultural or natural value for inclusion in a World Heritage List and to protect those sites. The US has also ratified the 1983 International Tropical Timber Agreement, entered into force in the USA on 25 May 1990. This agreement provides for cooperation between producing and consuming nations to promote conservation of tropical timber species and preservation of tropical forest ecosystems. In addition, the USA has entered into a series of international agreements to promote preservation of Arctic and Antarctic species and ecosystems and agreements to protect ocean ecosystems and oceanic species. The USA agreed to an international regime for environmental protection in the North American Free Trade Agreement (NAFTA).

Thus, the refusal of the USA to ratify the Convention on Biodiversity might seem surprising. The USA, under the Administration of President George H.W. Bush, refused to sign the Convention on Biological Diversity for two primary reasons. The Administration cited concerns, first about inadequate protection of intellectual-property rights and secondly that the financial assistance mechanism under the Convention was vague and inadequate. The Administration of President William J. Clinton reversed this position, signed the Convention on 4 June 1993 and transmitted it to the Senate for ratification. Before exploring the reasons for the Clinton Administration's reversal of position and the Senate's refusal to ratify the Convention, it is useful to discuss what the Convention seeks to accomplish.

Objectives of the convention on biological diversity

The three primary objectives of the Convention on Biological Diversity are conservation of biological diversity, promotion of the use of biological diversity through access to genetic resources and equitable sharing of the benefits through adequate funding (Convention on Biological Diversity, Arts. 1, 21). Article 3 of the Convention recognizes the principle that nations have the sovereign right to exploit their own resources and the obligation to ensure that the environments of other nations and areas beyond their national boundaries are not damaged. Article 6 obligates nations to develop systems for preserving biological diversity. Article 7 provides for identification and monitoring of the attributes that constitute biological diversity. Article 8 requires conservation of biological diversity within the natural habitat of the species to be protected. While the Convention favours preservation of the components of biological diversity in their native habitat, Article 9 recognizes that this is not always possible and provides for conservation outside the natural habitat of the species preserved. Article 10 encourages the sustainable use of biological diversity, and Article 11 encourages the use of economic incentives to promote biological diversity. Article 12 encourages research and training, and Article 13 provides for public education and awareness. Article 14 provides for impact assessment and minimization of adverse impacts.

Article 15, which encourages the sharing of genetic resources and the benefits derived therefrom through mutually agreed terms, is critically important to the development of biotechnology because it provides access to genetic resources. Article 19 reinforces the requirement that the benefits of the use of genetic resources be shared, particularly with source nations. Nations must ensure that their nationals and private organizations under their jurisdiction transfer technology developed as a result of the use of biological diversity on a fair, equitable and mutually agreed basis. The source nation must give informed consent prior to the use of its genetic resources. Article 16 provides developing nations with a right to technology, including biotechnology, on favourable terms 'consistent with the adequate protection of intellectual property rights'. Article 16, section 5, recognizes that patents and other intellectual-property rights may influence the operation of the Convention and therefore requires, consistent with domestic and international law, that nations ensure that intellectual-property rights promote, rather than derogate, the objectives of the Convention.

Articles 20 and 21 establish an obligation on the part of developed nations and the means to fund compliance by developing nations with the terms of the Convention. Article 20, section 4, could be interpreted to relieve developing nations of their obligations under the Convention should developed nations fail to fund the financial mechanism established under the Convention. Perhaps a better interpretation is that this section recognizes the practical reality that developing nations will not be able to meet their obligations under the Convention without financial and technical assistance from developed nations. Under Article 21, a Conference of the Parties (established in Article 23) is charged with developing the details of the financial mechanism established under the Convention.

Disputes under the Convention are to be settled by arbitration (Art. 27). A procedure was established for amending the Convention (Art. 29), and Article 38 provides a means for withdrawal from the Convention. Article 37 prohibits reservations to the Convention. Article 22 defines the relationship of this Convention to other international agreements. The Convention is not to modify or interfere with rights or obligations arising under other international agreements, except where those rights and obligations 'would cause a serious damage or threat to biological diversity' (Art. 22, § 1).

The US position on the Convention on Biological Diversity

In reversing the US position on the Convention, the Clinton Administration determined that no implementing legislation was needed because US law already adequately implemented the provisions of the Convention. In his letter of transmittal, dated 19 November 1993, President Clinton pledged to 'continue to pursue a vigorous policy in respect of the adequate and effective protection of intellectual property rights' (Clinton, 1993). The Letter of Transmittal of the Department of State that accompanied President Clinton's letter of transmittal articulated seven understandings to be included in the instrument of ratification (Department of State, 1993).

The first understanding applies to Article 3, which states the principle that all nations have 'the sovereign right to exploit their own resources pursuant to their own environmental policies' together with the responsibility to avoid consequences outside their borders. Article 3 is taken verbatim from Principle 21 of the Stockholm Declaration from the 1972 United Nations Conference on the Human Environment, a codification of long-standing customary international law. Noting that Article 3 lacks language to place this principle within the context of the Convention, the Department of State's Letter of Transmittal recommended the following understanding: 'The Government of the United States of America understands that Article 3 references a principle to be taken into account in the implementation of the Convention' (Department of State, 1993, p. VIII).

The United States had considerable concern about the impact of Article 16, Access to and Transfer of Technology, on intellectual-property protection. The Department of State's interpretation of the 'fair and most favourable terms' clause of Article 16, section 2, is that such agreements are to be entered voluntarily by all parties. The Convention has no provision for compulsory licensing laws that might force private companies to transfer technology. The Department noted that a country with an inhospitable climate for investment could not claim that the subsequent refusal of private companies to transfer technology constitutes a violation of the Convention. Article 16, section 4, of the Convention on Biological Diversity is interpreted to apply to governments and the incentives that governments may create to encourage the private sector to transfer technology; it does not apply directly to the private sector. The Department of State further determined that the Convention is consistent with existing intellectual-property law. The implication of this position is that nothing in the Convention is inconsistent with TRIPS. In the light of these concerns, the second understanding recommended for inclusion in the instrument of ratification states that the technology-transfer provisions of the Convention are voluntary.

The third understanding applies to the conduct and location of research based on genetic resources. It focuses on Articles 15 and 19, discussed above, and clarifies the effect of the Convention on public and private research. The fourth understanding applies to Article 20, Financial Resources, and states that financial resources provided by developed countries for developing countries are for the purposes of helping them meet their obligations under the Convention.

The fifth and sixth understandings relate to the financial mechanism of Article 21. The fifth states that the authority of the Conference of the Parties should be used to determine 'the policy, strategy, program priorities and eligibility criteria relating to the access to and utilization of such resources'. The sixth understanding is essential to avoid any conflict with the appropriations process of the Constitution. It notes that Articles 20 and 21 of the Convention do not authorize the Conference of the Parties to make decisions about the 'amount, nature, frequency or size' of the contributions of Parties. The US Constitution requires an annual appropriations process with all appropriations bills originating in the House. Any financial mechanism that transferred that authority to a Conference of the Parties or resulted in multi-year obligations would be in violation of the Constitution. Thus this understanding is critical to ratification.

Article 22, section 1 states that nothing in the Convention shall adversely affect any existing right or obligation under any existing international agreement unless that right or obligation causes serious damage or threat to biological diversity. The Department of State determined that no international agreement to which the USA is a party, including international agreements related to intellectual property, would cause serious damage or threat to biological diversity. This confirms that it is the view of the Department of State, as implied in the second understanding, that there is no conflict between the Convention and TRIPS.

During negotiations, the USA proposed a sovereign immunity clause. Upon recognition by many delegates that sovereign immunity is a principle of customary international law, the USA withdrew its proposal as unnecessary. The seventh understanding expresses this principle.

The Department of State Letter of Transmittal (1993) states that no additional federal legislation is required to implement the Convention on Biological Diversity and lists the many federal statutes through which the requirements of the Convention are already carried out. The Letter notes that the USA cannot compel the transfer of technology that is privately owned. As to federally owned technology, the Letter discusses the complex federal statutes and regulations that govern its use and transfer, including the Federal Technology Transfer Act of 1986 (FTTA). FTTA requires that biological materials collected under the National Genetic

Resources Program be made available to all who request materials without charge and without regard to the country from which the request comes. The Letter concludes that ratification of the Convention 'is consistent with US foreign policy and economic and environmental interests' (Department of State, 1993, p. XIX).

Despite this interpretation and the understandings proposed by the Department of State, the Convention ran into serious opposition in the Senate and was not ratified. The Biotechnology Industry Organization (BIO), more than 500 companies, academic institutions, state biotechnology centres and other organizations involved in all phases of biotechnology, provided qualified support subject to conditions. First, it wanted the Senate in its understanding to make clear that the Convention does not conflict with TRIPS and that no legal right to property of US persons would be affected adversely by the Convention (Pell, 1994). Secondly, BIO opposed the development of any biological-safety protocol under the Convention, which would provide a forum for political agendas adverse to the US biotechnology industry rather than promoting legitimate science-based concerns about biological safety. BIO proposed that the Senate include clear conditions for withdrawal from the Convention should the Parties to the Convention reach an interpretation of the Convention contrary to these understandings. Merck & Co., Inc. strongly supported ratification and cited its ongoing agreement with the *Instituto Nacional de Biodiversidad* (INBio) in Costa Rica as an example of how mutually beneficial agreements might work under the Convention. Concerns, however, were raised by Senator Jesse Helms that the Convention unconstitutionally undermined US sovereignty by transferring constitutionally mandated federal functions to an unaccountable international body. This view was ultimately adopted by Senator Robert Dole, the Senate Minority Leader, who led a group of 35 Senators in opposition to the Convention. Following the 1994 election, leadership of the Senate shifted to the Republican Party, and Senator Helms assumed the Chairmanship of the US Senate Committee on Foreign Relations. There has been no further consideration of the Convention.

Enforcement of rights in genetic material in the light of the Convention on Biodiversity

From the foregoing, it is clear that substantial disagreement exists about the correct interpretation of the terms of the Convention on Biodiversity. The Department of State believed that the Convention creates neither a liability regime for damage to biological diversity nor a private right of action for such damage or for rights in genetic materials (Department of State, 1993, p. XI). If this interpretation is correct, and it is certainly a reasonable interpretation of the Convention, the question naturally arises as to how rights in genetic source materials together with related knowledge might be enforced.

Numerous commentators have noted the difficulty in enforcing rights in genetic source materials; biopiracy has been discussed widely as a major issue (RAFI, 1998). There are, however, several means by which these rights may be recognized and enforced. The first is through contract. The agreement between INBio and Merck, mentioned above, is an example of such a contract. Contracts have many advantages, including flexibility and independence from a particular nation's decision on ratification of agreements like the Convention. Dispute-resolution mechanisms such as mediation or arbitration can be included in the contract. Devices such as irrevocable letters of credit can be required to ensure parties' financial performance. Even where such terms are not included in the contract, the courts of most US states and countries will recognize and enforce each other's judgements under principles of comity or reciprocity. US federal courts may be obliged to recognize foreign judgments under the terms of various bilateral or multilateral treaties.

TRIPS permits enforcement of rights in genetic material

TRIPS recognizes the right and obligation of any Member to create an effective *sui generis* system for the protection of plant varieties (TRIPS, Art. 27(3)(b)). While TRIPS requires an inventive step as an essential condition for patent protection, no such inventive step is required for an effective *sui generis* system (Art. 27). Indeed, TRIPS does not define the term *sui generis*. A carefully drafted property-rights system for genetic stock could result in rights that would be enforceable in courts of the states of the USA. One interesting question is whether such a *sui generis* system could create property rights in genetic stock that was as yet undiscovered or unrecognized. The English common law as adopted by the various US states would suggest an answer in the affirmative. At common law, owners of real property were considered to own not only the surface of the land but everything above and below it. Ownership included the plants growing on the property even where the owner of the property did not know of the existence of such plants. One could not cut a tree from another's real property and claim ownership of it because the landowner did not know the tree was there. It does not seem a large step to extend this principle to the genes that are contained within the tree. Of course, the analysis is more complicated for animals and birds, because they are mobile and therefore subject to different rules at common law.

Assuming that a country can craft a law that creates property rights in genetic source material, could such a right be enforced in courts in the USA? The answer is almost definitely positive. The courts of US states routinely recognize the property rights of non-resident nationals under principles of comity or reciprocity. State courts recognize these rights under several legal theories. Conversion, the intentional wrongful interference with the property of another (Keeton, 1984), is an intentional tort that might provide a theory of recovery. Once confined only to tangible personal property, it is now generally recognized as applying to intangible personal property as well. A successful plaintiff may be awarded punitive damages as well as actual damages. Where a bio-prospector obtains genetic material through misrepresentation or non-disclosure of his true motives and objectives, no property right need exist to support a recovery; most courts of US states recognize misrepresentation, fraud and related wrongdoing as torts in their own right (Keeton, 1984). It is beyond the scope of this chapter to explore these common-law torts in depth; however, it may well be possible to recover from US-based operators who misappropriate genetic source material through legal actions in the courts of US states. Such a result is neither inconsistent with TRIPS nor dependent upon the US Senate's ratification of the Convention on Biological Diversity. In addition, most US states have adopted unfair or deceptive-trade-practices acts that may have application. These laws often have the advantage to plaintiffs of offering the opportunity to recover treble damages and attorney fees.

There is one last note to add to this discussion as it relates not to the genetic source material itself but to indigenous knowledge related to the property expressed by such source material in native species. To the extent to which indigenous peoples have maintained the secrecy of that information, claims to trade secrets may exist. Under TRIPS, Member countries must protect such undisclosed information (TRIPS, Art. 39). TRIPS lists three requirements that must be satisfied, if protection is to be available. First, the information must not be generally accessible to those who normally deal with the type of information in question. Secondly, the information must have commercial value. Thirdly, those lawfully in control of the information must have taken steps to protect the information (Art. 39). While this provision of TRIPS was clearly drafted with more traditional industrial trade secrets in mind, it is not at all inconceivable that nations could adopt legislation to help indigenous peoples protect their traditional information, consistent with the provisions of TRIPS.

Conclusion

TRIPS moved national intellectual-property systems substantially towards the goal of harmonization; however, much remains to be done. TRIPS provides the powerful hammer of trade sanctions to force those nations with weak intellectual-property systems to strengthen them. As to biotechnology, much is excluded from TRIPS. In particular the rights of nations and individuals in genetic source materials remains problematic.

None the less, the US system provides considerable protection for the developers of biotechnology.

Notes

[1] Sections of the chapter covering US domestic intellectual-property law are excerpted from Feitshans (2001) © *Drake Journal of Agricultural Law* and used with permission. Excerpts have been updated to reflect developments in the law, and more complete references are included in the published article.

[2] The USA granted a patent on a living organism to Louis Pasteur in 1873. Despite this precedent, the US Patent and Trademark Office developed the practice of not granting claims to living organisms, which resulted in the *Diamond* v. *Chakrabarty* litigation.

[3] A common practice among researchers whereby one or more third parties sign and date the researchers' notebook pages. The practice increases the weight of evidence accorded researchers' notebooks by reducing the likelihood of fraud.

[4] Estoppel may be rebutted by a showing that the claim amendment was not made to narrow the scope of the claim to exclude the claimed equivalent (*Festo Corp.* v. *Shoketsu Kinzoku Kogyo Kabushiki Co., Ltd.* (2002)).

Cases

Animal Legal Defense Fund v. *Quigg*, 932 F.2d 920 (Fed. Cir. 1991).
Asgrow Seed Company v. *Winterboer*, 513 US 179 (1995).
Barton v. *Adang*, 162 F.3d 1140 (Fed. Cir. 1998).
Brenner v. *Manson*, 383 US 519 (1966).
Burroughs Welcome v. *Barr Laboratories, Incorporated*, 40 F.3d 1223 (Fed. Cir. 1994).
Diamond v. *Chakrabarty*, 447 US 303 (1980).
Dickinson v. *Zurko*, 527 US 150 (1999).
Elekta Instrument v. *O.U.R. Scientific Int.*, 214 F.3d 1302 (2000), pet. reh'g den., 2000 US App. LEXIS 17777 (2000).
Enzo Biochem, Incorporated v. *Calgene, Incorporated*, 188 F.3d 1362 (Fed. Cir. 1999).
Festo Corporation v. *Shoketsu Kinzoku Kogyo Kabushiki Company, Limited*, 534 US 1053 (2002).
Fiers v. *Revel*, 984 F.2d 1164 (Fed. Cir. 1993).
Finnigan Corporation v. *International Trade Commission*, 180 F.3d 1354 (Fed. Cir. 1999).
Florida Prepaid Postsecondary Education Expense Board v. *College Savings Bank*, 527 US 627 (1999).
Genentech, Incorporated v. *Chiron Corporation*, 112 F.3d 495 (Fed. Cir. 1997).
Genentech, Incorporated v. *Novo Nordisk A/S*, 108 F.3d 1361 (Fed. Cir. 1997), cert. denied, 1997 US LEXIS 6493 (1997).
Imazio Nursery, Incorporated v. *Dania Greenhouses*, 69 F.3d 1560 (Fed. Cir. 1995).
In re Alton, 76 F.3d 1168 (Fed. Cir. 1996).
In re Frank R. Bonczyk, No. 01-1061 (unpublished), 10 Fed. Appx. 908, 2001 US App. LEXIS 9256 (Fed. Cir.), cert. denied, 534 US 895 (2001).

In re Hiniker Company, 150 F.3d 1362 (Fed. Cir. 1998).
In re Mayne, 104 F.3d 1339 (Fed. Cir. 1997).
In re Ziegler, 992 F.2d 1197 (Fed. Cir. 1993).
J.E.M. Ag Supply Incorporated v. *Pioneer Hi-Bred International Incorporated*, 534 US 124 (2001), reh'g. den., 122 S. Ct. 1600 (2002).
Juicy Whip, Incorporated v. *Orange Bang, Incorporated*, 185 F.3d 1364 (Fed. Cir. 1999).
Juicy Whip, Incorporated v. *Orange Bang, Incorporated*, 292 F.3d 728 (Fed. Cir. 2002).
Kridl v. *McCormick*, 105 F.3d 1446 (Fed. Cir. 1997).
Life Technologies v. *Clontech Laboratories, Incorporated*, 224 F.3d 1320 (Fed. Cir. 2000).
Markman v. *Westview Instruments, Inc.*, 517 US 370 (1996).
Perseptive Biosystems v. *Pharmacia Biotech*, 225 F.3d 1315 (Fed. Cir. 2000).
Pfaff v. *Wells Electronics, Incorporated*, 525 US 55, reh'g den., 525 US 1094 (1999).
Pickholz v. *Rainbow Technologies, Incorporated*, 284 F.3d 1365 (Fed. Cir. 2002).
Regents of the University of California v. *Eli Lilly & Company*, 119 F.3d 1559 (Fed. Cir. 1997).
Rexnord Corporation v. *The Laitram Corporation*, 274 F.3d 1336 (Fed. Cir. 2001).
Sibia Neurosciences v. *Cadus Pharmaceutical*, 225 F.3d 1349 (Fed. Cir. 2000).
Singh v. *Brake*, 222 F.3d 1362 (Fed. Cir. 2000).
The Gentry Gallery, Incorporated v. *The Berkline Corporation*, 134 F.3d 1437 (Fed. Cir. 1998).
The Holmes Group, Incorporated v. *Vornado Air Circulation Systems, Incorporated*, 122 S. Ct. 1889 (2002).
Warner-Jenkinson Company v. *Hilton Davis Chemical*, 520 US 17 (1997).

References

Agreement Establishing the World Trade Organization, Annex 1A Multilateral Agreements on Trade in Goods, GATT 1994 http://www.wto.org/english/docs_e/legal_e/legal_e.htm#wtoagreement

Agreement Establishing the World Trade Organization, Annex 1C Trade-Related Aspects of Intellectual Property Rights (TRIPS) http://www.wto.org/english/docs_e/legal_e/legal_e.htm#TRIPs

Agreement Establishing the World Trade Organization, Understanding on Rules and Procedures Governing the Settlement of Disputes (DSU), http://www.wto.org/english/docs_e/legal_e/28-dsu_e.htm

Berne Convention for the Protection of Literary and Artistic Works (Berne Convention (1971)) http://www.wipo.org/treaties/ip/berne/index.html

Budapest Treaty on the International Recognition of the Deposit of Microorganisms for the Purposes of Patent Procedure (1977) http://www.wipo.org/treaties/registration/budapest/index.html

Clinton, W.J. (1993) Letter of Transmittal, *Congressional Record*, S16572.

Convention for the Protection of the World Cultural and Natural Heritage (1972) http://sedac.ciesin.org/entri/register/reg-066.rrr.html

Convention on Biological Diversity (1992) http://www.biodiv.org/convention/articles.asp

Convention on Nature Protection and Wildlife Preservation in the Western Hemisphere (1940) http://sedac.ciesin.org/entri/register/reg-003.rrr.html

Convention on Wetlands of International Importance Especially as Waterfowl Habitat (1971) (Ramsar Convention) http://sedac.ciesin.org/entri/register/reg-056.rrr.html

Department of State (1993) Letter of Transmittal, Treaty Document 103–20, Senate, 103d Congress, First Session.

Feitshans, T.A. (2001) A review of US intellectual property law applicable to inventions in biotechnology: US intellectual property law continues to demonstrate its adaptability to new technology. *Drake Journal of Agricultural Law* 61, 7–29.

Felgueroso, J. (2002) TRIPS and the Dispute Settlement Understanding: the first six years. *American Intellectual Property Law Association Quarterly Journal* 30, 165–231.

Gonzalez, J.C. (2000) The on-sale bar to patentability: the US Supreme Court sheds some light. *PTC Research Foundation of Franklin Pierce Law Center IDEA: the Journal of Law and Technology* 40, 83–104.

Grubb, P.W. (1999) *Patents for Chemicals, Pharmaceuticals and Biotechnology*, 3rd edn. Clarendon Press, Oxford, 448 pp.

Guidelines for Examination of Patent Applications Under the 35 U.S.C. Sec. 112, para. 1 'Written Description' Requirement, 66 Federal Register 1099 (2001)

International Tropical Timber Agreement (1983) http://sedac.ciesin.org/entri/register/reg-130.rrr.html

Keeton, W.P. (1984) *Prosser and Keeton on Torts*, 5th edn. West, St Paul, Minnesota, 1286 pp.

Manual of Patenting Examination Procedure, 7th edn (MPEP) (1998–2001) US Patent and Trademark Office, Washington, DC. http://www.uspto.gov/web/offices/pac/mpep/index.html

Myers, S.L. (1999) South Africa and US end dispute over drugs. *New York Times*, 18 September, A8.

North American Free Trade Agreement (NAFTA) (1992) http://www nafta-sec-alena.org/english/index.htm

Paris Convention (1967) Stockholm Act of the Paris Convention for the Protection of Industrial Property (14 July) http://www.wipo.org/treaties/ip/paris/index.html

Pell, C. (1994) *Congressional Record*, S14046–S14053.

Revised Interim Guidelines for Examination of Patent Applications Under the 35 USC Sec 112, para. 1 'Written Description' Requirement; Request for Comments, 64 Fed. Reg. 71427, 71430 (1999).

Revised Utility Examination Guidelines; Request for Comments, 64 Fed. Reg. 71440, 71441 (1999).

Rural Advancement Fund International (RAFI) (1998) Plant Breeders Wrongs, http://www.rafi.org/

Stockholm Declaration of the United Nations Conference on the Human Environment (1972) http://environment.harvard.edu/guides/intenvpol/indexes/treaties/STOCK.html

Title 7 United States Code (USC) §§ 2321–2582 (2002) (Plant Variety Protection Act of 1970, as amended) http://www.loislaw.com

Title 15 United States Code (USC) §§ 3701–3717 (2002) (Federal Technology Transfer Act of 1986) http://www.loislaw.com

Title 17 United States Code (USC) §§ 1–1332 (2002) (Copyright Act) http://www.loislaw.com
Title 28 United States Code (USC) §§ 1–4001 (2002) (Judiciary and Judicial Procedure) http://www.loislaw.com
Title 35 United States Code (USC) §§ 1–376 (2002) (Patent Act) http://www.loislaw.com
Title 35 United States Code (USC) §§ 161–164 (2002) (Plant Patent Act of 1930) http://www.loislaw.com
Utility Examination Guidelines, 66 Fed. Reg. 1092 (2001)
United Nations Conference on Environment and Development (1992) Convention on Biological Diversity, reprinted in *International Legal Materials* 31, 818.
Wigley, D.E. (2000) Note: evolution of the concept of non-obviousness of the novel invention: from flash of genius to the trilogy. *Arizona Law Review* 7, 581–606.

8

TRIPS, Biotechnology and the Public Domain: What Role will World Trade Law Play?

John Linarelli*

Introduction

It would be difficult to challenge the argument that no other category of legal rules affects the distribution of wealth more than property rules. In the 18th century, Hume postulated as his central reason why people engage in society that it is for stability in the possession of property (Hume, 1739). Property rights and agriculture share a longstanding historical relationship. The endowments of any society are vitally connected to agriculture because agriculture is about food production. The political economics of British agriculture in the 17th and 18th centuries produced the so-called first enclosure movement, in which commons in agricultural land areas were enclosed and the rights of small farmers in estates such as copyholds were expropriated (Smith, 2000; Travis, 2000). While the conflict in the first enclosure movement was over rights in real property, the conflict in the second enclosure movement is over rights in intellectual property (Boyle, 2001). Similes and metaphors abound in the literature. We are in the process of the 'enclosure of the intangible commons of the

* Associate Professor of Law, University of La Verne College of Law, Ontario, California, USA. Funding provided by the US Department of Agriculture and the University of Arkansas is gratefully acknowledged. I would like to thank all those who commented on my presentation of the chapter at the Symposium on International Agricultural Trade held on 14–15 May 2001 at the National Agricultural Library of the US Department of Agriculture. I am also grateful to Laurence Helfer for his comments on a near final draft. The usual disclaimers apply. Substantial parts of this article are drawn from Linarelli (2002). Publisher permissions are gratefully acknowledged.

mind' and the 'intellectual land grab' (Boyle, 2001). The battle for rights in intellectual property is 'an information arms race with multiple sides battling for larger shares of the global knowledge pool' (Hess and Ostrom, 2001). The enclosure of the intellectual commons is occurring in various disciplines of science and technology, including in information technology and cyberspace, and in biotechnology relating to pharmaceuticals, medicine and human genetics. The focus is on agriculture here, with its continued place in the forefront of commons debates.

Innovations in agriculture are not resistant to the trends towards commodification, and whether and how that commodification will continue will affect various stakeholders, from farmers to scientists. Technological innovation, in the language of economics, pushes the production frontier outwards, allowing the production of more food with the expenditure of the same or fewer resources. But improvement in allocative efficiency is not the whole story. The institutions for allocating rights in technology have to be examined, are those include domestic and international laws that set standards for intellectual-property protection. Intellectual-property rights matter because they affect both efficiency and distribution. From the standpoint of efficiency, they affect the ability of agricultural production to continue to outpace population growth. Rights in intellectual property affect the choices that people make to invest or not to invest in innovation in agricultural science, which in turn affects agricultural productivity. With the allocation of rights comes the allocation of rents. The issue of distribution has become contentious, as biological resources that for centuries had no commercial value and were treated as a common resource now have significant commercial value. What has attracted international attention to distributional concerns is that key biotechnologies seem to concentrate in a few large multinational firms headquartered in North America and Western Europe.

As this book is on international agricultural trade law, policy and economics, this chapter necessarily focuses on the World Trade Organization (WTO) Agreement on Trade-related Aspects of Intellectual Property Rights (TRIPS). TRIPS was concluded on 15 April 1994 and entered into force on 1 January 1995. It is a multilateral agreement in the WTO regime and thus all WTO members must adhere to it. Although an international trade agreement and not a domestic intellectual-property law, TRIPS is relevant to the question of ownership of rights in biotechnology. It specifies standards for the intellectual-property laws of the WTO Members. It is unlike any other trade agreement preceding it, unlike anything produced in the General Agreement on Tariff and Trade (GATT)/WTO framework since the GATT's humble beginnings as an Anglo-American-inspired agreement to regulate tariffs. TRIPS harmonizes intellectual-property protection at a high level of protection for rights holders, and this is one of its controversial characteristics (Brenner, 1998).

This chapter examines the intellectual-property rights of relevance to agricultural biotechnology: patents and *sui generis* rights in plant varieties, and how these rights are dealt with in TRIPS.[1] Patents and *sui generis* protection remain the core methods of protecting technological innovation.

The chapter is in three parts, excluding the introduction and the conclusion. The first part explains the TRIPS provisions relevant to agriculture. The relevant TRIPS provision for biotechnology is Article 27, dealing with patentability, and in particular Article 27.3(b). The obligations set forth in Article 27.3(b) will be analysed and how TRIPS promotes a partly open science- and technology-rights regime will be explained. TRIPS promotes a commodified and privatized system of intellectual-property rights that benefits WTO Members with endowments in science and technology industries. It also promotes a commons regime that is disadvantageous to WTO Members with endowments in biodiversity and traditional knowledge. This has led to controversy between Europe, America and the Quad group of countries – Canada, the European Union (EU), Japan and the USA – and the developing countries.

The effects of this controversy cannot be overstated. It almost stalled the latest round of multilateral trade negotiations, which made an auspicious start in Doha. The second part of the chapter examines whether the Doha Round of multilateral trade negotiations will affect rights in agricultural biotechnology. It explores the controversy surrounding TRIPS Article 27.3(b), as it surfaced in the aftermath of the Uruguay Round, in the WTO TRIPS Council, the Seattle Ministerial Conference and thereafter. It is difficult to say whether the Doha Round will produce changes in the substance of TRIPS, and it is doubtful that any radical changes will emerge. It is unlikely that 27.3(b) will be altered much, and also unlikely that WTO Members will amend TRIPS to promote a more open science and technology regime.

The third part of the chapter examines European law and policy, focusing on the substantive obligations of the European countries, mainly those who are EU Member States, relevant to TRIPS.[2] It focuses on the principal legal regimes in Europe, the European Patent Convention, the EU Biotechnology Directive and EU Plant Varieties Regulation. It concludes that, while Europe has been more cautious in the development of patent protection for plant and animal varieties, the trend towards overlapping forms of intellectual-property protection is still evident in Europe as it is in the USA.

TRIPS and Biotechnology: Patents, *Sui Generis* Rights and Commons

The key TRIPS provision relevant to intellectual-property rights in biotechnology is Article 27, entitled 'Patentable Subject Matter'. Article 27 provides that patents 'shall be available for any inventions, whether products or processes, in all fields of technology, provided they are new, involve an inventive step and are capable of industrial application' (TRIPS Art. 27.1). WTO Members must make patents rights available in their territories 'without discrimination . . . as to the field of technology' (TRIPS Art. 27.1). Three exceptions exist to this 'any technology' standard for patentability. Article 27.2 provides that WTO Members may exclude from patentability inventions:

> the prevention within their territory of the commercial exploitation of which is necessary to protect *ordre public* or morality, including to protect human, animal or plant life or health or to avoid serious prejudice to the environment, provided that such exclusion is not made merely because the exploitation is prohibited by their law.

Article 27.3 provides that Members may exclude from patentability: (a) 'diagnostic, therapeutic and surgical methods for the treatment of humans or animals'; and (b) 'plants and animals other than micro-organisms, and essentially biological processes for the production of plants or animals other than non-biological and microbiological processes'. Further, under subparagraph (b), WTO Members may provide for the protection of plant varieties either by patents or by an 'effective *sui generis* system' or by a combination of the two methods.

One of the basic distinctions between *sui generis* protection and patents is that *sui generis* rights tend to be subject to a farmer's exemption and a research exemption. TRIPS does not explicitly permit either exemption. Such exemptions would arguably fall, at least implicitly, within the *sui generis* category of protection permitted under Article 27.3(b). In addition, TRIPS Article 30 provides that:

> Members may provide limited exceptions to the exclusive rights conferred by a patent, provided that such exceptions do not unreasonably conflict with normal exploitation of the patent and do not unreasonably prejudice the legitimate interests of the patent owner, taking account of the legitimate interests of third parties.

Thus, Article 30 could be interpreted to permit such exemptions even for patented seeds and plants (Llewelyn, 2001).

Article 27 reflects a compromise between the approaches that Europe and America have traditionally taken on intellectual-property rights in plant and animal varieties. The original US text of TRIPS did not include the exceptions found in Articles 27.3 and 27.2, and opted only for patents for the protection of plant varieties (Hamilton, 1993). Article 27 was negotiated in the Uruguay Round to substantially adopt the wording of Article 53(b) of the European Patent Convention, but important differences maintain.[3] Convention Article 53(b) uses the term 'shall' when referring to exclusions from patentability of plant varieties, but TRIPS Article 27 uses the term 'may'. The European Patent Convention excludes from eligibility for European patents rights in plant and animal varieties and 'essentially biological processes' for plant and animal production, but leaves the door open for rights in microbiological processes. In contrast, Article 27 leaves the policy choice to WTO Members to use patents, *sui generis* rights or both to protect rights in plants and animal varieties. WTO Members, in addition, may exclude from patentability 'essentially biological processes' for plant and animal production, but not for the production of microorganisms or rights in non-biological and microbiological processes. TRIPS and the European Patent Convention are consistent with each other.

Article 27.3(b) is controversial. The controversy is not between the USA and the EU nor is it among the members of so-called 'Quad' group of WTO Members. Rather, it is between the Quad group and the developing countries. Article 27, as with TRIPS generally, provides a strategic bargaining problem for WTO Members as to which members, or the industries and communities they represent, capture the appropriable rents from intellectual property. The interests of the EU and the USA, technology exporters, are in protecting the interests of powerful lobbying groups with influence in the political process, namely firms that produce biotechnological innovation, which have an interest in strong intellectual-property laws. Representatives of these industries were involved in the preparation of the language currently found in Article 27. The clash with developing country interests is stark. Developing countries have traditionally been importers of innovation. With the growth of biotechnology, there is a growing perception that developing countries are exporters of biological resources and traditional knowledge, but that TRIPS fails to offer meaningful protection that would lead to compensation for such exports. During the negotiation of TRIPS in the Uruguay Round, some TRIPS negotiators for developing countries asserted that TRIPS should either declare such resources as common heritage and incompatible with a private intellectual property rights regime, or that TRIPS should recognize or accommodate traditional knowledge as valuable by granting some sort of rights, as yet undefined, of excludability.

The core of the controversy is that biotechnological innovation often has origins in property held in common and found in developing countries. TRIPS says nothing about traditional knowledge or preservation of biodiversity. As a result of advances in science and technology, seeds and plants that were once of value only to isolated communities in the developing world now have significant commercial value in transnational markets. The result, predictably, is claims of 'bioprospecting' and, worse, 'biopiracy', as interested parties seek to obtain private property rights in the biodiversity commons.[4]

As a result of the divergence of interests between the North and the South, many developing countries opposed the inclusion of Article 27.3(b) in TRIPS. In addition, they resist the implementation of the provision. As will be explained below, one of the major points of contention in any future WTO negotiating round will be whether the substantive obligations in TRIPS should be reopened or whether the focus should be on the 'built-in agenda'. The concept of the built-in agenda focuses attention on implementation, not renegotiation. Predictably, the USA and, to some extent, the EU favour a focus on implementation and the built-in agenda, while the developing countries favour a focus on TRIPS obligations.

TRIPS is one of several institutions implementing American and European public policy favouring the privatization of biotechnological innovation. In the USA, public investment in agricultural biotechnology has not increased since 1990, while the private sector has increased research and development expenditures substantially (Wolf and Zilberman, 1999). The US Congress has provided incentives to remove publicly funded discoveries from the public domain. The Bayh Dole Patent Policy Act, promulgated in 1980, allows individuals and institutions to obtain patents on federally funded research. The Federal Technology Transfer Act, promulgated in 1986 and which amends the Stevensen Wydler Technology Innovation Act, authorizes the creation of cooperative research and development agreements (CRADAs) as mechanisms for public–private research collaboration. The Federal Technology Transfer Act requires the public sector to transfer the rights to the private sector to exploit the commercial potential of discoveries that emanate from CRADAs. The trend to privatize seems to apply to Europe as well, although the European Commission seems interested in public funding to promote a European Research Area. European governments have funded science and technology, but not at the level that the US government has done so. The UK has privatized all of its significant agricultural research (Wolf and Zilberman, 1999). The EU has funded agriculutural biotechnological initiatives in its Framework Programmes, which the European Commission Research Directorate administers. The Fourth Framework Programme included funding for agricultural biotechnology. The focus of the newest Framework Programme, the Sixth, is to promote more intensely a European Research Area, intended ostensibly to improve European 'competitiveness', particularly given the science and technology competitiveness of North America and Japan (europa.eu.int/comm/research/why.htm).

The Built-in Agenda: Article 27.3(b) and Doha

Article 27.3(b) provides that 'the provisions of this subparagraph shall be reviewed four years after the date of entry into force of the WTO Agreement'. TRIPS entered into force on 1 January 1995, which means that the provisions on patentability of biotechnology were to be reviewed in 1999, as part of the WTO's so-called built-in agenda. In addition, TRIPS by its terms required a review of its implementation in 2000 (TRIPS Art. 71). Developing countries, moreover, were required to bring their domestic laws into compliance with TRIPS on or before 1 January 2000, and least developed countries have until 2006. All WTO Members, however, were required to bring their existing laws into compliance with the TRIPS most-favoured-nation and national treatment provisions by 1 January 1996 (TRIPS Art. 65.2). The review of Article 27.b(3) was thus to take place 1 year before developing countries were required to implement it, and the review of TRIPs generally was to occur in the same year that developing countries were required to implement it.

Tracing the progress of the built-in agenda under these provisions brings home the distinction well known by lawyers between the words of the contract and the actual conduct of the parties in performing it. Article 27.3(b) has been under scrutiny at least since December 1998, when the TRIPS Council initiated an information-gathering exercise on how WTO Members were implementing the provisions (WTO, 2000a). In the TRIPS Council meeting in December 1998, the developed countries wanted to focus on implementation while the developing countries wanted to focus on substance. Developing-country representatives argued that the language of Article 27.3(b) supported their position because it says that 'the provisions' of Article 27.3(b) were to be reviewed 4 years after the date of TRIPS's entry into force. The TRIPS Council decided that the WTO Secretariat would collect information about implementation from WTO Members, in response to a questionnaire the Council would furnish.

Throughout 1999, WTO Member responses to the questionnaire trickled in to the TRIPS Council. WTO Members, including the EU, submitted responses to questionnaires that the TRIPS Council circulated, and these responses were collated and available for general circulation. Not all WTO Members responded. As of April 1999, only 30 countries had submitted information on implementation.

In summer 1999, preparations started to take on some seriousness for the WTO's Third Ministerial Conference, to be held in Seattle. It was widely contemplated that Seattle would result in the launching of a new WTO negotiating round, the so-called Millennium Round. Developing countries expressed more of an interest in the preparations for Seattle than in the TRIPS Council assessment of implementation. About 100 developing countries agreed to almost a dozen proposals, to be tabled in Seattle, to reform TRIPS. The gist of these proposals was that TRIPS failed to deal adequately with biodiversity and traditional knowledge. Perhaps the most influential of these proposals was that of the Africa Group, which Kenya led, submitted to the WTO General Council on 6 August 1999 (WTO, 1999b). This submission proposed to extend the deadline to implement Article 27.3(b) in the developing countries and to ban patents on 'life', including those on microbiological organisms, and sought clarification of some of the TRIPS provisions.

Throughout late summer and autumn of 1999, discussions continued in the TRIPS Council, with, as the developing countries sought initially, a focus on the substance of TRIPS rather than on its implementation. The USA and the EU responded. They agreed that granting intellectual-property rights in biotechnology was vital to providing proper incentives to innovate, and that the International Convention for the Protection of New Varieties of Plants (UPOV Convention) provided an acceptable *sui generis* system for protecting rights in plant varieties. The EU urged all WTO Members to promulgate laws that complied with the UPOV Convention. The EU was prepared to address the ethics of biotechnology patenting and to consider the sorts of protection traditional knowledge might require (WTO, 1999a).

The Seattle Ministerial Conference took place in late November–early December 1999. The differences in the positions of the WTO Members at the Conference were significant. In Seattle, the USA did not offer a proposal on TRIPS. Instead, the USA wanted to work on the built-in agenda, primarily to get developing-country Members to meet existing obligations when TRIPS transition periods expire (WTO, 1999c; Abbott, 2000). The EU focused substantially on trying to persuade the USA to adopt a 'first-to-file' system for patents. The USA alone uses a 'first-to-invent' system, while virtually the rest of the world uses first-to-file (Abbott, 2000). The EU was in substantial agreement with the USA on the need for compliance with the TRIPS built-in agenda. The EU position was that a new round would offer an opportunity to examine areas of TRIPS for possible amendment, but by the time a new round came into operation, the transition periods for developing-country implementation of TRIPS would have expired (WTO, 1999a, c). The EU Communication to the WTO General Council on TRIPS states, among other things, that:

> [i]t should of course be kept in mind that the TRIPS *acquis* is a basis from which to seek further improvements in the protection of IPR. There should therefore be no question, in future negotiations, of lowering of standards or granting of further transitional periods.
>
> (WTO, 1999)

Contrast these positions with those of the developing countries in Seattle. The Africa Group, led by Kenya, reiterated that Article 27.3(b) by its terms provides for a review of its provisions, and that implementation is covered under TRIPS Article 71.1. Here is a summary by the WTO Secretariat of the position of the Africa Group on the problems with Article 27.3(b):

> Artificial distinctions between biological and microbiological organisms and processes:
> (a) The review of the substantive provisions of Article 27.3(b) should clarify the following:
> – Why the option of exclusion of patentability of plants and animals does not extend to micro-organisms as there is no scientific basis for the distinction.

- Why the option of exclusion of patentability of 'essentially biological processes' does not extend to 'microbiological processes' as the latter are also biological processes.

(b) The review process should clarify that plants and animals as well as microorganisms and all other living organisms and their parts cannot be patented, and that natural processes that produce plants, animals and other living organisms should also not be patentable.

Clarifying the option of a *sui generis* system for plant varieties: After the sentence on plant variety protection in Article 27.3(b), a footnote should be inserted stating that any *sui generis* law for plant variety protection can provide for:

(i) the protection of the innovations of indigenous and local farming communities in developing countries, consistent with the Convention on Biological Diversity [CBD] and the International Undertaking on Plant Genetic Resources;
(ii) the continuation of the traditional farming practices including the right to save, exchange and save seeds, and sell their harvest;
(iii) preventing anti-competitive rights or practices which will threaten food sovereignty of people in developing countries, as is permitted by Article 31 of the TRIPS Agreement.

Relation between Article 27.3(b) and CBD and the International Undertaking on Plant Genetic Resources: The review process should seek to harmonize Article 27.3(b) with the provisions of the CBD and the International Undertaking, in which the conservation and sustainable use of biological diversity, the protection of the rights and knowledge of indigenous and local communities, and the promotion of farmers' rights, are fully taken into account.

(WTO, 1999c)

WTO Members not in the Africa Group, particularly those from Latin America, shared the concerns of the Africa Group. The USA and the EU both offered various 'trade-off' initiatives to assist developing countries in building institutional capacity and improving governance structures to implement TRIPS, but these did not go to the core of what the developing countries wanted, which was a major revision of the substantive obligations in TRIPS.

The result of Seattle was that the WTO Members took no significant decisions on TRIPS. Towards the conclusion of the Conference, the US Trade Representative (the Chairperson of the Seattle Ministerial Conference) and the WTO Director-General both declared the Conference to be 'suspended', although the import of such language and its effects on the Seattle proposals are unclear.

Post-Seattle, the Quad countries held various meetings about the future of a new WTO round. In various statements to the press, the EU has strongly supported the idea of a comprehensive new round. A good segment of the post-Seattle discussions have focused on 'confidence building' measures intended for least-developed countries. In March 2000, the Quad countries proposed a plan to improve the confidence of the least-developed countries. The plan included four elements: (i) zero tariffs and zero quota access to developed-country markets; (ii) mechanisms for addressing the implementation problems of developing countries; (iii) enhanced technical assistance for least-developed countries, and (iv) increased transparency in WTO decision-making. As part of the package, extensions requested of TRIPS implementation deadlines would be considered on a country-specific basis. The WTO Director-General expressed disappointment in the package, contending that it did not go far enough, and it is unclear what the result of this exercise was. In June 2000, however, the momentum seemed to be in favour of focusing on implementation issues. In the 22 June 2000 meeting of the WTO General Council, a programme of meetings on implementation was agreed with the goal of concluding discussions before the next ministerial conference, to be held in 2001. The developing countries again baulked, contending that they faced considerable institutional and financial problems in achieving compliance with existing WTO obligations. These battles continued throughout 2000.

In a 27 November–1 December 2000 meeting of the TRIPS Council, Brazil and India renewed efforts to seek a review of TRIPS to avoid conflict with the Convention on Biological Diversity. The Brazilian and Indian efforts represent an attempt at a major refocus on substantive policy and legal obligations. Brazil's Communication to the General Council is telling. It includes as issues to be considered technical issues relating to patent protection under Article 27.3(b) and *sui generis* protection of plant varieties, ethical issues relating to patentability of life-forms, the relationship between conservation and sustainable use of genetic material and the relevance of traditional knowledge and farmers' rights (WTO, 2000b).

As of 16 March 2001, the WTO Chair of the General Council declared the discussions over TRIPS with developing countries to be stalled. The most recent event was a 27 March 2001 meeting, attended by delegates of 20 WTO Members, including delegates from the EU, Japan and Canada. The USA did not participate. From the press reports of this meeting, the EU and Japan said that they would take a harder line on implementation at the upcoming Ministerial Conference in Doha, Qatar, and in any future negotiating round. The distinction made at the meeting was between countries that want to implement but cannot because of a lack of institutional capacity and those who want a different obligation. The former problem is one of 'capacity building' while the latter is one of 'negotiation'. Negotiation, or more properly renegotiation, entails a change in a treaty obligation and possibly also in domestic implementing legislation. The EU and Japanese delegates took a legalistic position to the effect that the developing-country delegates knew or should have known what they were agreeing to in the Uruguay Round.

The frictions between the Quad group of countries and developing countries continued in the Doha Ministerial Conference in November 2001. The Quad group of developed countries, particularly the USA, sought to focus the Conference on implementation but also want some movement in the Conference, so as to avoid the label 'failure' that has been so persistently affixed to the Seattle Ministerial Conference. The implementation question was one of the most difficult of the Conference, bordering on intractable. The actual negotiating positions of the EU and the other Quad members did not offer substantial revisions of TRIPS obligations. The positions of the Quad versus the developing countries stood in stark contrast for most of the Conference. The Declaration of the African, Caribbean and Pacific (ACP) countries on the Conference, submitted by Kenya to the WTO before the Conference, urged WTO Members to develop mechanisms to 'allow for the disclosure of the sources of traditional knowledge and genetic resources used in inventions in order to achieve a fair and equitable sharing of benefits' and sought a review of the TRIPS Agreement to 'clarify that all living organisms including plants, animals, and part of plants and animals, including gene sequencing and biological and other natural processes for the production of plants and animals and their parts shall not be patented' (WTO, 2001a).

India took the lead in stridently opposing the TRIPS built-in agenda. The Indian Minister of Commerce and Industry, in a fourth session of the Doha Conference agreed that, '[a]fter the setback at Seattle, all of us want Doha to be a success,' but warned that '[w]e cannot be held hostage to unreasonable demands that concessions be made for carrying forward what are already mandated negotiations' (WTO, 2001b). The Indian position is that there 'should . . . be no misappropriation of the biological and genetic resources and traditional knowledge of the developing countries' (WTO, 2001b).

It is difficult to say whether Doha produced any concrete agreement on the way forward on TRIPS and biotechnology. The two relevant provisions of the Doha Ministerial Declaration are the following:

> 12. We attach the utmost importance to the implementation-related issues and concerns raised by Members and are determined to find appropriate solutions to them. In this connection, and having regard to the General Council Decisions of 3 May and 15 December 2000, we further adopt the Decision on Implementation-Related Issues and Concerns in document WT/MIN(01)/W/10 to address a number of implementation problems faced by Members. We agree that negotiations on outstanding implementation issues shall be an integral part of the Work Programme we are establishing, and that agreements reached at an early stage in these negotiations shall be treated in accordance with the provisions of paragraph 47 below. In this regard, we shall proceed as follows: (a) where we provide a specific negotiating mandate in this Declaration, the relevant implementation issues shall be addressed under that mandate; (b) the other outstanding implementation

> issues shall be addressed as a matter of priority by the relevant WTO bodies, which shall report to the Trade Negotiations Committee, established under paragraph 46 below, by the end of 2002 for appropriate action.
>
> 19. We instruct the Council for TRIPS, in pursuing its work programme including under the review of Article 27.3(b), the review of the implementation of the TRIPS Agreement under Article 71.1 and the work foreseen pursuant to paragraph 12 of this Declaration, to examine, *inter alia*, the relationship between the TRIPS Agreement and the Convention on Biological Diversity, the protection of traditional knowledge and folklore, and other relevant new developments raised by Members pursuant to Article 71.1. In undertaking this work, the TRIPS Council shall be guided by the objectives and principles set out in Articles 7 and 8 of the TRIPS Agreement and shall take fully into account the development dimension. (WTO, 2001c)

Paragraph 12 seems wholly concerned with implementation of existing WTO obligations, including TRIPS. Paragraph 19 seems more directly on point in the dichotomy between developed- and developing-country interests in how TRIPS should or should not regulate biotechnological innovation. Paragraph 19 directs a review of Article 27.3(b) in the context of Article 71.1. Article 71.1 has to do with the review of transitional periods for developing countries to comply with TRIPS, since they had longer periods of time in which to achieve compliance with the TRIPS obligations, and hence paragraph 19 focuses the work of the TRIPS Council on the built-in agenda and implementation. Article 71.1 provides, however, that the TRIPS Council 'may . . . undertake reviews in the light of any relevant new developments which might warrant modification or amendment' of the TRIPS Agreement. Moreover, paragraph 19 of the Ministerial Declaration refers the TRIPS Council to TRIPS Agreement Articles 7 and 8. TRIPS Article 7 provides that:

> [t]he protection and enforcement of intellectual property rights should contribute to the promotion of technological innovation and to the transfer and dissemination of technology, to the mutual advantage of producers and users of technological knowledge and in a manner conducive to social and economic welfare, and to a balance of rights and obligations.

TRIPS Article 8 provides that WTO Members may adopt measures necessary 'to promote the public interest in sectors of vital importance to their socio-economic and technological development' and that 'may be needed to prevent the abuse of intellectual property rights by right holders or the resort to practices which unreasonably restrain trade or adversely affect the international transfer of technology', provided that such measures are consistent with the provisions of the TRIPS Agreement. As of the time of this writing, it remains to be seen whether the Quad and the developing countries will reach a suitable compromise in any future work emanating from the Doha Round.

European Law and Policy on TRIPS and Biotechnology

The European regime of intellectual-property rights in biotechnology has undergone significant development in the past decade. This development has been contentious in Europe, with public hostility continuing unabated against genetically modified foods and with concerns expressed about the ethics and morality of 'patenting life'. *The Economist* analogizes adverse British public opinion to a genetically modified organism: 'it seems to resist anything that might kill it, from scientific evidence to official reassurance'.[5]

Although not so clearly embracing dual forms of protection as has the USA, the trend to dual protection is evident in Europe. The European regime of patent and *sui generis* protection is governed by four legal instruments at the multilateral or European level and a host of national laws to implement these instruments.[6]

The Biotechnology Directive

One of the principal European laws on the patenting of biotechnological inventions is Council Directive 98/44 on the Legal Protection of Biotechnological Inventions. The Directive came into force on 30 July 1998 and EU Member States had until 30 July 2000 to implement it. The Biotechnology Directive took almost 10 years to bring to completion, with the European Parliament rejecting a prior draft in 1995. There have been no rigorous studies of whether the delay in implementing the Directive has affected the growth of the biotechnology industry in Europe, although it has been suggested (Perdue, 1999). Policy-makers have expressed concern that European laws and institutions tend to lag behind their American counterparts, thus losing 'competitive advantage' for the European biotechnology industry. Whether this is actually the case is unknown. It is difficult to assess the effect of laws and institutions on the growth of a particular industry sector and thus it is difficult to verify these claims.

The Biotechnology Directive came into existence after TRIPS. The Recitals in the Directive say that it is in part designed to implement TRIPS. Recitals are important in EU law for purposes of interpreting legislative language. One of the 56 Recitals in the Directive states, as one of the reasons for the adoption of the Directive, that TRIPS, 'signed by the European Community and the Member States, has entered into force and provides that patent protection must be guaranteed for products and processes in all areas of technology'.

TRIPS Article 27.1 provides that patents may be obtained on 'any technology'. Directive Article 3 implements TRIPS Article 27.1. Article 3 provides:

> 1. For the purposes of this Directive, inventions which are new, which involve an inventive step and which are susceptible of industrial application shall be patentable even if they concern a product consisting of or containing biological material or a process by means of which biological material is produced, processed or used.
> 2. Biological material which is isolated from its natural environment or produced by means of a technical process may be the subject of an invention even if it previously occurred in nature.

The additional requirements imposed on patentability under Article 3, stated in the alternative, are reproducibility or the presence of a technical process in the invention. Under Article 3.1, inventions are patentable even if they concern a product or process that uses 'biological material'. The Directive defines 'biological material' as 'any material containing genetic information and capable of reproducing itself or being reproduced in a biological system' (Article 2.1(a)). Under Article 3.2, even biological material that occurs in nature may be patented if it is isolated from nature or a technical process is used to produce it.

TRIPS Article 27.3(b) contains four rules on the patentability of biotechnology. First, WTO Members may exclude from patentability plants and animals other than microorganisms. Secondly, they may exclude from patentability essentially biological processes for the production of plants or animals. Thirdly, they may exclude plant varieties from patentability if they provide *sui generis* protection for plant varieties. Finally, WTO Members may not exclude from patentability non-biological and microbiological processes for the production of plants and animals. The Biotechnology Directive complies with all four principles found in Article 27.3(b). First, under Directive Article 4.2, inventions relating to plants or animals are patentable 'if the technical feasibility of the invention is not confined to a particular plant or animal variety'. Secondly, under Directive Article 4.1(b), 'essentially biological processes for the production of plants or animals' are not patentable. Thirdly, under Directive Article 4.1(a), plant and animal varieties *per se* are not patentable, but plant varieties receive protection under Council Regulation 2100/94/EC, which implements the UPOV Convention. Finally, Directive Article 4.3 provides that the ban on the patenting of essentially biological processes for the production of plants or animals is 'without prejudice to the patentability of inventions which concern a microbiological or other technical process or a product obtained by means of such a process'.

Directive Article 6 implements TRIPS Article 27.2. It provides that '[i]nventions shall be considered unpatentable where their commercial exploitation would be contrary to *ordre public* or morality; however, exploitation shall not be deemed to be so contrary merely because it is prohibited by law or regulation'. Directive Article 6 provides more detail than the TRIPS provisions by adding four specific patentability exclusions to which EU Member States must adhere. All but one of these exclusions are irrelevant to agriculture, as they exclude patenting of processes for cloning of human beings, processes for modifying the germ-line genetic identity of human beings and uses of human embryos for industrial or commercial purposes. The one exclusion potentially relevant to agriculture is in Article 6.2(d), the exclusion from patentability of 'processes for modifying the genetic identity of animals which are likely to cause them suffering without any substantial medical benefit to man or animal, and also animals resulting from such processes'. Article 6.2(d) imposes a morality requirement similar to that found in TRIPS Article 27.2 and European Patent Convention Article 53(a), but it is different in that it expressly sets forth a utilitarian calculus to assess the morality of animal varieties.

The European Patent Convention

The Convention on the Grant of European Patents, popularly known as the European Patent Convention, came into existence on 5 October 1973 in Munich. It was thus in existence long before TRIPS, and, as explained in the first part of this chapter, it in part formed the basis for language used in TRIPS Article 27.3(b). The members of the Convention include the EU Member States in their capacities outside the EU system. The Convention is not part of the EU legal system. In addition, five other European states not in the EU are members of the Convention: Cyprus, Liechtenstein, Monaco, Switzerland and Turkey. With a single application to the European Patent Office (EPO) established under the Convention, an inventor can obtain patent protection in all countries that are members of the Convention. The European-level Convention and its registration system coexist with national patent laws and national registration systems. The European patent is valid in the countries that are members of the Convention, but the interpretation and enforcement of the patents are issues for national law (Convention Article 64).

As explained in the first part of this chapter, Convention Article 53(b) provides that 'European patents shall not be granted in respect of . . . plant or animal varieties or essentially biological processes for the production of plants or animals; this provision does not apply to microbiological processes or the products thereof.' The term 'varieties' is not defined in the Convention. The *travaux préparatoires* of the Convention indicate that plant varieties were excluded from patent protection principally because *sui generis* protection existed under the UPOV Convention, then in its 1971 version, and under national laws implementing the UPOV Convention.

The European Patent Convention thus excludes plant varieties from patentability. This is consistent with TRIPS Article 27.3(b). The Convention interpretation of Article 53(b) has evolved over the years, and there would seem to be a trend in the law towards harmonization with US law. The US approach has been to permit three kinds of intellectual-property rights in plant varieties: (i) patents under general patent law, available for all inventions that meet patentability criteria, also known as utility patents; (ii) plant patents under the Plant Protection Act 1930, available for patents on asexually producing plants; and (iii) PBRs under the Plant Variety Protection Act 1970, for sexually producing varieties. In Europe, the approach has been to exclude patent protection for plant varieties, but the trend is towards permitting both kinds of protection. When the European Patent Convention was drafted in the early 1970s, the UPOV Convention required signatories to use plant-variety rights as the exclusive means of protection of rights in plant varieties. The UPOV Convention was amended in 1991 to freely permit countries to use patents, plant-variety rights or both to protect rights in plant varieties. The Council Regulation on Community Plant Variety Rights still provides that, in the EU, plant-variety rights are 'the sole and exclusive form of Community industrial property rights for plant varieties' (Council Regulation Article 1). When the European Patent Convention was drafted, plants and animals were not patentable because breeding did not result in plants and animals that could be reproduced. Genetic engineering has advanced since the early 1970s to the point where the reproducibility objection no longer exists, and patenting would seem to be a feasible option for protection of inventions in plant and animal varieties (Nott, 1999). The Biotechnology Directive underscores this conclusion. Recital 15 of the Directive states, 'no prohibition or exclusion exists in national or European patent law (Munich Convention) which precludes *a priori* the patentability of biological matter'.

The Convention contains grounds for denying patent applications on moral grounds substantially similar to the standard found in TRIPS Article 27.2. Convention Article 53(a) prohibits the grant of patents in Europe whose invention, publication and exploitation would be contrary to *ordre public* or morality.

A significant event towards patent protection for plant and animal varieties in Europe was the issuance of the decision of the EPO Enlarged Board of Appeal in *Novartis Transgenic Plant*, G01/98, on 20 December 1999. The case concerned the patentability of plants containing foreign genes inserted into their genomes (Blöchlinger, 2000). The transgenic plants produced with the claimed inventions would have characteristics that inhibit the growth of plant pathogens. One of the questions that the Technical Board of Appeal asked the Enlarged Board of Appeal was whether a patent claim relating to plants but in which specific plant varieties are not individually claimed avoids the prohibition in European Patent Convention Article 53(b) even though the patent embraces plant varieties. The Enlarged Board of Appeal ruled that Convention Article 53(b) prohibits patents for specific plant varieties but patents can be granted if varieties fall within the scope of the claims of the patent. The Enlarged Board looked to the UPOV Convention for guidance on what constitutes a plant variety, and found that the plant-variety concept embraces 'the entire constitution of a plant or a set of genetic information' (Blöchlinger, 2000). Plant-variety rights were designed at a time when varieties were the result of breeding processes, as in the use of selection and crossing to produce hybrids (Nott, 1999). This was in contrast to the patent claim in issue in *Novartis*, which involved a plant into which a piece of recombinant DNA was inserted, which, according to the Enlarged Board, was 'not a concrete living being but an abstract and open definition embracing an indefinite number of individual entities defined by a part of its genotype or by a property bestowed on it by that part' (Nott, 1999). According to the Enlarged Board, the subject-matter of the Novartis patent claim was ineligible for protection under the UPOV Convention. Plant-variety rights are granted only for specific plant varieties and not for 'technical teachings' that can be implemented in any number of different plant varieties (Nott, 1999).

The Enlarged Board of Appeal issued *Novartis* under the European patent law system that the European Patent Convention establishes. It is not EU law. *Novartis* is consistent with the EU Biotechnology Directive, which provides in Article 4.2 that inventions relating to plants or animals are patentable 'if the technical feasibility of the invention is not confined to a particular plant or animal variety'. While *Novartis* and the Biotechnology Directive go a long way towards expanding patent protection in Europe, pure plant varieties remain ineligible for European patent protection.

The European Patent Convention and *Novartis* are consistent with TRIPS Article 27.3(b), which permits patent or *sui generis* protection or a combination of the two. As explained in the first part of this chapter, the TRIPS exclusions in Article 27.3(b) are permissive, not mandatory. Under Article 27.3(b), WTO members 'may' exclude from patentability 'plants and animals other than microorganisms, and essentially biological processes for the production of plants or animals other than non-biological and microbiological processes'. *Novartis* does not implicate this provision, because it permits patentability of a process that is not essentially biological. The transgenic technology involved in *Novartis* was biotechnological, not merely biological.

The UPOV Convention and implementing Council Regulation

The European countries are all members of the International Convention for the Protection of New Varieties of Plants (UPOV Convention), signed in Paris in 1961 and revised in 1972, 1978 and 1991. The Convention established the International Union for the Protection of New Varieties of Plants (Millett, 1999). The USA is also a signatory to the UPOV Convention. The EU has implemented the UPOV Convention with

Council Regulation 2100/94 on Community Plant Variety Rights. The 1991 Convention represents a substantial revision, in that it freely permits dual protection of plant varieties, by patents, *sui generis* rights or both (UPOV Convention Article 2). The original 1961 Convention required signatories to choose one form of protection for 'one and the same botanical genus or species', either patents or *sui generis* rights. The 1978 Convention relaxed this restriction to permit countries such as the USA and Japan to continue to provide dual protection if they had provided it before 31 October 1979. The 1978 revision facilitated the accession of the USA and Japan to the Convention. In the EU Member States, however, the Council Regulation provides that plant varieties are entitled only to protection as Community plant-variety rights (Council Regulation Article 1).

Conclusion

The above examinations of world trade law on intellectual property and substantive European intellectual-property law, and the examination of American intellectual-property law in Professor Fietshans's chapter (Chapter 7), are compelling evidence that the law is developing around a proprietary model with overlapping forms of intellectual-property protection for biotechnological innovation. While the law in the USA and the EU exhibit such a trend, it is far from clear that other WTO Members, outside the Quad, will uncritically accept the proprietary model. The WTO Members do not have an agreed framework for determining what is acceptable for patenting in the biotechnology field. It is by no means clear that more or stronger intellectual-property rights are always better, from the standpoint of economic efficiency or ethics. When multiple layers of property rights are coupled with the ability of firms to take effective technological measures to protect intellectual property, the balance that intellectual-property law is supposed to produce, between the public interest in dissemination of knowledge and the private interest in gain from one's invention, is likely to come undone. The tragedy here is not one of a commons, but of an anti-commons, in which the only or primary goal that intellectual-property law seems to serve is the production and maintenance of monopolies, legally contrived, regardless of whether these monopolies provide net benefits for society (Heller and Eisenberg, 1998).

What can Europe offer to promote consensus? At the WTO level, it is doubtful that the EU and its Member States would take positions differing radically from those of the USA on matters involving intellectual-property rights in biotechnology. Europe and the USA share a similar stake. The European Commission Directorate for Trade has offered some proposals in various forums for assuaging developing-country concerns. In September 2002, the Commission submitted a 'concept paper' to the

WTO advocating the disclosure of geographical origin of biological material in patents involving biotechnology (European Commission, 2002). It is unlikely that these proposals will result in negotiating positions antagonistic to those of the USA, although the USA does not support the disclosure-of-origin approach, at least at the time of the writing of this chapter. Europe's contribution is more likely from its patent-law regime, which exists separate from the EU, and in its Biotechnology Directive, which sets standards for biotechnology law in the EU Member States. European patent law takes a narrower and more cautious approach than US law in offering patent protection for plant and animal varieties, but Europe may well succumb to more closely following US law, in the name of 'competitiveness'. The EPO has used the European morality requirements discussed above in its jurisprudence, but to date has offered little other than an application of a utilitarian standard, comparing the suffering of the animal, for example, in the context of a patent involving an animal, with the benefits to humanity of the invention. Patent examiners are ill-equipped to make public policy balancing ethics and economics (Grubb, 1999). The teachings that will inform policy are more likely to emanate from the various European think-tanks that have devoted substantial resources to investigating the important ethical questions (Nuffield, 1999, 2002). Time will tell whether they succeed in influencing public policy in a way that will ultimately affect the content of TRIPS.

Notes

1 Other matters of relevance to agriculture in TRIPs are the provisions on agricultural chemicals and geographical indications. Geographical indications are relevant to agriculture because European countries have substantial interests in the branding of agricultural and food products, such as cheese, wines and spirits. A discussion of geographical indications is omitted, however, because the focus of the chapter is on rights in technology and not rights in commercial property. Geographical indications are of increasing relevance in identifying the source of genetic material in patents involving biotechnology, although it is unclear what sorts of positive rights, if any, result from such identification. A discussion of agricultural chemicals is omitted as well, as it presents no contentious legal issues at the present time. Biotechnology innovation in bioinformatics databases will make copyright an important concern in the biotechnology sector. Trade-secret laws are important to the extent that the biotechnology sector uses these laws to protect innovative ideas that they wish to keep confidential. Trade marks are important too, as companies begin to market products derived from biotechnological innovation. Examples are Monsanto's Roundup Ready®, Aventis's Liberty® and Libertylink technologies® (Binenbaum *et al.*, 2000).

2 The 'European Community' (EC) has international legal personality and is thus a WTO Member. The 'European Union' (EU) has no international legal personality and is not a WTO Member (Bourgeois, 1999). The 'Union' terminology is nevertheless used throughout this chapter, as it is now standard usage.

3 The European Patent Convention Article 53 provides:

European patents shall not be granted in respect of:
(a) inventions the publication or exploitation of which would be contrary to ordre public or morality, provided that the exploitation shall not be deemed to be so contrary merely because it is prohibited by law or regulation in some or all of the Contracting States;
(b) plant or animal varieties or essentially biological processes for the production of plants or animals; this provision does not apply to microbiological processes or the products thereof.

4 For example, the US Patent and Trademark Office (PTO) granted and subsequently cancelled a patent for an invention based on the pharmacological properties of the ayahuasca vine, a plant found in the Amazon rain forest. The PTO based its cancellation on the fact that publications describing the pharmacological aspects of the ayahuasca vine were known and available prior to the filing of the patent application. US patent law denies patentability to inventions described in printed publications more than 1 year prior to the date of the patent application. The pharmacological properties of the ayahuasca vine have been known for centuries and are part of the traditional knowledge of indigenous communities in Brazil. Patents have been claimed on inventions relating to turmeric, karela and the neem tree, plants found in India. The European Patent Office (EPO) has revoked a patent granted jointly to W.R. Grace and the US Department of Agriculture for an insecticide and fungicide derived from the seed of the neem tree. The EPO revoked the patent on grounds similar to those that the US PTO used to cancel the patent on the invention relating to the ayahuasca vine. The EPO ruled that the patent claims were not novel and there was prior public use.

5 Who's afraid? *The Economist*, 17 July 1999.

6 It is only possible to provide an overview here in the context of examining the relationship between European law on intellectual-property rights in biotechnological inventions and TRIPS Article 27. For detailed discussion of other major aspects of European patent law, see (Grubb (1999), Muir *et al.* (1999) and Paterson (2000).

References

Abbott, F. (2000) TRIPS in Seattle: the not-so-surprising failure and the future of the TRIPS agenda. *Berkeley Journal of International Law* 18, 165–179.

Binenbaum, E., Nottenburg, C., Pardey, P.G., Wright, B.D. and Zambrano, P. (2000) *North–South Trade, Intellectual Property Jurisdictions, and Freedom to Operate in Agricultural Research in Stable Crops*. EPTD Discussion Paper No. 20, www.apec.umn.edu/faculty/ppardey/EPT_DP70.pdf.

Blöchlinger, K. (2000) A variety of interpretations of 'plant variety'. *CASRIP Newsletter* 7, 10–11. www.law.washington.edu/Casrip/newsletter/.

Bourgeois, J.H.J. (1999) External relations powers of the European Community. *Fordham International Law Journal* 22, S149–S173.

Boyle, J. (2001) *The Second Enclosure Movement and the Construction of the Public Domain*. www.law.duke.edu/pd/papers/boyle.pdf.

Brenner, C. (1998) *Intellectual Property Rights and Technology Transfer in Developing Country Agriculture: Rhetoric and Reality*. Technical Papers No. 133, OECD.

European Commission (2002) *Communication by the European Communities and their Member States to the TRIPs Council on the Review of Article 27.3(b) of the TRIPs Agreement, and the Relationship between the TRIPs Agreement and the Convention on Biological Diversity (CBD) and the Protection of Traditional Knowledge and Folklore*. 12 September.

Grubb, P.W. (1999) *Patents for Chemicals, Pharmaceuticals and Biotechnology: Fundamentals of Global Law, Practice and Strategy*. Oxford University Press, Oxford.

Hamilton, N.D. (1993) 'Who owns dinner? Evolving legal mechanisms for ownership of plant and genetic resources. *Tulsa Law Journal* 28, 587–665.

Michael A., Heller, M.A. and Eisenberg, R.S. (1998) Can patents deter innovation? The anti-commons in biomedical research. *Science* 280, 698.

Hess, C. and Ostrom, E. (2001) *Artifacts, Facilities, and Content: Information as a Common Pool Resource*. www.law.duke.edu/pd/papers/ostromhess.pdf.

Hume, D. (1739) *A Treatise on Human Nature*. Reprinted 1990, Oxford University Press, Oxford.

Llewelyn, M. (2001) PIP project and legislative overview. In: *Conference on Plant Intellectual Property within Europe and the Wider Global Community*. www.shef.ac.uk/~pip/session1.doc.

Linarelli, J. (2002) Trade Related Aspects of Intellectual Property Rights and Biotechnology: European Aspects. *Singapore Journal of International and Comparative Law* 6, 406–435.

Millett, T. (1999) The community system of plant variety rights. *European Law Review* 24, 231–258.

Muir, I., Brandi-Dohrn, M. and Gruber, S. (1999) *European Patent Law*. Oxford University Press, Oxford.

Nott, R. (1999) The Novartis case in the EPO. *European Intellectual Property Review* 21, 33–36.

Nuffield Council on Bioethics (1999) *Genetically Modified Crops: the Ethical and Social Issues*. London.

Nuffield Council on Bioethics (2000) *The Ethics of Patenting DNA: a Discussion Paper*. London.

Paterson, G. (2000) *The European Patent System*, 2nd edn. Sweet & Maxwell, London.

Perdue, D.O. (1999) The changing landscape of patenting transgenic plants in Europe. *CASRIP Newsletter* 6. www.law.washington.edu/Casrip/newsletter/news6i1Perdue.html.

Smith, H. (2000) Semicommons property rights and scattering in the open fields. *Journal of Legal Studies* 29, 131–169.

Travis, H. (2000) Pirates of the information infrastructure: Blackstonian copyright and the First Amendment. *Berkeley Technology Law Journal* 15, 777–862.

Wolf, S.A. and Zilberman, D. (1999) Public science, biotechnology and the industrial organization of agro-food systems. *AgBioForum* 2, 37. www.agbioforum.missouri.edu.

WTO (1999a) *EC Approach to Trade-Related Aspects to Intellectual Property in the New Round*. WT/GC/W/193, 2 June.

WTO (1999b) *Communication from Kenya on behalf of the African Group* WT/GC/W/302, 6 August.

WTO (1999c) *Preparations for the 1999 Ministerial Conference*. Job(99)/4797/Rev. 3, 18 November.

WTO (2000a) *The Relationship Between the Convention on Biological Diversity (CBD) and the Agreement on Trade-Related Aspects of Intellectual Property Rights (TRIPs)*. IP/C/W/175, 11 May.

WTO (2000b) *Communication from Brazil: Review of Article 27.3(b)*. IP/C/W/228, 24 November.

WTO (2001a) *Communication from Kenya: ACP Declaration on the Fourth Ministerial Conference*. WT/L/430, 9 November.

WTO (2001b) *Statement by the Honourable Murasoli Maran, Minister of Commerce and Industry for India*. WTO/MIN(01)/ST/10, 10 November.

WTO (2001c) *Doha Ministerial Declaration*. WT/MIN(01)DEC/1, 20 November.

9

Regulating GM Products in the EU: Risk, Precaution and International Trade

Chris Hilson and Duncan French

Introduction

The potential conflict between international free trade and national environmental policies is one that has attracted significant comment. The first round of cases involving the General Agreement on Tariffs and Trade (GATT) and the World Trade Organization (WTO) concerned 'pure' environmental disputes. With both *United States – Restrictions on Imports of Tuna* (*Tuna/Dolphin Dispute*) (1991) and *United States – Import Prohibition of Certain Shrimp and Shrimp Products* (1998) and (2001) (*Shrimp/Turtles I* and *II*), for example, national policies were seeking to protect known environmental threats to marine creatures, at the expense of free trade. However, recent environmental trade cases or potential disputes, such as *EC Measures Concerning Meat and Meat Products (Hormones)* (*Beef Hormones Dispute*) (1998), *EC – Measures Affecting Asbestos and Asbestos-containing Products* (*Asbestos Dispute*) (2001) and the recently commenced Community/USA dispute over genetically modified (GM) products, are markedly different in two respects. First, they involve human-health questions rather than just environmental ones. And, secondly, they involve the question of risk to human health and/or the environment rather than currently observable environmental effects. The question thus becomes how far states can maintain their own national preferences on risk and where these have to yield to international rules on free trade.

The first part of this chapter will explore how Community regulation of GM products addresses the question of risk. It will become clear that the Community is in the process of tightening its risk regulation in relation to GM produce. Not only has the range of potential risks assessed within risk assessment been broadened considerably, but also the risk-management response by Member States to these risks has increasingly been one based on precaution. The chapter will then move on to address some of the US objections to the Community system of regulation, which could conceivably form the basis of a trade dispute between the two regions. In the final part, some tentative suggestions will be offered as to how international trade law might approach any such dispute and on the likely impact of the 2000 Cartagena Protocol on Biosafety.

Community Regulation of GM Products

The term GM products is used in this chapter to refer to both living, viable GM organisms (GMOs), such as tomatoes, and non-viable products derived or produced from GMOs, such as tomato paste. For trade purposes, the most significant items of existing or proposed Community legislation relating to GM products are the Deliberate Release Directive 2001/18/EC (OJ 2001, L106/1), the Novel Foods Regulation (Council Regulation (EC) 258/97 (OJ 1997, L43/1)) and its proposed replacement covering GM food and feed (Commission of the European Communities (CEC), 2001b, 2002b), and the proposed Regulation concerning traceability and labelling (CEC, 2001a, 2002a).

Council Directive 2001/18

Council Directive 2001/18/EC, which regulates only living GMOs, repealed and replaced Council Directive 90/220/EEC (OJ 1990, L117/15) on 17 October 2002. It deals in its part B with 'deliberate releases' and in its Part C with the 'placing on the [Community] market' of GMOs. Deliberate releases are experimental releases, which have as their direct purpose research rather than commercial gain. Placing on the market covers not only Community products containing or consisting of GMOs but also imports. In agricultural terms, Part C therefore covers both the authorization of seeds for the commercial-scale cultivation of crops within the Community and also the import of GMOs that are then processed for use in the food or animal-feed industries.

Under the standard procedure for Part B and under Part C, 'notifiers' are obliged to submit a notification with detailed information (including, in particular, an environmental risk assessment) to the Member State Competent Authority where the deliberate release is to take place or where the GMO is to be first marketed or imported into the Community. From there, the procedures for the two Parts differ. Under Part C – the more relevant Part for international trade purposes – the initial

Competent Authority then sends the notification to other Member States and the Commission and itself examines it for compliance with the Directive. On the basis of this examination, it prepares an assessment report. If the report is against granting consent, then the notification will be rejected. If it is in favour and neither the Commission nor any other Member States object, consent must be granted. However, if another Member State or the Commission objects, then the matter proceeds to a regulatory committee procedure, which is discussed in more detail later in this chapter.

Risk

Council Directive 2001/18/EC requires deliberate releases to conform with the 'step-by-step' principle. Under this principle the containment of GMOs is reduced and the corresponding scale of release increased gradually, step by step, only if risk assessment at each stage proves satisfactory. This is beneficial in risk terms because potential risks can be easily monitored and controlled on a small scale. If, in contrast, large-scale deliberate releases were allowed right from the start, any undesirable effects would be geographically much more widespread and thus more difficult to control. However, there is obviously a balance to be drawn here. If the area is too small, then the data provided on risks to, for example, biodiversity are likely to be of limited application and hence value. It is for this reason that the UK government, for example, decided to give the go-ahead to 'farm-scale' evaluations or trials at sites across the UK.

Although Council Directive 90/220/EEC required risk assessment, there was no detailed guidance on how, precisely, notifiers and Member State Competent Authorities were to approach this, which led to problems of inconsistency in practice across the Community. Council Directive 2001/18/EC has sought to address this problem by harmonizing in its Annex II the objective to be achieved, the elements to be considered and the general principles and methodology to be followed in carrying out environmental risk assessments. Further guidance notes will in time supplement this information in Annex II.

Besides providing a greater level of detail on risk assessment, Council Directive 2001/18/EC also broadens the effects that must be considered under the assessment – to include not only the direct and immediate effects of the release of the proposed GMO on human health or the environment, but also indirect and delayed effects. Direct effects are defined as 'primary effects on human health or the environment which are a result of the GMO itself and which do not occur through a causal chain of events' (Annex II, para. 3), while indirect effects refer to effects 'occurring through a causal chain of events, through mechanisms such as interactions with other organisms, transfer of genetic material, or changes in use or management' (Annex II, para. 4). Indirect effects, the Directive notes, are likely to be delayed. An example of an indirect effect associated with the management of the GMO rather than the GMO itself is the effect that a herbicide-tolerant GMO crop variety may have on weed management, which may itself affect the diet of certain types of farmland wildlife (DETR, 1999a). Immediate effects are those which are observed during the period of release of the GMO and may be direct or indirect. Delayed effects are those which become apparent as direct or indirect effects at a later stage and will include any cumulative effects. The inclusion of cumulative long-term effects is significant because applications for authorization under the Directive are dealt with on a 'case-by-case' basis. While this has its advantages in risk terms (in that GMOs, with their own potentially distinct risks, are considered individually rather than *en bloc*), it also has the potential to ignore the cumulative effect that increasing numbers of separate GMOs in the environment may be having.

The revised Directive also introduces, for GMOs placed on the market, an obligation on notifiers to implement a monitoring plan in order to trace and identify any of the above effects. This is because, as the UK Advisory Committee on Releases to the Environment has noted, 'however detailed the scientific evidence which indicates that a product will not cause adverse effects, field observations are important to confirm that this is in fact the case' (DETR, 1999b). The reason for this is that, where ecological systems are concerned, one finds oneself more in the realm of uncertainty than in the case of risk proper, which presupposes not only knowledge of a particular hazard but also the probabilistic likelihood of its occurrence (DETR, 1999a). In other words, the scientific evidence may be detailed but, if it is inevitably uncertain, it is worth employing monitoring, both during and after the release of the GMO, to confirm whether those initial, uncertain scientific predictions were correct (DETR, 1999a).

Closely linked to the issue of monitoring is the requirement for time-limited consents imposed by Council Directive 2001/18/EC: under the new Council Directive, consents will run for a maximum period of 10 years. This time limitation means that problems highlighted by the monitoring data or any new information on risk that comes along should be acted upon. Under the old Council Directive 90/220/EEC, consents for placing on the market were granted for an indefinite period and the onus was on notifiers to come forward with any new information on risk, which obviously has its limitations in regulatory terms. With Council Directive 2001/18/EC in contrast, consents must contain monitoring plans and must include obligations to report the results of this monitoring to the Commission and Member States. And, since notifiers are bound to comply with these reporting duties, this means that, via the monitoring data, Member States are much more likely to be alerted to emerging risks than under the old, essentially self-regulatory approach (though the latter still remains in place as an additional mechanism). However, even with this system in place, the relevant authorities could still fail to take much notice of the new data (though this seems unlikely given that the relevant authorities consist of both the Commission and all of the Member States). The fail-safe is thus provided by the time-limited nature of the consent. On an application for renewal after 10 years, the applicant will be expected to show that the monitoring data from the previous 10 years and any other new information on risk do not stand in the way of renewing the authorization. There is, in other words, a positive obligation here that cannot be avoided. This obligation falls both on the notifier and on the original Member State Competent Authority, which will be obliged actively to engage in a detailed re-evaluation of the GMO consent in the light of the monitoring data and any new information on risk.

Council Directive 2001/18/EC, like Council Directive 90/220/EEC, contains a 'safeguard clause' that enables a Member State provisionally to restrict or prohibit the use and/or sale of an already authorized GMO where risks posed by that GMO have become apparent since consent was granted. Needless to say, this provision has the capacity to interfere not only with international trade but also with intra-Community trade. A number of Member States have invoked the safeguard clause (Article 16) of Council Directive 90/220/EEC to impose a temporary ban on the import of GM maize and oilseed-rape products into their countries (European Community, 2002). There are nine ongoing Article 16 cases involving six Member States – Austria, Luxembourg, France, Greece, Germany and the UK (European Community, 2002). Eight of those cases have been sent to the Scientific Committee on Plants for an opinion and, in all eight cases, the Committee was of the view that the information submitted by the Member States concerned did not justify their import bans (European Community, 2002). However, under Council Directive 90/220/EEC – as under the new Council Directive 2001/18/EC – the decision on the particular Member State's action is taken not by the relevant scientific committee, but under the regulatory committee procedure, which will be examined further below.

The precautionary principle

The precautionary principle is one of the key principles of Community environmental law (Article 174(2) EC; CEC, 1999b). It has also recently been applied more generally in other Community policy areas concerning human, animal and plant health, such as food safety (CEC, 1999a; *Pfizer Animal Health SA* v. *Council* (2002) and *Alpharma Inc.* v. *Council* (2002)). The position of Community GMO regulation in relation to the precautionary principle is a complicated one.

Recital 8 of Council Directive 2001/18/EC notes that 'the precautionary principle has been taken into account in the drafting of this Directive and must be taken into account when implementing it'. Articles 1 and 4 then go on to give expression to this statement in the main text of the

Directive. Article 1 mirrors the first part of the recital in stating that, '[i]n accordance with the precautionary principle', the objective of the Directive is a dual one of single-market harmonization and protecting human health and the environment in relation to the deliberate release of GMOs and the placing on the Community market of GMOs. Article 4 then concerns the latter 'implementation' part of the recital, in stating that:

> Member States (which are responsible for implementing Directives within their national laws) shall, in accordance with the precautionary principle, ensure that all appropriate measures are taken to avoid adverse effects on human health and the environment which might arise from the deliberate release or the placing on the market of GMOs. GMOs may only be deliberately released or placed on the market in conformity with Part B or Part C respectively.

What is being claimed here, in other words, is that the regulatory controls set out in the Directive are themselves an application of the precautionary principle in relation to the risk posed by GMOs. At this stage, it is necessary to distinguish between risk assessment and risk management. In so far as the Directive contains a stringent regime for risk assessment, it can be seen as an application of the principle. However, the precautionary principle is perhaps more often associated with the risk-management decision adopted in the light of an uncertain scientific risk assessment. The new regulatory committee procedure in Council Directive 2001/18/EC allows expression of the principle in this latter sense. With applications for GMOs to be placed on the market, if another Member State Competent Authority or the Commission raises and maintains an objection to the notification, then the committee procedure in Article 30(2) comes into play. While Council Directive 90/220/EEC also contained a committee procedure in these circumstances, Council Directive 2001/18/EC has made some key changes, which will have a significant bearing on individual Member State preferences on risk. The committee under Council Directive 90/220/EEC was what is known as a 'type IIIa' regulatory committee under the Council 'comitology' Decision 87/373, whereas the committee under the new Council Directive 2001/18 is a new-style regulatory committee under Council Decision 1999/468. Committees consist of representatives of the Member States, chaired by a non-voting representative from the Commission. Under both the old and new systems, the Commission puts a proposal forward to the regulatory committee. If the proposal is to grant marketing consent and a qualified majority in the regulatory committee is in favour of the proposal, then a decision to grant consent will be adopted. However, if there is no qualified majority in favour, the Commission has to submit a proposal to the Council. This is where IIIa and the new-style committee procedures part company. Under IIIa – the procedure used in Council Directive 90/220/EEC – the Council could only reject or amend the proposal by unanimity. With the new-style committee, in contrast, the Council can reject the proposal by a qualified majority (though it seems that unanimity is still required for the Council to amend the proposal), in which case the Commission is likely to submit an amended proposal to the Council.

The old IIIa procedure placed considerable power with the Commission. In a number of cases, there was no qualified majority in favour of consent in the regulatory committee and thus the matter would proceed to the Council. There, the original Competent Authority would invariably remain in favour of granting consent and thus a unanimous vote in the Council to reject consent was a most unlikely event. This meant that the Commission's proposal to grant consent would be adopted, despite potentially considerable Member State opposition. IIIa was, as a result, a firmly science-based procedure: the original Competent Authority that carried out the initial scientific risk assessment and the pro-science Commission (which tends to follow the advice of its scientific committees) were in the driving seat. The new-style committee, in contrast, has politicized matters to a much greater extent. Now, if a qualified majority in the Council wants to prevent a release on precautionary rather than strictly scientific grounds, there would appear to be little to stop it from doing so. Risk-management decisions by the Council, in other words, may be much more cautious than the already cautious scientific risk assessment suggests is necessary. And politically this may well make sense: if citizens in a particular Member State perceive a risk even where none is thought by scientists to exist, then a Member State government will be better served politically by taking action that reflects its citizens' perceptions on risk. Of course, whether a qualified majority vote in the Council can be achieved is another matter. As will be seen below, there are currently six Member States that remain somewhat anti-GM. They, however, would have a total of only 34 votes in the Council, whereas a qualified majority requires 62 votes out of a possible 87. A precautionary vote on their part against granting consent would thus have to be accompanied by other Member States also voting on precautionary grounds or by some votes against on scientific grounds. There is also, of course, the question of whether the Court of Justice would uphold such an approach to the precautionary principle as legally permissible. However, the recent rulings in the *Pfizer* and *Alpharma* cases suggest that it would – with the Court in those cases revealing its unwillingness to second-guess risk-management decisions so long as they are premised on a scientific risk assessment that is as full as is possible in the circumstances.

The precautionary principle has also been cited in connection with the *de facto* moratorium on new consents for placing GMOs on the Community market, which was officially declared in June 1999 and remains in place today. In the June 1999 Environmental Council, two groups of Member States issued Declarations on GMOs that made reference to the principle. On the one hand, Denmark, Greece, France, Italy and

Luxembourg cited the precautionary principle in imposing a *de facto* moratorium on new consents for growing and placing on the market of GMOs until rules on labelling and traceability for GMOs and GMO-derived products were adopted; and, on the other, Austria, Belgium, Finland, Germany, the Netherlands and Sweden stressed their intention to 'take a thoroughly precautionary approach in dealing with notifications and authorizations for the placing on the market of GMOs' and stated, *inter alia*, that they would not 'authorize the placing on the market of any GMOs until it is demonstrated that there is no adverse effect on the environment and human health' (Council, 1999).

The first usage, by Denmark and others, implies that, once traceability and labelling are in place, then consumers will be able to make their own precautionary decisions, but that, until then, the Member States concerned will adopt a precautionary approach on their behalf in banning what may, in scientific terms, be products that pose no proved risk. At first sight, the second usage appears to reverse the burden of proof – a very strict, environmentalist interpretation of the precautionary principle. In other words, rather than a presumption in favour of trade unless Member States can adduce scientific evidence to demonstrate that there is evidence of risk posed by a particular GMO, it implies that a notifier must provide evidence that there is no risk. However, there are two points to be made in this regard. First, the Court of Justice has recently ruled against such a zero-risk interpretation of the precautionary principle in *Pfizer* and *Alpharma*; and, secondly, in any event, it seems most unlikely that Austria or any of the other Member States in the second camp were seeking to adopt that interpretation of the precautionary principle. In fact, their stance is weaker than that of the first camp in that they do not go as far as officially to declare a moratorium. For these Member States, adopting a 'thoroughly precautionary approach' appears to mean little more than applying the improved risk-assessment provisions of the new Council Directive.

In this connection, it is worth mentioning the Commission's attempts to lift the moratorium. Instead of waiting until the official, final date for implementation of the new Council Directive by Member States in October 2002, the Commission was proposing to apply its provisions straight away on a voluntary basis (European Community, 2000). However, six Member States (Denmark, Greece, France, Italy, Austria and Luxembourg) signalled that they were unwilling to drop the moratorium on new authorizations until effective rules on traceability and labelling were adopted (Council, 2001).

Novel foods

The first thing to be said about the Novel Foods Regulation (Council Regulation (EC) 258/97 (OJ 1997, L43/1)) is that there is a proposal currently going through the Community legislative process which will, in time, lead to the repeal of its provisions concerning GM food. It will therefore not be discussed in detail here.

The existing Regulation contains rules on the regulatory control and labelling of both foods and food ingredients containing or consisting of GMOs and of foods and ingredients produced from but not containing GMOs. Labelling requirements of the Regulation will be analysed in the next section and thus will not be dealt with at this stage. As for the stringency of regulatory control imposed by the Regulation, this depends on the type of food product involved and the degree of risk associated with it. Living GMOs, which pose the greatest risk, require prior authorization. This involves a food-'safety' assessment and an environmental-risk assessment, followed by a series of procedures that have much in common with the old Council Directive 90/220/EEC. In contrast, derived products that are 'substantially equivalent to existing foods or food ingredients' as regards their composition, nutritional value, metabolism, intended use and the level of undesirable substances contained in them, require only notification and a much simplified set of procedures is involved. The approach taken by the UK authorities and all other Member States is that substantial equivalence will only be demonstrated where neither novel protein nor DNA is detectable in the final product (for example, with highly refined oils). Where, in contrast, traces of DNA or protein are detectable, the authorization procedure must be followed (MAFF, 1999).

The new proposal is for a Regulation on GM food and feed (CEC, 2001b, 2002b). As the title of the proposal suggests, one of the key changes this will make is to bring the regulation of food and feed together within one instrument. Furthermore, under the 'one door–one key' principle, a single authorization under the Regulation will cover both the deliberate release of a GMO and its subsequent use as food or feed, avoiding the need for a separate application under Council Directive 2001/18/EC. Another key change is that the Regulation will abandon the simplified notification procedure for substantially equivalent foods on the grounds that this 'regulatory shortcut' has been 'very controversial in the Community in recent years' and also because 'whilst substantial equivalence is a key step in the safety assessment process of genetically modified foods, it is not a safety assessment in itself' (CEC, 2001b). The proposal thus covers all products produced from a GMO in the sense that material from the GMO is present in the end-product (whether or not novel DNA or protein is detectable). Products such as highly refined oils will therefore require authorization. Products to be excluded will be those produced with GMOs but where no material from that GMO is present in the end-product (such as cheese made with a GM enzyme that does not remain in the final product, and milk products, meat and eggs from animals fed on GM feed). Finally, the presence of *de minimis* levels of adventitious contamination by material that contains, consists of or is produced from GMOs will not require authorization, so long as the relevant Scientific Committee or European Food Safety Authority has approved the particular GMO concerned after a scientific risk assessment.

Traceability and labelling

It will be recalled that a number of Member States have refused to drop the existing moratorium on the placing on the market of new GMOs until detailed legislation on traceability and labelling has entered into force.

Labelling

At the core of the Community system of GM regulation is the idea of prior authorization based on sound, scientific risk assessment. However, in recent years a series of food crises (including bovine spongiform encephalopathy (BSE)) mean that consumers are much more sceptical about the claims of science and scientists and of government regulatory choices based on that science. Labelling has therefore become an attractive proposition in Europe because it devolves decision-making on risk to individual consumers. State regulation may declare that there is no scientific evidence that GM products pose any risk to human health or the environment, but labelling enables consumers to adopt a precautionary approach if they so wish. Of course, mention of human health and the environment points to one of the limitations of labelling as a regulatory addition or alternative and that is that it only affords real protection to human concerns about health. If I am concerned about health risks, then I can choose not to purchase a labelled GM product. But, with the environment, my not choosing a labelled GM product will not stop others from doing so. If this product poses a risk to the environment, then state regulation is needed to tackle this.

Some labelling provisions are found in Council Directive 2001/18/EC, which requires labels for all products consisting of or containing GMOs to state that 'this product contains genetically modified organisms'. However, the duty to label in this way is only imposed on the initial, consent-holding notifier, and not on other operators who subsequently place the GMO product on the market (CEC, 2001a). Nevertheless, once the proposed Regulation concerning traceability and labelling (CEC, 2001a) is implemented, labelling will be required at all stages of the placing on the market for prepackaged products consisting of or containing GMOs. Finally, it should be noted that Council Directive 2001/18/EC provides for the possibility of excluding from its labelling requirements products that contain adventitious or technically unavoidable traces of authorized, live GMOs (through, for example, cultivation, harvest, transport or processing).

Under the current Novel Foods Regulation, food products that consist of or contain GMOs must be labelled as such. As for products produced from GMOs, the labelling requirements turn on whether the newly proposed food or ingredient can be regarded as 'no longer equivalent' to an existing food or ingredient. For guidance on when a food or ingredient will be no longer equivalent, one turns to Council Regulation (EC) 1139/98 (OJ 1998, L159/4), as amended by Commission Regulation (EC) 49/2000 (OJ 2000, L6/13). In theory, this sets out labelling requirements only for products derived from certain varieties of maize and soybean that were authorized under Council Directive 90/220/EEC before the labelling requirements for food in the Novel Foods Regulation were introduced. The requirement of labelling under Council Regulation (EC) 1139/98 is dependent on whether novel protein or DNA is detectable in the food or ingredient. Later revisions to the Regulation mean that food will still not require labelling if adventitious contamination with these GM materials is detectable at levels of less than 1% (Commission Regulation (EC) 49/2000). In practice, this latter Regulation and its revision currently serve as a model for all Community labelling of food products derived from GMOs (MAFF, 1999; European Community, 2002). Thus for all other products not covered by Council Regulation (EC) 1139/98, if either protein or DNA is detectable, other than at adventitious levels, then the product will be regarded as 'no longer equivalent' for the purposes of the Novel Foods Regulation and labelling will be required.

As noted earlier, there is, however, a proposal on GM food and feed which will lead to the repeal of the Novel Foods Regulation (CEC, 2001b, 2002b). As far as labelling is concerned, there are two significant elements of the proposal, both of which have attracted controversy.

The first is its intention to extend the labelling requirement to all GM food and feed, irrespective of the detectability of DNA or protein. The Commission was essentially confronted by four possible options for labelling of products produced using GMOs (CEC, 2000). First, it could have maintained the existing system, whereby labelling is based on the detectability of DNA or protein. Secondly, it could have kept this system, but introduced alongside it a voluntary 'GM-free' scheme, with producers vouching that no gene technology had been used in production. The third option – and the one chosen in the proposal – is to label all foods and feed produced from GMOs, thus informing the consumer about the production process irrespective of whether DNA or protein is detectable in the final product. The final option would have been to extend labelling still further, requiring not only products produced from GMOs to be labelled, but also products produced with GMOs.

The Commission's choice of the third option has proved controversial. Much of the controversy has revolved around the accompanying traceability regime which the option chosen will necessarily require. Under the existing arrangements, labelling claims are easily tested in the laboratory: if DNA or protein is detected in the relevant product above adventitious levels, then labelling is required. However, under the new, proposed system, the only way of controlling and verifying labelling claims for products with no detectable traces of GM material in them (such as highly refined oils) is to have a traceable audit trail in place. Critics claim that such a traceability regime will be unreliable and open to fraud. They also believe that it places unacceptable costs on consumers who are not concerned about GM-derived food, and with no apparent benefits in terms of food safety (House of Lords Select Committee on the European Union, 2002). Critics thus support something like the second option above, whereby those who are concerned can pay a premium for GM-free (or non-GM) products produced without any use of gene technology. Supporters, for their part, point to similar regimes in other contexts which appear to work (such as fair-trade foods). And they in turn criticize the second option because it means that, as with organic foods, the poor are unable to exercise an effective choice. Rather than GM-free being a more expensive niche market, their view is that GM foods should be in that position (House of Lords Select Committee on the European Union, 2002).

The second controversial labelling element of the proposal involves adventitious contamination. As seen above, existing GM labelling rules provide for a 1% adventitious-contamination threshold, below which labelling is not required. The Commission's proposal was that the threshold for labelling under the new Regulation should be set via comitology – a position it maintained in its revised proposal despite pressure from the European Parliament to lower the threshold in the Regulation itself to 0.5%.

Traceability

Traceability can be defined as the ability to establish the identity and history of a product, including its components and ingredients, back through each stage of the distribution and production chains, so that the source of a particular GMO can be traced. The benefits of establishing a traceability system are numerous (CEC, 2000). First, it facilitates the targeted withdrawal of products where a risk to human health or the environment is revealed. Secondly, it enables one to identify and monitor effects (both positive and negative) on human health and the environment. Thirdly, as seen above, it facilitates labelling and the control of labelling claims. And, finally, it can be seen as a prerequisite for the creation of an effective liability regime. In risk terms, therefore, traceability is principally concerned with the *ex post* regulation of risk, with the exception of labelling, where it facilitates *ex ante* choice on risk by individuals.

At present, only Council Directive 2001/18/EC contains any provisions on traceability for GMOs or derived products. However, although Article 4(6) of the Directive requires Member States to ensure traceability at all stages of the placing on the market for GMOs, it contains little detail on exactly what is required and it only applies to products containing or consisting of GMOs (CEC, 2001a). The scope of the Directive does not, in other words, extend to products produced from GMOs. With this in mind, the Commission issued a working document on traceability and labelling (CEC, 2000), which was followed by a proposal for a Regulation on the subject (CEC, 2001a, 2002a). When the proposal is adopted, the traceability requirements in Council Directive 2001/18/EC will be repealed and a new detailed, documentary, traceability regime will apply to both GMOs and products produced from GMOs. In essence, the proposal requires food and feed businesses to be able to identify other operators from whom products have been received and to transmit information concerning products to the next businesses in the chain (CEC, 2001a).

United States Objections to the Community Regulatory System

The regulatory approach of the Community is contested by the USA, which argues that certain aspects are incompatible with the rules of the international trading system. The final part of this chapter outlines which aspects of the Community's regulatory approach the USA argues are in breach of international trade law. Some consideration will also be given as to the impact of the 2000 Cartagena Biosafety Protocol on these issues.

While in danger of oversimplifying the matter, the USA has three principal concerns. First, there is the inclusion within the regulatory system of provisions that entitle Member States to take decisions not on the basis of the available scientific evidence but in an attempt to adopt a more precautionary approach for reasons of domestic politics. The two key provisions in this regard are, first, the regulatory committee procedure that entitles a qualified majority of the Council to reject a proposed individual authorization – potentially on grounds of precaution rather than scientific assessment – and, secondly, that which allows individual Member States provisionally to restrict the use and/or sale of an already authorized GMO (the 'safeguard clause' under both Council Directive 90/220/EEC and Council Directive 2001/18/EC).

The second US concern is the continuing political moratorium adopted by a number of Member States on the granting of any authorization until such time as the Community has adopted adequate labelling and traceability legislation. To what extent can the Member States implement a 'blanket' ban on authorizations, regardless of the individual circumstances of each application?

And, thirdly, Community requirements on labelling and tracing of GMOs and GMO-derived products are significantly stricter than in the USA, forcing US exporters to change their market behaviour if they are to gain access to the Community market. The USA argues that genetic modification *per se* is an insufficient reason to justify mandatory labels.

Compatibility of the Community System with WTO Rules

Any dispute between the USA and the Community over the latter's regulatory approach to GMOs and GM products would ultimately be decided according to the rules of the WTO, which determine the legality of national and regional barriers to international trade. The three most important treaties are the Agreement on the Application of Sanitary and Phytosanitary Measures (SPS Agreement), the Agreement on Technical Barriers to Trade (TBT Agreement) and the GATT. While the GATT is in many ways the 'parent' treaty, any trade dispute as regards GMOs is much more likely to involve the SPS and TBT Agreements.

The SPS Agreement

The purpose of the SPS Agreement is to regulate when – and how – a Member State can adopt sanitary and phytosanitary measures. The SPS Agreement seeks to elaborate upon the GATT, noting that 'measures which conform to the relevant provisions of this Agreement shall be presumed to be in accordance with' the GATT. The key obligation in the SPS Agreement is Article 2.2:

> Members shall ensure that any sanitary or phytosanitary measure is applied only to the extent necessary to protect human, animal or plant life or health, is based on scientific principles and is not maintained without sufficient scientific evidence.

It is not sufficient for a Member State to say that a measure is non-discriminatory; a Member State must also be able to prove that the measure itself is scientifically justifiable. And, whereas sanitary or phytosanitary measures that conform to 'international standards, guidelines or recommendations' will be 'presumed to be consistent' with the SPS Agreement, measures adopted by a Member State that achieve 'a higher level of . . . protection' must be the result of a risk assessment as required under Article 5 of the SPS Agreement. Failure to conduct a properly constituted risk assessment will invalidate the measure. Of central importance is the requirement that a 'rational or objective relationship' must be proved between the contested measure and the scientific evidence, as made clear by the Appellate Body in *Japan – Measures Affecting Agricultural*

Products (1999). This, of course, does not mean there has to be consensus as to the nature and degree of the risk, as a 'divergent opinion coming from qualified and respected sources' may be sufficient. However, the scientific research relied upon must be more than 'relevant': it must be 'sufficiently specific to the case at hand', as stated by the Appellate Body in the *Beef Hormones Dispute* (1998).

As regards the level of risk that must be shown under a risk assessment, in the *Beef Hormones Dispute* the Appellate Body drew a distinction between 'ascertainable' risks and 'theoretical uncertain[ties]', with only the former satisfying a risk assessment under Article 5.1. This, of course, has significant implications for the application of a precautionary approach to scientific uncertainty within the SPS Agreement. Both the original panel and the Appellate Body in that case were asked by the Community to respect the precautionary principle in its application of the SPS Agreement. In its decision, the Appellate Body held that whatever its status in international law, the precautionary principle could not overrule the explicit obligations placed upon Member States by the SPS Agreement. Nevertheless, the Appellate Body did note that the 'precautionary principle . . . finds reflection in Article 5.7', which permits the adoption of provisional measures where 'relevant scientific evidence is insufficient'. However, as Article 5.7 states, such provisional measures must be based on available pertinent information, and can only be maintained if the Member State that adopts the measure 'seek[s] to obtain the additional information necessary for a more objective assessment of risk' and 'review[s] the . . . measure . . . within a reasonable period of time'. Article 5.7 is therefore only a temporary derogation from the usual rule.

While there remains some uncertainty as to whether or not the scope of the SPS Agreement includes GMOs (Eggers and Mackenzie, 2000), it is clear that any Member State that intends to adopt permanent measures to regulate GMOs and GM foods on the grounds of sanitary and phytosanitary protection will be under an obligation to prepare a risk assessment that is sufficiently scientifically robust to withstand international scrutiny. This may prove to be a significant hurdle for those Member States concerned about the health and/or phytosanitary implications of GMOs and GM products. Any measures adopted will need to be supported by at least some credible scientific research. Of particular interest is how a dispute settlement panel will interpret the view taken by earlier Appellate Body decisions that 'divergent opinion coming from qualified and respected sources' may be sufficient to establish an actionable risk. Finally, one must not forget that undertaking a risk assessment is only one aspect of the SPS Agreement. In addition '[o]nce a . . . measure has passed the science test . . . it would still need to comply with [the requirements that it] is not more trade restrictive than necessary, is consistent with comparable regulations, and shall be taken without undue delay' (Eggers and Mackenzie, 2000).

Particular features of the Community regulatory system, including the safeguard clause, as well as the current moratorium, may well be in violation of the SPS Agreement. First, the ability of Member States to refuse an authorization on grounds of precaution, rather than 'hard science', appears to be strictly limited under the SPS Agreement. Allowing the final decision within the regulatory committee procedure to be taken by political representatives of Member States within the European Union Council introduces – the USA would suggest – an unnecessary political element into what should be a purely scientific decision-making process. The politicization of the procedure is therefore, according to this view, a fundamental flaw in the Community's approach. In response, the Community may argue that, while science can determine the risk, 'judging what is an "acceptable" level of risk for society is a political responsibility' (CEC, 1999b). Any WTO dispute settlement panel examining the compatibility of Community legislation with the SPS Agreement would have to decide whether the Community reliance on political participation is valid under international trade law.

Similar problems also arise when individual Member States utilize the 'safeguard clause' to restrict the marketing of an approved GMO within their territory where the scientific assessment has not shown there to be a risk but, for political reasons, Member States nevertheless adopt restrictive measures. As noted above, there is some evidence that the Community's own scientific committees dispute the basis of many of the measures undertaken under this provision. In itself, the 'safeguard clause' is not a violation of the SPS Agreement, as both concern temporary derogations from the normal risk-assessment procedure. What is more contentious is where Member States have relied upon the safeguard clause to adopt precautionary measures that are more permanent in nature. As the WTO Appellate Body has emphasized in *Japan – Measures Affecting Agricultural Products* (1999), Article 5.7 requires such measures to be subject to a 'more objective risk assessment' within a reasonable period of time in the light of any new additional information. The failure of the Community to comply with its own legislation by ensuring that the regulatory committee procedure meets to determine whether individual derogatory measures are compatible with the available science and, if not, to require their withdrawal is a serious issue. It would certainly seem unlikely that the present situation where a number of individual derogations have become *de facto* permanent are compatible with the SPS Agreement.

A second likely infringement of the SPS Agreement is the current moratorium on processing applications for marketing authorizations. With no formal scientific basis, the moratorium runs counter to all the basic precepts of the SPS Agreement. Moreover, a 'blanket' ban – as this moratorium is – infringes the jurisprudence of the Appellate Body. Thus, in *Japan – Measures Affecting Agricultural Products* (1999) it required a case-by-case analysis of individual sanitary and phytosanitary risks.

There seems little that the Community can say to justify the moratorium imposed by a majority of its Members, particularly as it is, in addition, a violation of the Community legislation, which similarly requires a case-by-case analysis. Arguments justifying the moratorium based on the potential risk GMOs pose are weakened by the lack of risk assessment undertaken and the blatant political nature of the moratorium.

Labelling and the WTO rules

In trade circles, there are often two positions on the role of labelling as regards the promotion of non-trade objectives, such as environmental protection. First, there are those who consider labelling, if not an innocuous form of national regulation, to be 'less trade-restrictive' than more direct forms of state intervention. The alternative viewpoint is that national labelling schemes are a 'disguised restriction on trade', which are intrinsically more likely to promote the sale of domestic goods and which disadvantage foreign producers, who are less likely to have been consulted during the process of devising the labelling scheme. As regards GM labelling, much will depend upon the reason for the labelling. If the labels are allegedly required for reasons of sanitary and phytosanitary protection, the question of their validity will be decided by the SPS Agreement. As was noted above, vague and unproved assertions about the potential negative effects of GMOs and GM foods will be insufficient. Only labels that highlight 'ascertainable' physical risks (such as causing allergic reactions in susceptible people) are likely to prove acceptable.

Most of the debate as regards GM labelling, however, takes place within the scope of the TBT Agreement. The purpose of the TBT Agreement is to prevent the preparation, adoption or application of regulatory standards and technical specifications so as to create unnecessary obstacles to trade. As Article 2.2 states, 'technical regulations shall not be more trade-restrictive than necessary to fulfil a legitimate objective'. Labelling is considered a form of technical regulation or standard. The compatibility of Community labelling legislation with the TBT Agreement requires the surmounting of numerous 'hurdles'.

First, to what extent are the reasons most often espoused for adopting labelling of GMOs and GMO derived products (namely, environmental protection and allowing consumers to make an informed choice) compatible with the TBT Agreement? The key issue is how the concept of 'legitimate objective' is defined. Article 2.2 of the TBT Agreement provides a non-exhaustive list of possible legitimate objectives, including 'national security . . . protection of public health . . . or the environment'. The TBT Agreement does, therefore, allow national measures on the grounds of environmental protection, though the level of proof required is unclear as there is no explicit requirement for Member States under the TBT Agreement to undertake a risk assessment.

An alternative basis for justifying labelling is that a consumer's right to information on the risk of GMOs and GMO-derived products is a legitimate objective under the TBT Agreement. There is no mention of such a right in the TBT Agreement list of legitimate objectives and, while Article 2.2 is a non-exhaustive list and may include such an objective, the USA has said that, if there is no scientific risk posed by GMOs and GMO-derived products that is distinct from the conventional counterparts, '[w]e therefore have questions about what the EC's legitimate objectives are in respect of providing "proper information to the final consumer"' (WTO, 1998).

The second issue relating to the legality of GM labelling is the requirement under the TBT Agreement that any national measure (including labelling) must satisfy the twin non-discrimination principles of 'most favoured nation' and 'national treatment'. If national measures are discriminatory, they are subject to the more restrictive requirements of Article XX of the GATT. Discrimination will have occurred if 'like products' have been treated differently. This concept of 'like product' is not defined within the WTO agreements, and its meaning has largely been the result of judicial interpretation. While not universally accepted, one of the most understood aspects of the concept is that the 'like'-ness of any two products must be determined by their characteristics as products *per se*, and not by any differences in their manufacture or production. This was the clear message from the *Tuna/Dolphin Dispute* (1991), a pre-WTO case, where the panel held that '[r]egulations governing the taking of dolphins incidental to the taking of tuna could not possibly affect tuna as a product'. Such process and production methods (PPMs), according to this interpretation of the WTO disciplines, are to be disregarded.

Are GMOs and GMO-derived products 'like' their conventional counterparts? There is a significant difference in opinion between the USA and the Community. Whereas, as noted above, current Community law considers GMOs and GMO-derived products not to be 'equivalent' in their characteristics or food property if they contain novel DNA or protein (so justifying labelling *per se*), the USA argues that labelling should be restricted to cases where physical harm may occur (for instance, where a GM food contains a new allergen). As the USA notes:

> [t]he mere presence of protein or DNA resulting from genetic modification is not sufficient to establish that a food is no longer equivalent to its conventional counterpart in terms of its 'composition, nutritional value or nutritional effects or the intended use of the food'.
>
> (WTO, 1998)

Critics of the US approach argue that they are interpreting 'like product' too narrowly and note that numerous panels and the Appellate Body have pointed out that 'like'-ness should be determined on a 'case-by-case basis', taking into account both consumers' tastes and habits, which change from country to country, and the product's 'properties, nature and quality' (as in *Japan – Taxes on Alcoholic Beverages* (1996)), as well as 'those physical properties of products that are likely to influence the competitive relationship between products in the marketplace' (as in the *Asbestos Dispute* (2001)). From this broader perspective, it is argued that it is difficult to consider GMOs and GMO-derived products as 'like' conventional counterparts.

Moreover, US concerns over Community labelling are only likely to increase, as the Community is currently revising its legislation so that the determining factor as regards GMO-derived products is no longer the existence of novel DNA or protein, but rather that the product has, somewhere along the production line, been subject to or affected by genetic modification. Such a change in approach would only serve to emphasize, in the view of the USA, that the Community system is a blatant violation of the WTO rules. Of particular concern would be the Community reliance on genetic modification as a distinguishing factor, because, as noted above, PPMs have traditionally been considered matters outside the remit of the notion of 'like product'.

A related aspect is the matter of traceability, whereby GMO and GMO-derived product manufacturers will be required to prove whether their goods have been subject to or derived from genetic modification. In terms of international trade law, the issue is how far the Community can impose such an obligation upon non-Community importers. The documents issued by the Commission on traceability have so far been ambivalent on this question. On the one hand, the Commission notes that '[a]n EU [European Union] traceabilty system cannot be imposed on trading partners' (CEC, 2000). However, in a more recent document it states that '[o]perators importing products into the Community would . . . have to follow the requirements . . . and transmit the information that the product is derived from GMO(s)' (CEC, 2001b). Of itself, requiring documentation prior to the importation and/or authorization of a product is not unlawful *per se*. It becomes controversial, however, when the request for documentation imposes a heavier burden upon an importer than upon a domestic supplier, or where the information required falls outside the exemptions permitted by the WTO. So, once again, the GM debate returns to the issue of science and risk – why is such information necessary when, according to the USA, GMOs and GMO-derived products are comparable to conventional material? How the WTO resolves the issue of risk will go a long way in determining the legality of more specific issues within the genetic-modification trade debate.

Biosafety Protocol and its Relationship with the WTO

The negotiation and adoption of the 2000 Cartagena Biosafety Protocol has introduced a further element into this debate. The objective of the Protocol may be environmental in focus, but it is very definitely trade-related in its application. Any environmental agreement that seeks to regulate imports is inevitably going to be sharing legal territory with the WTO disciplines. The possibility for conflict is therefore very real, as a brief summary of the provisions of the Protocol will show.

There are a number of misconceptions about the Biosafety Protocol. One of the most important is that of its coverage. First, it only covers living GMOs. It therefore has nothing to say on GMO-derived products. Moreover, further distinctions are made between what it terms living modified organisms (LMOs) that are intended for 'intentional introduction into the environment' (e.g. seeds) and LMOs intended for direct use as food or feed or for processing (LMO-FFPs) (for example, GM tomatoes). Both types of LMOs require, under the Protocol, the prior informed consent of the importing state, though the procedure for LMO-FFPs is somewhat simpler as the potential environmental risks posed by LMO-FFPs are fewer and the commercial importance of trade in LMO-FFPs meant that a more stringent regime was unlikely. LMOs intended for intentional introduction into the environment are subject to Advance Informed Agreement (AIA), whereby before the 'first intentional transboundary movement' of such an LMO into the importing state, the national authorities must decide whether to permit importation on the basis of a risk assessment 'carried out in a scientifically sound and transparent manner' in accordance with Annex III to the Protocol. As one commentary notes, the 'SPS [Agreement] does not spell out what a risk assessment entails, but the Protocol does so in detail' in Annex III (Cosbey and Burgiel, 2000).

On one level, therefore, the AIA procedure seems to 'flesh out' the WTO rules and may provide assistance in future interpretations of what a risk assessment should involve. On another level, however, the AIA procedure raises significant tensions between the Protocol and WTO rules, in particular the fact that the Biosafety Protocol explicitly allows, under the risk-assessment procedure, a Party to adopt a precautionary approach in relation to scientific uncertainty. Unlike the SPS Agreement, which permits such action only as a temporary derogation, Articles 10.6 and 11.8 of the Protocol state:

> [l]ack of scientific certainty due to insufficient relevant scientific information and knowledge regarding the extent of the potential adverse effect of a living modified organism on the conservation and sustainable use of biological diversity in the Party of import, taking also into account risks to human health, shall not prevent that Party from taking a decision, as appropriate, with regard to the import of the living modified organism in question . . . in order to avoid or minimize such potential adverse effects.

The explicit inclusion of such a precautionary approach is in direct contrast with the emphasis on 'sound science' in the WTO. How the WTO will deal with the existence of inconsistent international – as compared with national – law remains uncertain.

Another provision of the Protocol which will be of concern is Article 18 on labelling. Of particular significance is the requirement that shipments of LMO-FFPs are labelled with the phrase that they 'may contain' LMOs and that they are not intended for intentional introduction into the environment. When the Biosafety Protocol enters into force, Article 18.2(a) also requires the Parties to consider the necessity of further rules relating to 'specification of their identity and any unique identification', namely, some form of traceability scheme. Legally, there is the issue as to whether the Protocol's provisions, in particular the 'may contain' clause, are compatible with either the SPS or TBT Agreements. Whereas the Protocol suggests that genetic modification is a sufficiently distinguishing factor in itself to justify labelling, the WTO rules appear to require more objective justification before labelling is permitted.

Once again, the Community will seek to argue that the Protocol justifies its own position on this issue. In particular, the Community will welcome the general thrust of the Protocol, which treats labelling as an acceptable – and legally valid – approach to the management of GMOs. Of some concern, however, may be the Protocol's rules on LMO-FFPs. Under Community law, the marketing of live GM products (e.g. tomatoes) currently requires a label stating that such material contains GMOs (which presupposes segregation from non-GMOs). The Protocol, however, is less restrictive, permitting the label 'may contain', thus not requiring segregation and thereby reducing transaction costs. It may become more difficult for the Community to justify its current approach if it is shown to go beyond that required by the Protocol.

Notwithstanding the potential inconsistencies that may exist between the Community legislation and the Protocol, an inevitably more important issue for the Community is whether the Protocol can be utilized to support its position in any trade dispute within the WTO. If there are differences between WTO rules and the Protocol as suggested, how will they be resolved? Both the WTO disciplines and the Protocol (when it enters into force) are valid treaties under international law. The Protocol recognizes this potential for conflict, though does little to mitigate its effect. The Preamble to the Protocol contains two recitals, which, when read together, merely highlight the political and legal tensions rather than doing anything to resolve them (www.iisd.org/pdf/biosafety.pdf.)

> Emphasizing that this Protocol shall not be interpreted as implying a change in the rights and obligations of a Party under any existing international agreements.

> Understanding that the above recital is not intended to subordinate this Protocol to other international agreements.

With no clear hierarchy, one is tempted to resort to traditional rules of treaty interpretation, namely, that, where there exist rules in different international agreements that are contradictory, one must first determine whether the Parties to the dispute are bound by both agreements. If not, the rule to which both are bound is the one that is legally binding. But, if the Parties have ratified both agreements, other rules of interpretation come into play, such as that the later in time prevails or the more specific rule prevails. As should become immediately apparent, these rules of interpretation are of little help in that they may point in different directions as to the correct result. To take the likely example of a dispute between the USA and the Community: whereas the Community has signed and ratified the Protocol, the USA cannot sign the Protocol because it has yet to ratify the Protocol's parent treaty, the 1992 Convention on Biological Diversity. Therefore, if the Community relies upon the Protocol in any international dispute, the USA could justifiably argue that the provisions of the Protocol are not binding upon it and therefore the two sides should be bound only by what are, namely the WTO disciplines.

But to approach the issue simply in terms of treaty interpretation undermines a number of aspects of the wider debate. As well as asking what is the applicable law in any dispute, one should also enquire as to how any dispute would be resolved. In a dispute between a Party and a non-Party to the Protocol, the dispute settlement procedure of the WTO seems the most appropriate forum as the WTO rules are the only legal obligations that both accept. However, even a dispute determined strictly under the SPS and/or TBT Agreements would inevitably have to recognize the importance of the Protocol as the most detailed international attempt yet to regulate GMOs. It would be extremely politically difficult for a WTO dispute settlement panel to rule that another agreement validly adopted under international law is incompatible with the rules of the WTO. Juridical subtlety would be the order of the day. A panel might find it useful to draw a distinction between the Protocol *per se* and a Party's implementation of the Protocol. It will be much easier for any WTO panel to rule, for example, on the legality of an individual state's decision as to whether or not to permit certain imports under the AIA procedure than to question the AIA procedure itself.

In this way, a WTO panel may well find synergies between the two systems, in particular the requirement of a risk assessment under both the SPS Agreement and the Protocol. However, the real issue is not where the systems are similar, but where they are different. In particular, the way the two systems deal with precautionary measures will be difficult for even the most imaginative WTO panel to reconcile. The Protocol and the WTO part company here; only the most optimistic would be able to find synergies on this issue. Many, including the Community, have welcomed the adoption of the Protocol, for they believe that its inclusion of a more precautionary approach to scientific uncertainty has fundamentally altered the balance between the WTO disciplines and environmental concerns, in favour of the latter. However, this remains a hypothesis that has yet to be tested.

Conclusion

The increasing commercial importance of GMOs and GMO-derived products, in particular the expanding international trade in such material, raises fundamental questions as to the relationship between the liberalization of the global economy and the regulation of risk. This chapter has examined the legal implications of a situation where different states have reacted differently to the same scientific development. The genetic-modification issue clearly highlights these differences and the consequent tensions between the imperative of global trade and the need for national regulation. Any 'legal' dispute arising from this would be decided according to the disciplines of the WTO. It is a strength of a rules-based international trading system that matters of great economic importance are resolved through political negotiation and judicial settlement, rather than conflict and unilateral sanctions. The benefits of such a rules-based system are to be commended and protected. However, the present system has yet to find an appropriate balance between the promotion of trade liberalization, on the one hand, and the protection of non-trade public policy objectives, on the other.

List of Cases

Case T-13/99, *Pfizer Animal Health SA* v. *Council* (Court of First Instance), judgment delivered 11 September 2002.

Case T-70/99, *Alpharma Inc.* v. *Council* (Court of First Instance), judgement delivered 11 September 2002.

EC Measures Concerning Meat and Meat Products (Hormones) (1998), WT/DS26/AB/R and WT/DS48/AB/R.

EC – Measures Affecting Asbestos and Asbestos-containing Products (2001), WT/DA/135/AB/R.

Japan – Taxes on Alcoholic Beverages (1996), WT/DS8/AB/R, WT/DS10/AB/R and WT/DS11/AB/R.

Japan – Measures Affecting Agricultural Products (1999), WT/DS76/AB/R.

United States – Restrictions on Imports of Tuna (1991) (30 ILM (1991) 1594).

United States – Import Prohibition of Certain Shrimp and Shrimp Products (1998), WT/DS58/AB/R.

United States – Import Prohibition of Certain Shrimp and Shrimp Products: Recourse to Article 21.5 of the DSU by Malaysia (2001), WT/DS58/AB/RW.

References

CEC (1999a) *White Paper on Food Safety*. COM (1999) 719.

CEC (1999b) *Communication for the Commission on the Precautionary Principle*. COM (2000) 1.

CEC (2000) *Working Document of the Commission Services on Traceability and Labelling of GMOs and Products Derived from GMOs*. ENV/620/2000.

CEC (2001a) *Proposal for a Regulation of the European Parliament and of the Council Concerning Traceability and Labelling of Genetically Modified Organisms and Traceability of Food and Feed Products Produced from Genetically Modified Organisms and Amending Directive 2001/18/EC*. COM (2001) 182.

CEC (2001b) *Proposal for a Regulation of the European Parliament and of the Council on Genetically Modified Food and Feed*. COM (2001) 425.

CEC (2002a) *Amended Proposal for a Regulation of the European Parliament and of the Council Concerning Traceability and Labelling of Genetically Modified Organisms and Traceability of Food and Feed Products Produced from Genetically Modified Organisms and Amending Directive 2001/18/EC*. COM (2002) 515.

CEC (2002b) *Amended Proposal for a Regulation of the European Parliament and of the Council on Genetically Modified Food and Feed*. COM (2002) 559.

Cosbey, A. and Burgiel, S. (2000) *The Cartagena Protocol on Biosafety: an Analysis of Results*. Briefing Paper for the International Institute for Sustainable Development Winnipeg, Manitoba. (On file with author).

Council of the European Union (1999) *Official Minutes of the Environmental Council of the EU (24 June 1999)*. Council, Brussels.

Council of the European Union (2001) *Statements for the Minutes (6068/01, 16 February 2001)*. Council, Brussels.

DETR (Department of the Environment, Transport and the Regions) (1999a) *Advisory Committee on Releases to the Environment: Risk Assessment for Release and Marketing of GMOs in the European Union*. DETR, London.

DETR (1999b) *Advisory Committee on Releases to the Environment: Proposal for a Directive of the European Parliament and of the Council Amending Council Directive 90/220/EEC on the Deliberate Release into the Environment of Genetically Modified Organisms: Key Issues and Developments*. DETR, London.

European Community (2000) *Commission Take Initiative to Restore Confidence in GMO Approval Process*. IP/00/778, Brussels, 13 July.

European Community (2002) *Questions and Answers on the Regulation of GMOs in the EU*. MEMO/02/160, Brussels, 15 October.

Eggers, B. and Mackenzie, R. (2000) The Cartagena Protocol on Biosafety. *Journal of International Economic Law* 3, 525–543.

House of Lords, Select Committee on the European Union (2002) *Twenty-second Report of the House of Lords Select Committee on the European Union: Labelling and Tracing of GM Food and Animal Feed: Informing the Consumer* Session 2001–2002, HL117.

MAFF (Ministry of Agriculture, Fisheries and Food) (1999) *Guidance Notes: Novel Foods and Novel Food Ingredients Legislation*. MAFF, London.

US Submission (1998) *European Council Regulation No. 1139/98 Compulsory Indication of the Labeling of Certain Foodstuffs Produced from Genetically Modified Organisms: Submission by the United States*. G/TBT/W/94, 16 October.

International Trade in Genetically Altered Agricultural Products: Impact of the Biosafety Protocol

10

Norman W. Thorson*

Introduction

Historically, agricultural trade policy focused largely on the extent to which agricultural products should be subject to international trade disciplines established by the General Agreement on Tariffs and Trade (GATT). International trade in agricultural commodities is a multibillion-dollar industry with highly visible benefits to the world community. At the same time, one of the most enduring and universal of economic phenomena has been the persistence of economic subsidies and trade barriers designed to protect domestic food producers. This fundamental conflict between agricultural protectionism and the benefits of free trade in agricultural products has long been a contentious issue, even before the complications brought about by the advent of modern biotechnology (Schoenbaum, 1993).

During the 1994 Uruguay Round, negotiators agreed to apply GATT free-trade principles to agricultural products within the context of the new World Trade Organization (WTO). While this was a seminal event in the law of international trade, trade in agricultural products remains highly regulated. Whether individual regulations serve a legitimate state interest or whether they are merely disguised barriers to trade is often an open question. Few interests are more fundamental to a nation's *raison d'être* than that of securing and distributing adequate supplies of healthful

* Judge Harry A. Spencer, Professor of Law, University of Nebraska. BS 1971; MS 1974; JD 1977; PhD 1979, Nebraska.

 Agriculture and International Trade (eds M.N. Cardwell, M.R. Grossman and C.P. Rodgers) and nutritious food. At the same time, however, international trade in agricultural commodities can be thwarted by the application of inconsistent health and safety standards at national borders. The WTO agreements incorporate an elaborate set of rules that attempt to reconcile the right of individual states to determine the level of protection that they wish to provide their citizens with the need of the international trading system for some level of uniformity. Applying the WTO rules to genetically modified (GM) agricultural products presents significant challenges. The difficulty is enhanced by the possibility that trade in GM agricultural products presents environmental risks wholly apart from any risks related to food safety.

Debate over the promise and peril of bioengineered agricultural products has been spirited (Krimsky and Wrubel, 1996). Proponents argue that genetic manipulation can increase yields and lower per-unit production costs, helping to alleviate world hunger (Buechle, 2001). Moreover, they argue, genetic modification has a potential to diminish negative environmental and energy impacts of production agriculture (Kershen, 2002). The need for fertilizers, herbicides and pesticides can be reduced; the conversion of environmentally sensitive lands to cropland can be arrested or slowed; and the scope of animal waste-disposal problems can be diminished. Finally, they argue, modifying organisms so they can thrive under hostile climatic conditions, such as short growing seasons or low water supplies, could make agricultural production more feasible in food-short areas of the globe.

Opponents of bioengineered products, on the other hand, raise two distinct safety concerns (Lambrecht, 2001; Teitel and Wilson, 2001; Hart, 2002; Mendelson, 2002). First, critics argue that consumption of GM organisms (GMOs) poses largely unquantified, but currently unacceptable health risks, including the possibility of allergenic or toxic reactions among susceptible portions of the population. They point out that insufficient time has passed to assess the long-term consequences of consumption of GMOs. Secondly, critics argue that potential environmental benefits are often overstated and unproved, and that release of GMOs into the environment poses serious threats to environmental quality. Among the concerns raised are the possible evolution of super-resistant pests, inadvertent harm to non-target species, transfer of herbicide resistance to weeds through natural cross-pollination and other events that have the potential to seriously affect global biological diversity.

The ethical and economic implications of genetic manipulation are also hotly debated (Coletta, 2000; Székely, 2001). Depending on one's perspective, genetic manipulation is morally reprehensible (it is immoral to transfer genes between species where natural cross-breeding is impossible), morally neutral (genetic manipulation is merely a consequence of normal scientific progress and is, by itself, devoid of moral content) or morally imperative (one is morally obligated to pursue genetic modification because it has the potential to alleviate the human suffering that flows from malnutrition and starvation in many parts of the globe). Clearly and inevitably, however, policy decisions taken in regard to genetic manipulation will produce economic winners and losers. Economic fallout from biotechnology will affect politically charged issues, including changes in structural patterns of agricultural production, accelerating trends toward specialization and changes in concentration and vertical integration.

Agricultural trade in the products of modern biotechnology has become a test case for determining whether established rules of international trade and emerging rules of environmental protection can coexist peacefully. The first step towards balancing the two regimes occurred on 29 January 2000, when the first extraordinary meeting of the Conference of the Parties to the Convention on Biological Diversity adopted the Cartagena Protocol on Biosafety (Cartagena, 2000). The Biosafety Protocol is designed to promote the safe transfer, handling and use of living modified organisms (LMOs). LMOs are defined as 'any living organism that possesses a novel combination of genetic material obtained through the use of modern biotechnology' (Cartagena, Art. 3(g)). Language in the Preamble to the Protocol recognizes the great potential of modern biotechnology to enhance human well-being (Preamble, ¶ 7), but it also recognizes that release of LMOs may have adverse effects on human health and on the conservation and sustainable utilization of biological diversity (¶ 6). Hence, the Protocol was drafted to apply to the 'transboundary movement, transit, handling and use of all living modified organisms that may have adverse effects on the conservation and sustainable use of biological diversity, taking also into account risks to human health' (Art. 4).

The Biosafety Protocol incorporates two emerging principles of international environmental law, the principle of prior informed consent and the precautionary principle. Prior informed consent is an extension of the generally recognized duty under international law to provide prior notification of actions in one state that might cause harm in another state, and the corollary obligation to consult in good faith with potentially affected parties (Hunter *et al.*, 2002, ch. 7(II)(N)). The precautionary principle, in contrast, addresses the complex question of how much evidence of harm is needed to justify legal policy responses following discovery of information that suggests that harm may be associated with particular acts, but before development of scientific consensus that harm, or significant risk of harm, necessarily follows from the acts (Hunter *et al.*, 2002, ch. 7(II)(I)).

As an agreement that contemplates precautionary limits on free trade in GMOs, the Protocol potentially conflicts with the rules of the WTO. Moreover, the agreement may add further weight to a contention that GM agricultural products and their conventional counterparts are not 'like products' for WTO purposes. If LMOs and conventionally produced organisms are not like products, discriminatory treatment of LMOs might not be inconsistent with international trade regimes. Precisely how these potential conflicts will be resolved is still an open question.

The lack of international consensus on many of the underlying principles, the infancy of the agricultural biotechnology industry and the concomitant level of uncertainty regarding the degree of the various risks perceived posed serious challenges for negotiators of the Biosafety Protocol. Incorporation of the precautionary principle into the agreement recognizes traditional sovereign rights to determine environmental and natural-resources policy within a nation's borders. On the other hand, the critical importance of international agricultural trade, the difficulty in certifying products as free of genetic manipulation and the long history of economic protectionism for domestic agriculture suggest that it might be difficult to separate legitimate precaution from illegitimate protectionism. Moreover, even apart from the question of overt protectionism, negotiations put into focus the question of how much latitude nations ought to have in demanding 'proof of safety' in contrast to 'evidence of harm' when setting domestic policies in respect of international trade in the products of modern biotechnology.

The completed Biosafety Protocol allows nation states some discretion to set their own standards for what constitutes an acceptable level of risk of exposure to imported bioengineered products. It also imposes obligations on exporters of such products, with the nature of the obligations varying depending on the intended use of the exported products. At the same time, WTO agreements, which recognize that ostensible health and safety measures may be used as disguised trade barriers, were not made subordinate to the Protocol. But neither was the Protocol subordinated to international trade agreements, raising questions of how operational conflicts between the two regimes might be resolved.

Genesis and Purpose of the Biosafety Protocol

The Biosafety Protocol is specifically authorized, though not mandated, by Article 19(3) of the Convention on Biological Diversity (United Nations, 1992a). The Biodiversity Convention attempted to address, comprehensively, all aspects of biodiversity, including conservation of biological diversity, the sustainable use of the components of biodiversity and the fair and equitable sharing of the benefits of biodiversity (Art. 1). Article 2 of the Convention defines biotechnology as any technological application that uses biological systems, living organisms or derivatives thereof to make or modify products or processes for specific use. Such a broad definition of biotechnology includes practices that have existed for thousands of years, such as selective breeding and the use of microorganisms to make wine, beer, bread and cheese. However, the definition also encompasses the methods of modern biotechnology, including tissue culture, cell fusion, embryo transfers, recombinant DNA and other sophisticated genetic-engineering techniques.

The Convention recognizes that biodiversity has intrinsic value and that conservation of biodiversity is a common concern of humankind. At the same time, however, the Convention states that nations have sovereign rights over their own biological resources, although these rights are tempered by an obligation to conserve biodiversity and use biological resources in a sustainable manner (United Nations, 1992a, Preamble).

Definitions in the Convention are critical. Biological diversity, that which is to be conserved, is defined as 'the variability among living organisms from all sources including, *inter alia*, terrestrial, marine and other aquatic ecosystems and the ecological complexes of which they are part; this includes diversity within species, between species and of ecosystems' (Art. 2). In other words, biological diversity is an attribute of life. Biological resources, the components of biodiversity that are to be used in a sustainable manner, are defined as 'genetic resources, organisms or parts thereof, populations, or any other biotic component of ecosystems with actual or potential use or value for humanity' (Art. 2). Examples include seeds, genes, plants, animals and parts of plants and animals. Finally, the benefits to be shared equitably include access to genetic resources, access to the benefits derived from the use of genetic resources and transfer of technology, including biotechnology (Arts. 15–19).

Biosafety, however, was not a central theme of the Biodiversity Convention. Article 19(4) contains the Convention's only direct biosafety mandate. It obligates exporters of LMOs to share information regarding use and safety regulations that apply in the exporting state. Exporting nations are also required to disclose any information on potential adverse impacts of the particular organism to be introduced. Thus, the protection provided by Article 19(4) is incomplete at best.

Given the limited protection provided by Article 19(4), negotiation of a Biosafety Protocol was seen by many nations, particularly developing nations, as an essential additional component of a mechanism for sharing the benefits of biotechnology among the suppliers of genetic material and the developers of biotechnology. Developing countries, in particular, were concerned that they might be used as the unwitting proving grounds for biotechnological innovations that might pose a significant risk to local environments. Specifically, developing countries argued that products tested in one environment might not be suitable for release in a different environment. Further, even if products were generally safe under similar environmental conditions, they might not be appropriate given local conditions. The lack of capacity to evaluate products independently for safety or to police or monitor purchases by farmers exacerbated the problem.

The Biosafety Protocol was enacted to address many of the perceived limitations of the Biodiversity Convention. The final negotiations, which took place in Montreal, Canada, capped a tumultuous period when the future of the Protocol was very much in doubt. Eleven months earlier, the original negotiations in Cartagena, Colombia, had broken down in a bitter dispute over the scope of the Protocol. Much of the disagreement that stymied the original negotiations concerned issues relating to trade and the environment (Weiskopf, 1999). Not surprisingly, the final agreement contains language designed to appease all interests and hence is not a model of clarity.

The primary objective of the Protocol is set forth in Article 1:

> In accordance with the precautionary approach contained in Principle 15 of the Rio Declaration on Environment and Development, the objective of this Protocol is to contribute to ensuring an adequate level of protection in the field of the safe transfer, handling and use of living modified organisms resulting from modern biotechnology that may have adverse effects on the conservation and sustainable use of biological diversity, taking also into account risks to human health, and specifically focusing on transboundary movements.
>
> (Cartagena, 2000, Art. 1)

Several observations are in order. First, the Protocol is concerned only with a subset of LMOs, those resulting from modern biotechnology, a term that excludes modification by traditional means such as plant breeding. Although the term living modified organism was not defined in the Convention on Biological Diversity, it was intended to have a broader reach than the term genetically modified organism and is broad enough to include living organisms modified by traditional means such as plant breeding or artificial insemination (Glowka *et al.*, 1994, p. 45). But the Protocol uses a three-step definition to limit the scope of the term. First, living modified organism is defined generally as 'any *living organism* that possesses a novel combination of genetic material obtained through the use of *modern biotechnology*' (Art. 3(g)). Second, a living organism is defined broadly as 'any biological entity capable of transferring or replicating genetic material, including sterile organisms, viruses and viroids' (Art. 3(h)). Finally, the broad scope of the term and the reach of the Protocol are limited by the definition of modern biotechnology. Modern biotechnology is limited to:

> *in vitro* nucleic acid techniques, including recombinant deoxyribonucleic acid (DNA) and direct injection of nucleic acid into cells or organelles, or fusion of cells beyond the taxonomic family, that overcome natural physiological reproductive or recombinant barriers and that are not techniques used in traditional breeding and selection.
>
> (Art. 3(i))

Secondly, the Protocol specifically references the precautionary approach as articulated in the Rio Declaration. The Preamble to the Convention on Biodiversity provides: '*Noting also* that where there is a threat of significant reduction or loss of biological diversity, lack of full scientific certainty should not be used as a reason for postponing measures to avoid or minimize such a threat' (Preamble, ¶ 9). Principle 15 of the Rio Declaration, referenced in the Biosafety Protocol, uses a slightly different articulation of the precautionary principle: 'Where there are threats of serious or irreversible damage, lack of full scientific certainty shall not be used as a reason for postponing *cost-effective* measures to prevent environmental degradation' (United Nations, 1992b, Principle 15, emphasis added).

Thirdly, to the extent that they are regulated by other international agreements and organizations, the Protocol does not apply to the transboundary movement of live modified organisms that are used as pharmaceuticals for humans (Art. 5).

Finally, the Protocol must be understood in the context of the Convention on Biological Diversity. A Protocol to a Convention is binding only on those parties to the Convention that ratify the Protocol. Thus, a party to the Convention is not subject to the Protocol until the Protocol is ratified separately. Furthermore, the Biosafety Protocol is open to signature only by the parties to the Convention on Biodiversity. The USA, with its huge interest in agricultural biotechnology, is currently ineligible to become a member of the Protocol because it is not a party to the Convention. Although the USA belatedly signed the Biodiversity Convention, it has not ratified it and therefore is not a party eligible to join the Protocol. A nation that has signed, but not ratified a treaty, is not bound by the terms of the document, but it is subject to an obligation not to act in ways that would defeat the object and purpose of the treaty (Vienna, 1969, Art. 18). Although the USA actively participated in the discussions that led to the adoption of the Protocol, its ability to participate meaningfully in future developments has been compromised by its failure to ratify the Biodiversity Convention.

Conceptually, the Biosafety Protocol distinguishes between LMOs that are to be introduced intentionally into the environment of the importing country and GM commodities that are intended for direct use as food or feed or for processing (Cartagena, 2000, Arts. 7–9, 10, 12). Those living organisms, such as seed or trees, intended for release into the environment are subject to a prior informed consent restriction, which the Protocol denominates the Advanced Informed Agreement Procedure (Art. 7). Commodities destined for food, feed or processing are not subject to prior informed consent requirements, but they are subject to labelling requirements and their import may be banned by countries in manners consistent with the agreement (Art. 11). Normally, an import ban would need to be supported by a science-based risk assessment.

The extent to which the Advanced Informed Agreement Procedure would apply to all transborder movements of LMOs was a matter of considerable debate at negotiating sessions (Pomerance, 2000; Schweizer, 2000). Some producing nations argued that the Biosafety Protocol should not apply to imports of products to be used for food, feed or processing.

Other nations, particularly developing nations, argued that all shipments should be subject to the same rules because there was no way of ensuring that commodities imported for food or feed might not be introduced into the environment, either intentionally or inadvertently. The eventual compromise was the dual-track system of regulation. The bifurcated approach recognizes that the most direct threat to biodiversity comes from the intentional introduction of LMOs into the environment; consuming the products of modern biotechnology raises a different set of issues.

The potential conflict between the trade-related provisions of the Biosafety Protocol and pre-existing international trade agreements was also a matter of great debate during negotiating sessions (Safrin, 2002). The compromise reached was incorporated into the final three paragraphs of the Preamble to the Protocol:

> *Recognizing* that trade and environmental agreements should be mutually supportive with a view to achieving sustainable development,
>
> *Emphasizing* that this Protocol should not be interpreted as implying a change in the rights and obligations of a Party under any existing international agreements,
>
> *Understanding* that the above recital is not intended to subordinate this Protocol to other international agreements.

In other words, trade and environmental agreements are both aimed at achieving sustainable development; the Biosafety Protocol does not amend pre-existing trade agreements; neither is it subordinated to pre-existing trade agreements. Without this rather convoluted set of savings clauses, the Biosafety Protocol would take precedence over conflicting WTO Agreements in a dispute between parties under two theories. First, in the event of a conflict, the agreement later in time usually receives preference. Secondly, in the event of a conflict, the most specifically applicable treaty is generally held to be controlling (Vienna, 1969, Art. 59(1)). In any event, whether or not the trade-related provisions of the Biosafety Protocol can be reconciled with the trade provisions of the WTO has important implications, not only for international trade in GMOs, but also for the economic future of GM agricultural products.

Essential Provisions of the Biosafety Protocol

General obligation to regulate

The Convention on Biodiversity imposes a general obligation to regulate, manage and control risks associated with the use and release of LMOs (United Nations, 1992a, Art. 8(g)). The Biosafety Protocol expands on this obligation by charging all parties with a duty to ensure that the development, handling, transport, use, transfer and release of any LMOs are undertaken in a manner that prevents or reduces the risks to biodiversity, also taking into account risks to human health (Cartagena, 2000, Arts. 2(2), 16). The Protocol also requires parties to take all appropriate measures to prevent unintentional transboundary movements of LMOs (Art. 16(3)).

Establishment of a Biosafety Clearing-house

The Protocol establishes a Biosafety Clearing-house to facilitate the exchange of scientific, technical, environmental and legal information on and experience with LMOs (Cartagena, 2000, Art. 20). Potentially, this exchange of information is particularly valuable for developing countries, which may lack the resources and capacity to develop information on their own. In addition to facilitating a general exchange of information, the Clearing-house will also play an important role in the implementation of the notice provisions of the Protocol. Parties that approve the domestic use or marketing of LMOs for food, feed or processing must inform the Clearing-house within 15 days of making the decision (Art. 11(1)). Similarly, parties must inform the Clearing-house of final decisions regarding the import or release of LMOs (Art. 20(d)). The Clearing-house will probably become a common source of shared knowledge regarding biotechnology and serve as a vehicle for harmonizing various domestic provisions relating to the use of GM products.

Procedures related to export for intentional release

Parties seeking to export LMOs for intentional release into the environment of an importing country are subject to the Advanced Informed Agreement Procedures of the Protocol (Cartagena, 2000, Art. 7(1)). This provision invokes the international principle of prior informed consent. The first multilateral environmental agreement to use prior informed consent as a tool for regulating environmental risks was the Basel Convention on the Control of Transboundary Movements of Hazardous Wastes and Their Disposal (Basel, 1989). The Basel Convention prevents parties from transporting hazardous waste through or disposing of hazardous waste in other Party States without their consent. Access to genetic resources under the Convention on Biological Diversity also requires the prior informed consent of the Party State providing the resources (United Nations, 1992a, Art. 15(5)). More recently, the principle of prior informed consent was extended to export of chemicals and pesticides under the Rotterdam Convention on the Prior Informed Consent (PIC) Procedure for Certain Hazardous Chemicals and Pesticides in International Trade (Rotterdam, 1998). Specific language included in the Biosafety Protocol is patterned on the Rotterdam Convention.

The scope of the advanced-consent provision was one of the major stumbling-blocks that prevented an agreement from being concluded in 1999 (Gupta, 2001). The so-called Miami Group, consisting of Canada, the USA, Australia, Argentina, Chile and Uruguay, insisted that GM commodities exported for food, feed or processing should be exempt from all Protocol procedures. The rest of the world disagreed with varying degrees of conviction, with most countries favouring Advanced Informed Agreement Procedures for all transborder movement of LMOs. Major producing countries continued to argue, however, that the prior informed consent rules were unworkable when applied to commodities. They pointed out that bulk grain is commonly commingled, making it difficult to determine whether a particular shipment contains specific LMOs or is entirely free of LMOs. Moreover, extensive documentation would be required to ensure that the consent of the importing party would be informed. Together, these two facts were deemed by the exporters to pose an almost insurmountable obstacle to international trade in agricultural commodities.

The final compromise achieved in Montreal distinguishes between commodity LMOs imported for food, feed or processing and LMOs imported for intentional release to the environment. The somewhat cumbersome and complex Advanced Informed Agreement Procedure, consequently, applies only to shipments of LMOs that are intended for release into the environment. Such intentional releases, of course, pose the most direct threats to biodiversity, particularly if the intended release is for large-scale commercial production.

When applicable, advanced informed agreement must be obtained prior to the first transboundary movement of organisms subject to the rule (Cartagena, 2000, Art. 7(3)). Limited excemptions from consent requirements are available for LMOs in transit (Art. 5(1)) and LMOs destined for contained uses that effectively limit their contact with and their impact on the external environment (Arts. 5(2), 3(b)).

In addition, the Conference of the Parties has the power to create categorical exclusions for LMOs that are not likely to pose significant risks to health or the environment (Art. 7(4)). The possibility of categorical exclusions is a trade-friendly provision. If experience proves that certain LMOs pose little or no risk to the environment, they can be excluded from the Advanced Informed Agreement Procedure notwithstanding the fact that they are intended for release into the environment. In other words, if the need for precaution is eliminated by future events, the agreement contemplates that restrictive measures could be eliminated as a matter of international consensus. Over time, this provision will probably play a role in determining how widely GM products are distributed.

The obligation to secure consent runs directly to the exporting state (Cartagena, 2000, Art. 8(1)). States, in turn, are free to delegate the consent duties to actual exporters, but only if the exporting state imposes legal requirements for the accuracy of information provided by the exporter (Art. 8(2)).

The first step in the Advanced Informed Agreement Procedure is for an exporter to provide written notice to a designated authority in the state of import (Art. 8(1)). A wide variety of information must be included in the notice, including contact details for the importer and exporter; name and identity of the LMOs; domestic safety classification of the LMO in the exporting state; intended dates of transborder transport; taxonomic information; centres of origin and centres of genetic diversity of the recipient organism or the parent organisms; description of the nucleic acid introduced, the technique used and the resulting characteristics of the LMOs; intended use of the LMOs; amount to be transferred; a previous and existing risk-assessment report; suggested methods for safe handling, transport, storage and use; regulatory status within the state of export; result and purpose of any exporter notification to other states; and a declaration that the notice contains factually correct information (Cartagena, 2000, Annex 1).

The importing party has 90 days to return a written acknowledgement of receipt of the notice (Art. 9(1)). The acknowledgement must contain the date that notice was received (Art. 9(2)(a)) and the result of a completeness review (Art. 9(2)(b)). The completeness review determines whether the documentation contains the required information. If the notice is obviously defective, presumably it will not receive any further consideration until the defect is cured. Finally, the written acknowledgement of receipt of notice must contain a statement of how to proceed, either under consistent domestic law or under the Protocol (Art. 9(2)(c)).

Significantly, the Protocol contains two options for granting or conditioning consent. A nation may condition consent on compliance with domestic legislation, but that legislation is subject to the traditional disciplines of international trade law. Alternatively, a nation may rely on default procedures contained in the Protocol, procedures that arguably might supersede conflicting principles of trade law. Not surprisingly, whether or not to include a default option in the Protocol was a contentious issue. Developing countries, in particular, sought the default option, because they often lack domestic regulatory schemes. Eventually, their view prevailed.

If proceeding under the default procedures of the Protocol, the importing state must inform the exporter that the shipment may proceed without written consent after a 90-day delay or that import will not be allowed until written consent is provided (Art. 10(2)(a),(b)). If written consent is required, the importing party has 270 days from the date of notice to approve the import, with or without conditions, including how the decision will apply to subsequent imports of the same LMO (Art. 10(3)(a)), to prohibit the import (Art. 10(3)(b)), to request additional information (Art. 10(3)(c)) or to extend the time period for consideration of the request (Art. 10(3)(d)). The Protocol thus imposes clear duties on importers to respond to any notice received in a timely manner. Failure to meet temporal deadlines is not, however, tantamount to consent (Arts. 9(4), 10(5)). Moreover, even if all parties comply with all deadlines, the Advanced Informed Agreement Procedures will be time-consuming.

Decisions whether to grant, deny or condition imports must be consistent with scientifically sound risk assessments (Art. 10(1)). Importers may require exporters to conduct or, alternatively, pay for the risk assessment (Art. 15(2),(3)). The risk-assessment requirement is an attempt to objectify the issue of whether restrictions on import are justified. Procedures for conducting acceptable risk assessments are set out in Appendix III of the Protocol. Unfortunately, a risk assessment will not necessarily produce a scientific consensus. In that event, the precautionary principle gives states a degree of political latitude in formulating policy. The Protocol itself provides little guidance. On the one hand, Annex III provides that lack of scientific knowledge or scientific consensus should not necessarily be interpreted as indicating a particular level of risk, an absence of risk or an acceptable risk (Annex III, ¶ 4). On the other hand, the body of the Protocol provides that lack of scientific certainty due to insufficient relevant scientific information does not preclude a party from restricting imports to avoid or minimize potential adverse effects (Art. 10(6)). Thus, under the Protocol, states applying the precautionary principle can prohibit or restrict imports of LMOs, even if the potential risk to biodiversity or human health is not conclusively established by the risk analysis. This, of course, does not imply that states are free to act arbitrarily in the face of a properly conducted risk analysis. One would expect that a particular dispute regarding the scientific justification for an import restriction might well turn on which party bears the burden of proof. Unfortunately, but perhaps intentionally, the burden of proof in the event of a conflict is a critical factor that the Protocol does not address.

Finally, as if the review process for import of LMOs for intentional release into the environment were not complicated enough, the Protocol provides that decisions are not final. Importers can review decisions or exporters can petition for a review of decisions, based on new information or changed circumstances (Art. 12).

Procedures related to export for direct use as food or feed or for processing

Parties that export commodity LMOs for direct use as food or feed or for processing are not subject to Advanced Informed Agreement Procedures. However, commodity exporters must communicate their intentions to the Clearing-house. For example, parties are required to inform the Clearing-house of any decision to approve LMOs for domestic use, including any placing of such goods on the market (Cartagena, 2000, Art. 11(1)). The notice must be provided within 15 days of the decision (Art. 11(1)) and must include prescribed information, including a risk assessment (Annex II, ¶ (j)). Any party may request additional information based on the Clearing-house filing (Art. 11(3)).

Although exporters are not subject to the Advanced Informed Agreement Procedures of the Protocol, parties may restrict the import of LMO commodities based on domestic law that is consistent with the Protocol (Art. 11(4)). Presumably, given the consistency requirement, import restrictions would have to be based on a scientific risk assessment, subject, of course, to the precautionary principle, which is restated in this article (Art. 11(8)). In any event, to ensure transparency, any restrictive laws, regulations or guidelines must be made available to the Clearing-house (Art. 11(5)). In theory, this information collected by the Clearing-house could simplify the task of exporters in attempting to comply with inconsistent domestic laws.

Finally, LMOs exported for food or feed or for processing must be labelled as 'may contain' LMOs and as 'not intended for intentional introduction into the environment' (Art. 18(2)(a)). The 'may contain' provision was an important compromise. Commodity exporters successfully argued that current handling technology made it impossible to determine whether a shipment of bulk grain was entirely free of LMOs.

Miscellaneous provisions

Additional provisions of the Protocol address protection of confidential information (Cartagena, 2000, Art. 21), public participation (Art. 23) and capacity building (Art. 23). Parties also are obligated to inform affected or potentially affected states and the Clearing-house whenever they learn of possible unintentional movements of LMOs (Art. 17). Moreover, parties must adopt domestic measures designed to prevent and penalize illegal transboundary movements of LMOs (Art. 25). If illegal shipments nevertheless occur, the affected party can request the party of origin to retrieve or destroy the LMO at its own expense (Art. 25(2)). Finally, the contentious issue of liability for damage resulting from release of LMOs was set aside for later consideration (Art. 27).

Article 26, Socio-Economic Considerations, is of particular interest. Article 26(1) reads as follows:

> The Parties, in reaching a decision on import under this Protocol or under its domestic measures implementing the Protocol, may take into account, *consistent with their international obligations*, socio-economic considerations arising from the impact of living modified organisms on the conservation and sustainable use of biological diversity, especially in regard to the value of biological diversity to indigenous and local communities. (emphasis added)

Although qualified by the term 'consistent with their international obligations' and directed especially towards 'the value of biodiversity to indigenous and local communities', the section is, nevertheless, quite a departure from the rules that normally govern environmental exceptions to international trade agreements.

At first glance, the provision suggests that social policy objectives might justify a departure from normal trade disciplines, at least where LMOs are at issue. The qualifying language, however, comes directly from the Convention on Biodiversity (United Nations, 1992a, Preamble, ¶ 12). Indigenous communities often possess unique knowledge of the possible beneficial uses of biological resources, knowledge that has been accumulated over hundreds of years (Glowka *et al.*, 1994, pp. 47–49, 60–61). Often, such communities are the only groups that actively cultivate and preserve certain resources. To the extent that industrialized agriculture replaces traditional practices of indigenous communities, biodiversity is threatened. Without the caretakers, knowledge of the value of particular biological resources and the resources themselves may be lost. Thus, on occasion, preservation of cultural diversity is tied inextricably to preservation of biological diversity. Consequently, in limited circumstances, the Protocol sanctions restrictions on import of LMOs to avoid displacing the agricultural practices of indigenous communities. If the language of Article 26 were more broadly interpreted, however, serious conflicts with international trade principles might arise.

Environmental Values and International Trade Agreements

The interaction between trade and the environment

International trade has both positive and negative environmental impacts (Thomas *et al.*, 2000, ch. 4). Evidence suggests that rising per capita wealth, facilitated in part by international trade, eventually leads to increased preferences for environmental amenities. On the other hand, increased production and consumption associated with economic growth will, *ceteris paribus*, increase the rate of depletion of renewable and non-renewable resources. Increased output of negative external environmental costs associated with increased production may or may not be offset by a wealth-driven increase in consumer preference for environmental amenities. Furthermore, environmental impacts, both positive and negative, may be exaggerated on a local level. Product specialization in conformity with principles of comparative advantage, facilitated by the existence of free international trade, may cause a redistribution of environmental costs across borders, at least in the short run. If a nation has a comparative advantage in producing goods with high negative environmental external costs, then local environments might suffer trade-induced deterioration.

Despite the clear linkages between trade and the environment, historically international trade agreements were little concerned with environmental impacts. With the evolution of international environmental law, however, international trade and the environment became a central focus of those concerned with protecting the global environment. Several international environmental agreements are directly concerned with trade in products regulated for environmental reasons. Examples include the Montreal Protocol on Substances that Deplete the Ozone Layer (1987) (bans trade in ozone-depleted substances and contains mechanisms to ban trade in products made using processes that contain ozone-depleting substances), the Convention on International Trade in Endangered Species of Wild Fauna and Flora (CITES) (1973) (bans international trade in listed species), the Basel Convention on the Control of Transboundary Wastes and Their Disposal (1989) (prohibits parties from importing or exporting waste to non-parties and implements a prior informed consent provision for transfer of hazardous waste among parties), the Rotterdam Convention on the Prior Informed Consent Procedure for Certain Hazardous Chemicals and Pesticides in International Trade (1998) (establishes a prior informed consent procedure to be applied to the export of chemicals that are banned or severely restricted by the domestic law of the exporting country, and to the export of severely hazardous pesticide formulations) and the Convention on Biological Diversity itself. Moreover, international trade in emission rights is an essential feature of the Kyoto Protocol to the United Nations Framework Convention on Climate Change (1997). International financial institutions, including the World Bank and the Regional Development Banks, have also adopted policies to mitigate the adverse environmental consequences of the projects they fund (Liebenthal, 2002).

The original 1947 GATT, however, made no specific reference to the environment. Not until the WTO was formed in 1994 was the link between trade and the environment formally acknowledged in a multilateral international trade agreement. The second paragraph of the Preamble to the Agreement Establishing the World Trade Organization (1994) contains the following statement:

> *Recognizing* that their relations in the field of trade and economic endeavor should be conducted with a view to raising standards of living, ensuring full employment and a large and steadily growing volume of real income and effective demand, and expanding the production of and trade in goods and services, while allowing for the optimal use of the world's resources in accordance with the objective of sustainable development, seeking both to protect and preserve the environment and to enhance the means for doing so in a manner consistent with the respective needs and concerns at different levels of economic development.

While environmental protection may not be a WTO mandate, it is at least a valid subject for discussion and consideration during trade negotiations. For example, the WTO now has an operating committee on Trade and the

Environment. In addition, environmental issues are increasingly involved in cases submitted to WTO dispute panels.

Legal regulation of international trade under the WTO Agreements

WTO Agreements include the GATT and a number of supporting agreements, including the Agreement on Technical Barriers to Trade (1994) (TBT Agreement) and the Agreement on the Application of Sanitary and Phytosanitary Measures (1994) (SPS Agreement). The GATT and the other WTO Agreements are designed to reduce existing barriers to free trade and to prevent new barriers from being established.

Under Article XI of the GATT, most non-tariff barriers to trade are presumptively forbidden. Moreover, the GATT prohibits members from discriminating against products on the basis on national origin. The non-discrimination provisions are found in two separate, but complementary, articles. First, Article III provides that imported products are entitled to the same treatment as domestically produced 'like products,' a principle known as national treatment. Secondly, Article I provides that no member nation is entitled to tariff concessions greater than any other member nation, a principle known as most-favoured-nation treatment.

Article XX sets forth a limited number of exceptions for policies that would otherwise violate GATT free-trade provisions. Among the environmentally relevant exceptions are measures necessary to protect human, animal or plant life or health, found in Article XX(b), and measures imposed for the conservation of exhaustible natural resources to the extent that they also are imposed on domestic production and consumption, found in Article XX(g). Biodiversity, for example, would be a natural resource within the meaning of Article XX(g). Finally, the so-called *chapeau* to Article XX sets forth a further requirement that the exceptions not be applied in a manner that would constitute a means of arbitrary or unjustifiable discrimination between countries where the same conditions prevail or a disguised restriction on international trade.

With respect of food-safety issues, whether a measure is 'necessary' to protect human, animal or plant life or health is determined by reference to the SPS Agreement. The Preamble to the SPS Agreement states that a purpose of the Agreement is to clarify how the provisions of GATT 1994 relate to sanitary and phytosanitary measures, in particular GATT Article XX(b). (Preamble, ¶ 8). Sanitary measures that conform to the relevant provisions of the SPS Agreement are presumed to be within the exceptions contemplated by GATT Article XX(b) (SPS Agreement, Art. 2(4)). Similarly, sanitary and phytosanitary

> measures which conform to international standards, guidelines or recommendations shall be deemed to be *necessary* to protect human,

> animal or plant life or health, and presumed to be consistent with the relevant provisions of this Agreement and of GATT 1994.
>
> (Art. 3(2) (emphasis added))

The term 'international standards, guidelines and recommendations' is defined in the SPS Agreement to include standards established by the Codex Alimentarius Commission (food safety), standards established by the International Office of Epizootics (animal health) and standards established under the auspices of the Secretariat of the International Plant Protection Convention (plant health) (Annex A(2)). Thus, compliance with international standards of food safety established by the Codex Alimentarius Commission presumptively satisfies both the SPS Agreement and GATT Article XX(b).

Although harmonization of standards is encouraged by the structure of the various trade agreements, a nation is free to establish health and safety standards that are more stringent than the international norm (SPS Agreement, Art. 3(3)). However, nations pursuing their own standards must support their regulatory regime with a risk assessment (Art. 5). Moreover, the measures may not be more trade-restrictive than necessary, given technical and economic feasibility (Art. 5(6)). A measure is not more trade-restrictive than required, however, unless there is another measure reasonably available, taking into account technical and economic feasibility, that achieves the appropriate level of sanitary or phytosanitary protection and is significantly less restrictive to trade (Art. 5(6), note 3).

Finally, if the state of scientific knowledge is incomplete or in dispute, the SPS Agreement incorporates a version of the precautionary principle:

> In cases where relevant scientific evidence is insufficient, a Member may provisionally adopt sanitary or phytosanitary measures on the basis of available pertinent information, including that from the relevant international organizations as well as from sanitary or phytosanitary measures applied by other Members. In such circumstances, Members shall seek to obtain the additional information necessary for a more objective assessment of risk and review the sanitary or phytosanitary measure accordingly within a reasonable period of time.
>
> (Art. 5(7))

Thus, states can adopt provisional sanitary standards in the face of scientific uncertainty without necessarily violating their trade obligations, but, if they do, they become obligated to undertake the research necessary to reduce the uncertainty. Moreover, there must be some scientific basis for concern; political expediency is not a justification for provisional standards. For example, while recognizing that the precautionary principle had been incorporated into the SPS Agreement, the recent opinion of the Appellate Panel in the *Beef Hormones* cases rejected an argument that the principle overrides the specific risk analysis provisions of the Agreement (WTO, 1998a, § 6).

Compliance with international standards is also a safe harbour under the terms of the TBT Agreement. The TBT Agreement governs other regulatory measures that might affect trade, such as labelling or packaging requirements. To the extent that measures are subject to the SPS Agreement, however, the TBT Agreement does not apply (TBT Agreement, Art. 1.5). If the TBT Agreement does apply, technical measures must not discriminate unjustifiably between products on account of their origin (Art. 2.1). Moreover, the measures must serve a legitimate purpose, such as national security, prevention of deceptive practices or protection of human health or safety, animal or plant life or health or the environment (Art. 2.2). In addition, technical measures must achieve their objectives in the least trade-restrictive manner (Art. 2.3). If relevant international standards exist, members should use them as a basis for their technical regulations unless they would be ineffective or inappropriate because of fundamentally different factors, such as climatic, geography or fundamental technological problems (Art. 2.4). While the TBT Agreement defines technical standards as mandatory technical requirements, Annex III of the Agreement creates a similar obligation to conform to voluntary standards, such as those promulgated by the International Organization for Standardization (ISO), the so-called ISO standards. In any event, technical regulations that are based on international standards are rebuttably presumed not to create an unnecessary obstacle to international trade (Art. 2.5). In contrast, regulations not based on international standards are subject to special transparency requirements (Art. 2.9). Members departing from international norms must justify the legitimacy of the departure when requested to do so by another Member (Art. 2.5). If disagreements under either the SPS Agreement or the TBT Agreements persist, the matter may be referred to a WTO dispute settlement panel (SPS Agreement, Art. 11; TBT Agreement, Art. 14).

Reconciling the Biosafety Protocol with the WTO Agreements

Conceptually, a suspect trade measure can be evaluated under the WTO Agreements by asking a series of questions. First, does a measure constitute a prima facie violation of one of the substantive provisions of the GATT found in Article I (Most Favoured Nation Treatment), Article III (National Treatment) or Article XI (Quantitative Restrictions)? In respect of Article I or Article III violations, this often involves a subsidiary question of whether the suspect measure applies to 'like products'. Secondly, if a measure violates a GATT provision, can an Article XX exception be invoked? In determining whether an Article XX exception exists for food-safety measures, the SPS Agreement and/or the TBT Agreement may be relevant and/or controlling.

The Biosafety Protocol articulates one set of rules that might be invoked to restrict entry of LMOs destined for intentional release into the environment and another set of rules that might be invoked to restrict entry of commodity LMOs destined for food, feed or further processing. The two sets of rules raise somewhat different questions in respect of international trade regimes.

Prima facie GATT violations and the 'like products' issue

The Biosafety Protocol defines circumstances when it is permissible to treat the products of modern biotechnology differently from similar conventional products. For purposes of GATT consistency, this raises the question of whether LMOs and similar conventional organisms should be considered to be 'like products'. GATT and WTO panel rulings have generally been unwilling to differentiate products whose only difference is that they were produced using different processes or production methods. For example, pre-WTO opinions by GATT panels refused to sanction the US practice of discriminating against tuna based on whether or not the tuna was harvested using dolphin-safe methods. Tuna was tuna irrespective of the harvesting method employed (*Tuna/Dolphin I* (1992); *Tuna/Dolphin II* (1994)). By analogy, the question posed is whether crops produced using the products of modern biotechnology should be viewed as generic commodities (e.g. cotton or maize) produced from different processes (genetically altered versus conventional seeds) or as different products (maize or cotton that contains *Bacillus thuringiensis (Bt)* genes and maize or cotton that does not). If LMOs and their conventional counterparts are not like products, then the non-discrimination rules of the GATT are simply irrelevant, and different domestic rules can be applied to LMOs from those that are applied to conventional organisms. Article XI prohibitions on quantitative restrictions, however, would still be relevant.

In the *Asbestos* case (WTO, 2001), the WTO Appellate Panel considered a Canadian challenge to a French regulation that prohibited the manufacture, processing, sale, import, marketing or transfer of all varieties of asbestos fibres, without regard to whether they had been incorporated into materials, products or devices. Canada argued that chrysotile asbestos fibres were 'like' polyvinyl alcohol, cellulose and glass fibres (PCG fibres) and that cement-based products containing chrysotile asbestos fibres are like cement-based products containing one of the PCG fibres. The Appellate Panel first stated that the fundamental factor determinative of likeness is the nature and extent of the competitive relationship between and among products. Stressing that the judgment must be made on a case-by-case basis, the Panel identified four factors that should be examined in searching for relevant evidence of likeness: (i) physical properties; (ii) similar end uses; (iii) consumer perception of products as alternatives to satisfy a particular demand; and (iv) the tariff classification of the products. Turning to the facts of the case before it, the Appellate Panel first rejected an argument that, if products have equivalent end uses, they necessarily have equivalent physical properties. In contrast, they found that the carcinogenicity or toxicity of asbestos fibres was a defining physical property that prevented a finding of likeness in respect of PCG fibres.

The product/process distinction was also a significant issue in both of the *Tuna/Dolphin* decisions (*Tuna/Dolphin I*, 1992; *Tuna/Dolphin II*, 1994) and was raised by the pleadings in the *Shrimp/Turtle* (WTO, 1998b) controversy. In both *Tuna/Dolphin* and *Shrimp/Turtle*, the effect of US legislation was to impose US conservation policy on other nations by imposing trade sanctions on those nations not adopting the US conservation policy as their own. In *Tuna/Dolphin I* and *II*, the USA unsuccessfully sought to establish that tuna harvested from the eastern tropical Pacific using purse-seine nets set on dolphins was a different product from tuna harvested in a dolphin-safe manner. The US Marine Mammal Protection Act prohibited the import of tuna from countries that did not require the use of dolphin-safe harvesting methods. Similarly, in *Shrimp/Turtle*, the USA sought to justify a ban on the import of shrimp from countries that did not require their fleets to use turtle extruder devices (TEDs) in their fishing nets. Unlike dolphins, however, which are numerically abundant, every species of sea turtle is listed as endangered under CITES. Although the issue of like products was not before the Appellate Panel because the challenged sanction involved an import ban prohibited under Article XI, it seems unlikely that shrimp harvested with and without TEDs would have been treated as separate products solely on the basis of the harvesting method employed.

Whether GM agricultural products and their conventional counterparts should be treated as like products remains a controversial question. In part, the answer may depend on the nature of the modification. Arguably, LMO commodities where the change is expressed as a pesticide found in all plant tissues, as in *Bt* corn or *Bt* potatoes, might be treated differently from LMO commodities where the change is merely an ability to break down glyphosates, as in Roundup-ready seeds. In part, the answer may depend on the end uses of the genetically altered products. Food products, for instance, may pose different questions from those regarding fibre products. As the WTO and prior GATT rulings have implied, relevant factors to consider include end uses of the product, consumer tastes and habits and the product's properties, nature and qualities.

Two products are generally considered to be like products if they are commercial substitutes for each other, unless the products differ in qualities that cause them to be treated differently in use, handling or disposal. A plant modified for genetic resistance to herbicides or insect pests may produce seeds that, apart from their genetic markers, are indistinguishable from other varieties of maize or soybeans. As long as LMO seeds are destined for consumption and not production, LMO commodities and their conventional counterparts would appear to be like products. On the other hand, a modification might be designed to affect the nutritional quality of the seeds by increasing the protein content or decreasing saturated fats. Even though the LMO commodity and the conventional commodity would have different characteristics, they would still seem to be like products for GATT purposes because they would be market substitutes and would not differ in use, handling or disposal. But suppose maize were genetically modified to serve as an economic poison or a component of a new medicine. In either case, LMO maize and conventional maize would no longer be like products for GATT purposes. Of course, one of the major concerns with GMO products is that they may be modified inadvertently in ways that express proteins that might trigger allergic reactions (FAO, 2001).

Ultimately, the extent to which LMO commodities are deemed to be like products may turn on how the Codex Alimentarius acts in determining international safety standards for the products of modern biotechnology. By definition, food-safety standards set by the Codex conform to the requirements of the SPS Agreement and therefore constitute a valid Article XX(b) exception to GATT non-discrimination rules. The Codex has adopted the same definition of modern biotechnology as is contained in the Biosafety Protocol (Codex, 2001, ¶ 23). Currently, the Codex seems likely to adopt a marginal impact analysis based on substantial equivalence as a method of determining the risk posed by foods derived from modern biotechnology (Codex, 2001, Appendix III, pp. 43–44). Such an analysis involves a comparison of LMO foods with their conventional counterparts in an attempt to determine whether the LMO food poses new or altered hazards taking into account both intended and unintended effects. Conventional counterpart is defined as a related organism/variety, its components and/or products for which there is experience of establishing safety based on common use as food (Codex, 2001, Appendix III, p. 40). This methodology recognizes that whole foods are not normally subject to risk analysis; instead, risk assessment has been applied to discrete chemical entities, such as pesticide residues or food additives, or to specific chemicals or microbial contaminants that have identifiable hazards and risks. In fact, many foods that are considered safe as a result of historical use contain substances that would be considered harmful if subjected to standard risk analysis. Consequently, the intent of the Codex is to focus on altered risks as a result of genetic manipulation and measures that may reasonably be imposed to respond to those altered risks.

At least in respect of food products, it seems unlikely that a WTO panel would hold that LMO commodities and their conventional counterparts are separate products in the absence of some action by an international standards group. While those who would make a separate-products argument may find some support in the language and structure of the Biosafety Protocol, the savings clause in the Protocol suggests that the Protocol itself should not be the basis for altering normal WTO understandings.

With respect to seeds and other LMOs to be introduced into the environment, however, the continuing vitality of the like-product status seems to be more problematic. Production of LMO crops engineered for pest resistance, for instance, may involve special handling in the form of spacing requirements to guard against the evolution of LMO-resistant pests through the process of natural selection. For example, the US Environmental Protection Agency (EPA) requires registrants of *Bt* maize and *Bt* cotton to ensure that growers of LMO crops devote a minimum percentage of their fields to a 'structured refuge' of non-*Bt* product (EPA, 2000). If LMO seeds and conventional seeds are not like products, then LMO regulation need not pose problems of GATT consistency (Werner, 2000). Whether a nation is a party to the Protocol is then of little or no importance. Even if LMOs and conventional organisms are considered like products, state regulation might still be justified under the Article XX exception.

The SPS Agreement and the precautionary principle

Products that satisfy the like-products test under GATT are subject to normal GATT antidiscrimination disciplines. Whether a domestic food-safety measure contravenes these disciplines is resolved by referring to the SPS Agreement. In a WTO challenge, discriminatory regulations must be shown to be within an Article XX exception. Article XX(b) authorizes trade measures necessary to protect human, animal or plant life or health. The SPS Agreement addresses the question of when a sanitary or phytosanitary measure is necessary. The *chapeau* of Article XX limits the scope of the exception, which does not apply if the challenged measure constitutes 'a means of arbitrary or unjustifiable discrimination between countries where the same conditions prevail, or a disguised restriction on international trade' (GATT, Art. XX, ¶ 1). In other words, bad motives as evidenced by subjective intent can doom a measure that might otherwise fit within a GATT exception.

The SPS Agreement generally requires that any deviation from international standards, such as those adopted for food by the Codex Alimentarius, be justified by a risk analysis (SPS Agreement, Art. 3.3). Subject to this requirement, nations are free to choose the level of protection they will impose (or the level of risk they are willing to accept), even if the level is more limiting than the one that underlies international standards. Furthermore, the SPS Agreement incorporates a version of the precautionary principle by recognizing that regulation may be justified even in the absence of clear scientific proof of harm or clear consensus on issues of risk.

Because the precautionary principle has been incorporated into the SPS Agreement, its inclusion in the Biosafety Protocol should not pose serious problems of GATT consistency (Gupta, 2001, pp. 275–278). On the other hand, the precautionary principle does not operate identically in the two agreements (SPS Agreement, Art. 5.7; Cartagena, 2000, Preamble ¶ 9, Arts. 10(6), 11(8)). Both agreements require that decisions be supported by a risk assessment, but the SPS Agreement is more clearly linked to a focus on avoiding disruptions in trade. Moreover, the application of precaution in the SPS Agreement is a temporary measure pending further study. The Biosafety Protocol does not link a precautionary standard to a duty to conduct further research. How closely the two principles will be harmonized will only be determined after the passage of time. The principal issue is how much evidence of risk will be required before a nation is justified in limiting trade in LMOs. Every effort should be made to harmonize the tests under the WTO Agreements and the Biosafety Protocol.

GATT consistency and the Protocol rules for intentional introduction of living modified organisms into the environment

The Advanced Informed Agreement Procedures of the Protocol constitute significant restrictions on the free flow of LMOs destined for introduction into the environment. The restrictions are both procedural and substantive. Substantively, the process could lead to a rejection of a proposed shipment of LMOs. Procedurally, the process significantly increases the cost of exporting LMOs for seed in the form of information to be supplied and the delay that the procedures contemplate. Final decisions are to be made within 1 year, but numerous opportunities for further procedural delays exist. Moreover, without any clear penalties for failure to meet temporal deadlines and with the understanding that silence is never deemed to be consent, procedural delays could block import indefinitely. Could such procedural delays be the basis for a WTO challenge? Given that LMO products undergo extensive testing and review in the developing country before receiving approval, it is at least a theoretical possibility. On the other hand, if the Biosafety Protocol results in the development of an international consensus for testing procedures and risk assessment, the likelihood of a successful WTO challenge to measures formulated using those procedures would be diminished. Furthermore, to the extent that LMO seeds fail to meet the like-products test, the basis for challenging a domestic regulation that discriminates against LMO products would evaporate.

Similarly, the likelihood of a measure being found substantively inconsistent with the GATT appears to be remote. First, there is the distinct possibility that LMOs subject to the Advanced Informed Agreement Procedures would be held not to be like products and therefore not subject to GATT antidiscrimination disciplines. In that event, even a ban on the import of LMOs for intentional introduction into the environment would not violate Article XI as long as LMO production was banned domestically (GATT, 1947, Annex I, Art. III). Secondly, even if the like-products test were met, the Protocol is clearly a multilateral agreement that addresses issues contemplated by the Article XX(b) exception to the GATT. Moreover, the inclusion of risk analysis in the Protocol as an essential element of the regulatory matrix strengthens the Article XX nexus. Thirdly, the intentional introduction of LMOs into the environment poses a clear hypothetical risk of potential harm to biodiversity, the resource whose protection is the primary subject of the Biosafety Protocol.

In the final analysis, conflicts between the intentional-introduction provisions of the Protocol and the free-trade provisions of the WTO Agreements ought to be resolved presumptively in favour of the party asserting the Protocol rules. Under WTO rules as interpreted in the *Beef Hormones* appellate decision (WTO, 1998a), the burden of proof lies with the challenger, even when a challenge is based on the SPS Agreement. Because an exporting nation would be mounting a challenge to an access rule adopted pursuant to the Protocol, the burden of demonstrating arbitrary or unjustified discrimination or a disguised barrier to trade would be on the exporting-party state. As a practical matter, such a burden would be very difficult to meet unless the importing party were acting in clear bad faith.

GATT consistency and the Protocol rules for the import of LMO commodities for use as food or feed or for further processing

Risks posed by the intentional introduction of LMOs into a nation's physical environment are different in kind from the risks posed by the importation and use of genetically manipulated products for food, feed or processing. The Biosafety Protocol recognizes this difference by exempting commodities destined for food, feed or processing from the Advanced Informed Agreement Procedures. Nevertheless, the Agreement subjects commodity LMOs to special labelling requirements and gives potential LMO importers the right to restrict imports if consistent with the objectives of the Protocol. As discussed above, substantive restrictions on the import of commodity LMOs are presumptively subject to GATT disciplines. Several commodity LMO issues, however, merit further discussion.

Procedural requirements

Exporting parties are required to submit substantial amounts of information to the Clearing-house. Parties may request additional information. Importing countries may regulate the import or use of commodity LMOs under domestic law. Developing-country and transition-country parties that lack domestic regulatory schemes may inform the Clearing-house that they reserve the right to make a decision on import within a reasonable time of a risk assessment being prepared. Although not insignificant burdens, the procedures applicable to commodity LMOs are not probably to conflict with WTO rules. Any conflict will probably be a consequence of substantive restrictions on the import of commodity LMOs. Moreover, there is at least some potential for the Clearing-house to assume a leadership role over time in respect of the development and negotiation of international harmonized standards for trade in LMOs and LMO commodities, perhaps similar to the role that the Codex has assumed in respect of food safety. When combined with the power of the Conference of the Parties to the Protocol to grant categorical exclusions (Cartagena, 2000, Art. 7(4)), one can easily envision the power and prestige of the Clearing-house being enhanced over time.

Labelling requirements

The nature of the labelling requirement that would apply to LMO commodities was a matter of considerable dispute among drafters of the Protocol until the final moments of negotiation. The language finally adopted requires exporters to include a statement that a shipment 'may contain' LMOs and is 'not intended for intentional introduction into the environment' (SPS Agreement, Art. 18.2(a)). Earlier drafts would have required exporters to identify specifically the presence of any LMOs in shipments. Commodity LMO producers argued that it was not possible under bulk distribution systems to certify that a shipment was entirely LMO-free or to identify LMO products that might make up a small percentage of a shipment. Some consumer advocates and environmental groups argue that failure clearly to identify products that contain LMOs threatens consumer sovereignty and exposes them to risks they should not be required to bear. On the other hand, zero-tolerance regulatory schemes that require complete tracing create serious potential problems with trade agreements. If it is impractical to isolate LMO commodities from traditional commodities completely in LMO-producing countries, zero tolerance effectively shields local producers from international competition. By opting for the 'may contain' labelling requirement, negotiators minimized the possibility that labelling requirements would be inconsistent with GATT.

Producers, of course, argue that labelling requirements perpetuate irrational prejudice against commodity LMOs, no matter how they are crafted. It is, of course, true that a combination of irrational prejudice fuelled by activist campaigns and legally mandated labelling requirements could diminish, if not destroy, the market for LMO products. Ultimately, however, consumers will decide the future of LMO products. While the potential for quick acceptance by not disclosing the presence of LMOs may be tempting, in the long run reluctance to provide information or submit to thorough tests can only be expected to result in a consumer backlash.

Socio-economic factors as a justification for import restriction

The possibility that the import of LMO commodities could be restricted on the basis of socio-economic impacts has the potential for serious conflict with GATT unless the language is interpreted restrictively. One of the major unifying principles of the Biosafety Protocol and the WTO Agreements is the use of science-based risk analysis to support the rationality of health- and safety-based regulations (SPS Agreement, Arts. 2.2, 5; Cartagena, 2000, art. 10(1)). Although risk analysis may not have the same outcome-determinative impact in both agreements, preserving the general applicability of the principle of risk analysis as a decision tool is important if conflicts between the WTO Agreements and the Biosafety Protocol are to be minimized. Trade restrictions based on socio-economic factors have the potential to disrupt this symmetry if not used judiciously. Restrictions reflecting socio-economic factors that are clearly linked to the preservation of biodiversity are justified, but restrictions designed to protect or preserve a particular farming methodology or agricultural economy are not. At a minimum, those states seeking to depart from risk-based trade restrictions should bear the burden of justifying a biodiversity-conservation purpose for their decision.

Categorical exclusions and flexible relief

The Biosafety Protocol provision on categorical exclusions can be an important tool to minimize conflicts between the WTO Agreements and the Protocol (Cartagena, 2000, Art. 7(4)). As a political matter, the judicious use of categorical exclusions would facilitate trade and enhance the authority of the Biosafety Protocol. If parties act in good faith to free some LMOs from the proscriptions of the Protocol, the likelihood that remaining Protocol restraints will be held to be GATT violations will be diminished significantly. The grant of exclusions thus helps to counter perceptions of a negative protectionist bias to Protocol restrictions.

Flexibility, if applied in good faith, could eliminate most GATT conflicts while accommodating individual national preferences for risk and protecting biodiversity resources. The Protocol is silent on procedures for obtaining a categorical exclusion, however, other than making it a decision of the Conference of the Parties (Art. 7(4)). Unfortunately, if a categorical exclusion may only be granted by consensus, as seems likely (Art. 29(5)), it may be almost impossible to achieve.

Unresolved Questions

Expansion of the precautionary principle

The Biosafety Protocol may move the world closer to accepting the precautionary principle as a customary rule of international law. The European Union argued unsuccessfully in *Beef Hormones* (WTO, 1998a) that the precautionary principle had independent force as customary international law and therefore should militate against a strict interpretation of the risk-analysis requirements in the SPS Agreement. Despite its rejection of the argument that the precautionary principle had independent force as customary international law, the opinion of the Appellate Panel in *Beef Hormones* seems more deferential to individual state choices based on precaution than does the opinion of the original dispute panel. Somewhat ironically, the very fact that negotiators included precautionary language in the Biosafety Protocol and perhaps even the Appellate Panel opinion in *Beef Hormones* may ultimately be viewed as events that help establish the precautionary principle as a customary rule of international law.

If the precautionary principle comes to be recognized and accepted as a customary rule of international law, it will probably become harder successfully to challenge an LMO trade-restrictive measure under the WTO Agreements. Of course, application of the precautionary principle must be predicated on the existence of some scientific evidence suggesting a need for precaution, a limitation that is sometimes ignored by proponents of expansion of the principle. In any event, to the extent that precautionary measures have an adverse impact on trade, the cost of precaution is increased.

Accounting for the costs of precaution

Properly understood, the precautionary principle is a necessary adjunct to pragmatic decision-making in the face of scientific uncertainty. The principle can be invoked either as a justification for acting when proof of harm is not conclusively established or it can be invoked to refute an argument that action may not proceed without clear scientific proof

(Dickson, 1999, pp. 213–215). The latter formulation of the principle is less proscriptive than the former and is the option incorporated into both the Biosafety Protocol and the WTO Agreements. A requirement that a state act in the face of uncertainty to establish trade-restrictive measures would pose serious problems of consistency with WTO Agreements. Authority to act with limited scientific proof is a much less threatening formulation.

The costs of precaution should not be underestimated. Delay in introducing new products, perpetuation of consumer bias and reduced incentives for continuing research and development are all costs of precaution. The cost of delay in product introduction is clearly illustrated by the introduction of new, potentially life-saving, pharmaceuticals. If risk tolerances are too low and approval processes are too protracted, the cost of precaution is unnecessary lives lost. Excepting pharmaceutical LMOs from the Biosafety Protocol avoided what would otherwise have been an increase in the costs of precaution for new medical products. Similarly, to the extent that genetic modification of agricultural products promises greater production per unit of resource expended, decreased reliance on chemical pesticides or fertilizers or more efficient use of water, measures that slow the introduction of LMO products come at a high environmental and human cost. The greatest threat to biodiversity is probably the conversion of natural habitat to economic use (Harte, 2001). If bioengineered products, for instance, can reduce conversion of tropical forests to cropland, then the introduction of LMOs into the environment may significantly enhance, rather than retard, efforts to conserve biodiversity. Of course, this is not to argue against stringent standards for introducing LMOs into the environment. In fact, a consequence of lax regulation may be consumer scepticism that hardens into anti-LMO bias, which could seriously threaten the near-term viability of the industry.

A second cost of precaution is the potential to distort consumer preferences, leading to economic inefficiencies. Precautionary measures validate the arguments of those urging precaution, whether the arguments are based on sound reasoning or pure emotion. Once economic preferences become incorporated into political belief systems, the resulting consumer biases are very difficult to overcome (Carbone and McLean, 2001). This is aptly illustrated by the French experience. France was one of the early proponents of GM agricultural products. Today, it is outspoken in opposition to LMOs. The change was a consequence of growing public opposition to the products, opposition that included attacks on farms where GM crops were grown. Today, although several GMO products are listed as authorized for production and a recent decision of the Conseil d'État upheld a 1998 decision of the Ministry of Agriculture authorizing production of *Bt* maize developed by Syngenta, GM maize is not cultivated because there is no demand for the product (Staff, 2000; Speer, 2001). Ironically, the LMO industry may exacerbate the problem if it promises too much and delivers too little. Without credibility, there is little to counter the emotional arguments of more extreme opponents of the technology.

A third and more subtle cost of precaution stems from the financial capacity of wealthier populations to eschew consumption of LMO products, essentially adopting a 'zero-risk' limitation on exposure to LMOs. Of course, the possibility exists that LMOs may, in fact, be safer than conventional commodities, particularly if the LMO commodities are free of pesticide residues. If markets for the products of modern technology are thwarted by consumer risk preference in developed countries, however, then basic and applied research and development will be threatened. Ironically, it is often developing nations that could benefit most from agricultural biotechnology, but they lack the financial capacity or the market potential to support basic research. Consumer resistance in developed countries could threaten economic incentives to develop LMO products that would benefit developing countries disproportionately. In the short run, lack of LMO products, even if perfectly safe and environmentally benign, would probably have a limited impact on consumer welfare in developed countries. Slowing the pace of development of LMO products, however, would have a much greater impact on food-short developing countries. According to Barun Mitra, managing trustee of the Liberty Institute in New Delhi, restricting availability of bioengineered crops because of potential but unknown risks ignores the existing and pressing problems of poverty and malnutrition. 'It is patently unfair that the Third World pay the price for the priorities of the richer world' (Safford, 2000).

The nature of risk assessments

Risk assessments are among the few tools available to provide some evidence that a trade-restrictive measure serves a legitimate, and not an illegitimate, purpose. Risk assessments are supposed to ensure that regulatory standards are linked logically to risks that a nation has chosen to guard against. Risk assessments, it is argued, should be technical decisions based on hard evidence and sound science, not mere political decisions. The SPS Agreement distinguishes between determining the level of risk that exists (risk assessment, a technical decision) and regulating to provide a level of protection that a particular society demands in the light of that risk (risk management, a political decision). By specifically extending the precautionary principle to transborder movements of LMOs, the Biosafety Protocol calls into question whether the risk assessment/risk management distinction is tenable. When regulating an infant industry with a seemingly unlimited potential to affect the lives of billions of people for better or worse, it seems that all decisions are inherently political.

Impacts on non-parties

By definition, there is no direct conflict between the Biosafety Protocol and the WTO Agreements for a state that is not a party to both agreements. If a state were a WTO Member, but not a party to the Protocol, Protocol language would not seem to affect a trade dispute. However, because the Protocol does not attempt to pre-empt trade agreements and recites that trade and environment agreements should be mutually supportive, resolution of trade disputes that arise from Protocol actions should not normally depend on the party status of the disputants. Given the savings clause in the Preamble to the Protocol, any trade-restrictive measure must stand on its own in a WTO dispute; the fact that the measure was adopted in accord with the Protocol should neither enhance nor detract from its chance of surviving a WTO challenge.

It now seems clear that future Conferences of the Parties under the Biosafety Protocol will be making decisions that have clear implications for international trade in biotechnology and for the economic future of biotechnology. For instance, parties will make future decisions regarding categorical exclusions and liability regimes. Non-parties to the Protocol were involved in negotiating the terms of the Protocol, but they will have no status once the agreement becomes operational. The matter is particularly important for the USA, which has signed, but not ratified, the Convention on Biodiversity. Unless and until it ratifies the Convention, the USA will have no ability to become a party to the Protocol. One implication of the Protocol, then, is to increase incentives for the USA to ratify the Biodiversity Convention, so that it can secure a place at the table when relevant issues are discussed.

Liability

One contentious issue that was not addressed in the Protocol was the question of liability for damage resulting from transboundary movements of LMOs. The Conference of the Parties will revisit the issue of liability, with a goal of completing a liability regime within 4 years (Cartagena, 2000, Art. 27). The implications of the precautionary principle should be evident in these negotiations. It is one thing to assign liability for a failed technology that causes environmental damage, and quite another to assign liability for inadvertently violating a zero-tolerance import standard, particularly if the liability is to be placed directly on a state. The level of precaution protected by liability systems and the measure of damages for various violations of precautionary rules will help determine the true scope of precautionary obligations.

One issue that has largely escaped attention is responsibility for the diversion of a product from its intended use. The bifurcated regulatory approach contained in the Biosafety Protocol assumes that commodity products imported for consumption will in fact be consumed. Although the StarLink débâcle in the USA raises questions regarding the ability of even the richest of nations to ensure the integrity of its regulatory process (Nelson, 2002; Uchtmann, 2002), third-world countries face a much more difficult problem. During negotiation of the Biosafety Protocol, one delegate noted that grain imported for consumption would be transported internally in trucks, which, because of age and condition, would inevitably lose a portion of the load to leakage during transport. Poor farmers, he said, will gather seeds wherever they can be found and use them to plant future crops (Schweizer, 2000, n. 112). Understanding the unique problems faced by developing nations will be key to achieving any type of acceptable liability scheme.

Traceability and labelling

Traceability is the ability to track a product back to its source. It is probably a necessary adjunct of some liability schemes and may be a legitimate component of some regulatory schemes. It is also a controversial issue because, if it is used indiscriminately, it will serve as a significant barrier to free trade. Traceability is currently a matter of discussion and negotiation in the Codex Alimentarius and in the ISO. It is also a component of Directive 2001/18, the European Union's recent directive on Deliberate Release into the Environment of Genetically Modified Organisms. The European Commission has also adopted a related proposal for regulation on labelling and traceability of GMOs (CEC, 2001). The USA has generally taken the position that, while traceability may be an appropriate requirement in certain circumstances, it should not generally be imposed on products that have been given pre-market approval. While the European Union (EU) generally believes that labelling and traceability are necessary to restore consumer confidence in the food regulatory system, the USA has opposed such measures if their only justification is to facilitate consumer choice.

The debate over labelling products is akin to the product/process distinction for permissible product discrimination under the GATT. The USA argues that labelling is appropriate only where the products of biotechnology differ in some material effect from their conventional counterparts. Thus, products should be labelled if they differ in nutritional content, allergenicity, potential use or composition, but not solely to satisfy a consumer desire for additional information. This position is based in part on industry concerns that labelling based on process alone is a form of product disparagement and in part on the cost of labelling, especially the cost of ensuring that a product conforms to a label (McGarity, 2002, pp. 499–504).

Perhaps more significant, however, is the potential impact of the label debate on the 'like-products' debate under WTO rules. If a combination of labelling and consumer preference results in separate markets for functionally identical LMO and non-LMO products and if special handling is required to prevent the commingling of LMO and non-LMO goods, then further credence is given to an argument that the LMO commodity and the conventional counterpart are not like products for purposes of GATT disciplines. This is particularly true if the market does not then treat the LMO product and the non-LMO product as product substitutes. If the products are not like products, then the basis for a WTO claim of unlawful discrimination is lost.

The potential to bring a credible WTO complaint against nations that discriminate against GMO products is an important bargaining chip in the overall debate. In fact, the recent decision of the EU to replace Directive 90/220 with Directive 2001/18 illustrates the importance of the WTO hammer. Under EU law, the directive must be transposed into national legislation within 18 months. But several EU nations that have strongly supported the EU *de facto* moratorium on the approval of GMO products indicated that they would not implement the new directive until acceptable labelling and complete traceability provisions are adopted. Although tracing and labelling regulations have now been proposed (CEC, 2001), the *de facto* moratorium has not been lifted as it requires action by individual Member States. The European Commission has been waiting to see whether the USA will pursue a WTO action before deciding how to proceed with its recalcitrant members (Kirwin, 2001). Meanwhile, political pressure to pursue a WTO action continues to mount in the USA, with three powerful US Senators recently urging the President to initiate a WTO challenge to the moratorium (Harkin *et al.*, 2002).

Maintaining flexibility to respond to new scientific developments

Modern biotechnology is still an infant industry. Any system must be flexible enough to incorporate new scientific findings. At the same time, the system must permit the rational assessment of new scientific information for its policy implications. The fact that *Bt* pollen may have adverse impacts on monarch butterflies may or may not justify additional restrictions on the production of *Bt* maize (Pew Initiative, 2002). As a minimum, policy-makers would have to consider whether the adverse impacts as a consequence of exposure to *Bt* pollen were greater or lesser than the adverse impacts that butterflies would face if conventional pesticides were used on the maize crop. Similarly, new research suggests that at least some LMO plants have no greater ability to survive without cultivation than their conventional counterparts and that neither thrives in the wild (Crawley *et al.*, 2001). Such findings lessen concerns that GMO traits might spread naturally to hybrids. On the other hand, it seems likely that increased use of recombinant DNA technology will raise questions of unintended consequences. Allergenic properties might be transferred from one species to another. Moreover, while biotechnology reduces the likelihood of introducing extraneous genetic material when compared with traditional hybridization, because breeders can introduce only genes of interest, it may enhance the likelihood of producing unintended effects through mutations (US FDA, 2001, p. 4710). More refined regulatory measures may be needed to address the impacts of more sophisticated genetic manipulations.

Summary of the United States Position on Regulating the Products of Modern Biotechnology

The USA has been actively involved in helping to fashion emerging international policy in respect of trade in the products of modern biotechnology. As a major exporter of both GM commodities and GM production inputs, the USA has a huge stake in the debate over trade in the products of modern biotechnology. The overall regulatory approach within the USA and in international negotiating has been deceptively simple: bioengineered products should be treated the same as substantially equivalent conventional counterparts unless a difference can be demonstrated by sound science. If the product that is produced as a consequence of a process that includes genetic manipulation is indistinguishable from a product that is produced by conventional means, then the market should not distinguish between the GMO product and the non-GMO product based solely on the differing process (US FDA, 1992; McGarity, 2002). If, however, product differences can be demonstrated, then whether additional standards or restrictions on use are required should be determined using pre-existing rules. Of course, an LMO intended for use in production may exhibit relevant differences, while the same LMO intended for consumption may be indistinguishable from its conventional counterpart – hence, the logic of the dual regulatory path incorporated into the Biosafety Protocol.

It is important to note that the USA does not assert that the products of modern biotechnology should be exempt from regulation. Domestically, regulation is shared among three agencies: the US Department of Agriculture (USDA), primarily through the Animal and Plant Health Inspection Service (APHIS); the EPA; and Health and Human Services through the Food and Drug Administration (FDA) (Grossman, 2002, pp. 223–227; McGarity, 2002, pp. 432–472). Developers must secure APHIS approval before field-testing a GMO crop and such tests are subject to inspection. Before commercial production can commence, an application for a 'determination of non-regulated status' must be filed with APHIS. Action on the petition is subject to notice and the comment procedures of the Administrative Procedure Act. If a GMO product contains a pesticidal substance, it is subject to further review by the EPA. The EPA exercises jurisdiction through the Federal Insecticide, Fungicide and Rodenticide Act (FIFRA) over GMO plants that contain pesticidal substances and over herbicides that are used in conjunction with herbicide-resistant plants. The FDA regulates the safety of new plant food products derived from the products of modern biotechnology. In 1992, the FDA determined that GMO foods are substantially equivalent to other foods and therefore do not require special labelling (US FDA, 1992). Since 1994, GMO foods have been evaluated in a voluntary consultation process. FDA has proposed a new rule that would require food developers to notify FDA at least 120 days in advance of their intent to market a food or animal feed developed through the use of modern biotechnology (US FDA, 2001). Developers would be required to demonstrate that the GMO product is as safe as its conventional counterpart, a review process that parallels the one making its way through the Codex Alimentarius. USDA's Food Safety and Inspection Service performs a similar regulatory function in respect of livestock or poultry consumed as food.

In international negotiations, the USA has insisted that restrictions on the production, consumption or distribution of GMO products be based on sound science, that the precautionary principle be employed only with extreme caution because of its potential to shield unlawful trade restrictions, that labelling requirements be restricted to information based on a conclusion that the GMO product differs from its conventional counterpart in some significant way, not merely that it is a product of the process of biotechnology, and that procedural requirements not be so onerous as to constitute a ban on trade in the products of modern biotechnology (Pomerance, 2000). While conceding that modern biotechnology raises legitimate regulatory issues in terms of both food safety and environmental protection, the USA is concerned that the promise of biotechnology not be sacrificed for political expediency.

Conclusion

The general direction of international efforts to address the issues raised by the products of modern biotechnology has been determined by the Biosafety Protocol. As gap-filling legislation, the Protocol poses little direct threat to established rules of international trade. In fact, the development of a Biosafety Clearing-house might actually enhance market access, especially if the exchange of information leads to a harmonization of rules and practices among states. At the same time, the Protocol addresses an issue largely beyond the competence or interest of the international trading community, namely the potential impact that unfettered trade in LMO products might have on conservation of biodiversity. In any event, much of the potential conflict between the WTO Agreements and the Biosafety Protocol was eliminated, or at least lessened, by the compromise that distinguishes for regulatory purposes between LMOs to be introduced intentionally into the environment and those to be used as food or feed or for further processing. In respect of the former, any conflicts between the language of the Protocol and the language of the WTO Agreements should be resolved presumptively in favour of the policies underlying the Protocol. In respect of the latter, any conflicts between the Protocol and the WTO Agreements should be resolved presumptively in favour of the policies underlying the WTO Agreements. The Agreements can then be viewed as complementary and not necessarily confrontational.

As with all international agreements, the final draft of the Biosafety Protocol was a product of compromise. Some critics will argue that the Agreement compromised the environment for the benefit of trade. Others may argue that excessive deference was paid to issues of national sovereignty at the expense of trade. Most, however, will understand that the Biosafety Protocol is an important development that builds on a growing consensus that trade and the environment are inextricably linked. Negotiators seem to have drafted an agreement that minimizes the potential for conflicts between the Protocol and the WTO Agreements, while meaningfully addressing a significant matter left open in the Convention on Biodiversity. Proper assignment of burdens of proof and burdens of persuasion could further harmonize the two agreements in ways that give each primacy in its dominant area of concern. Ultimately, however, the success of the Protocol will depend on the continuing willingness of parties to carry out their responsibilities in good faith and to avail themselves of some of the flexibility features of the Protocol.

References

Agreement Establishing the World Trade Organization (1994) Reprinted in *International Legal Materials* 33, 1125–1152.

Agreement on Technical Barriers to Trade, 15 April (1995) Marrakech Agreement Establishing the World Trade Organization, Annex 1A. Reprinted in *Australian Treaty Series* 3, 114–137.

Agreement on the Application of Sanitary and Phytosanitary Measures, 15 April (1994) Marrakech Agreement Establishing the World Trade Organization, Annex 1A. *United Nations Treaty Series* 1867, 493–507.

Basel Convention on the Control of Transboundary Wastes and Their Disposal (1989) Reprinted in *International Legal Materials* 28, 657–686.

Buechle, K. (2001) The great, global promise of genetically modified organisms: overcoming fear, misconceptions, and the Cartagena Protocol on Biosafety. *Indiana Journal of Global Legal Studies* 9, 283–324.

Carbone, J. and McLean, M. (2001) Genetically modified foods: the creation of trust and access to global markets. *Business and Professional Ethics Journal* 20, 79–104.

Cartagena Protocol on Biosafety to the Convention on Biological Diversity (2000) Reprinted in *International Legal Materials* 39, 1027–1046.

CEC (2001) *Proposal for a Regulation of the European Parliament and of the Council Concerning Traceability and Labelling of Genetically Modified Organisms and Traceability of Food and Feed Products Produced from Genetically Modified Organisms and Amending Directive 2001/18/EC*. COM(2001)182, amended by COM(2001)515 Brussels.

Codex Alimentarius Commission (2001) *Report of the Second Session of the Codex ad hoc Intergovernmental Task Force on Foods Derived from Biotechnology*. ALINORM 01/34A Rome.

Coletta, R. (2000) Biotechnology and the creation of ethics. *McGeorge Law Review* 32, 89–110.

Convention on International Trade in Endangered Species of Wild Fauna and Flora (1973) Reprinted in *International Legal Materials* 12, 1085–1104.

Crawley, M., Brown, S., Hails, R., Cohen, D. and Rees, M. (2001) Transgenic crops in natural habitats. *Nature* 409, 682–683.

Dickson, B. (1999) The precautionary principle in CITES: a critical assessment. *Natural Resources Journal* 39, 211–228.

EPA (2000) *Biopesticide Registration Action Document – Insect Resistance Management Section –* Bacillus thuringiensis *Plant-Pesticides*. United States Environmental Protection Agency, Washington, DC, 106 pp.

FAO (2001) *Evaluation of Allergenicity of Genetically Modified Foods – Report of a Joint FAO/WHO Expert Consultation*. Food and Agriculture Organization of the United Nations, Rome, 27 pp.

General Agreement on Tariffs and Trade (GATT) (1947) *United Nations Treaty Series* 55, 188.

Glowka, L., Burhenne-Guilmin, F. and Synge, H. (1994) *A Guide to the Convention on Biological Diversity*. IUCN – World Conservation Union, Gland, Switzerland, and Cambridge, UK, 161 pp.

Grossman, M.R. (2002) Biotechnology, property rights and the environment. *The American Journal of Comparative Law* 50, 215–248.

Gupta, A. (2001) Advanced Informed Agreement: a shared basis for governing trade in genetically modified organisms? *Indiana Journal of Global Legal Studies* 9, 265–281.

Harkin, T., Grassley, C. and Baucus, M. (2002) *Letter to the President of the United States, 20 December*. Distributed by the Office of International Information Programs, United States Department of State, Washington, DC.

Hart, K. (2002) *Eating in the Dark – America's Experiment with Genetically Engineered Food*. Pantheon Books, New York, 338 pp.

Harte, J. (2001) Land use, biodiversity, and ecosystem integrity: the challenge of preserving Earth's life support system. *Ecology Law Quarterly* 27, 929–965.

Hunter, D., Salzman, J. and Zaelke, D. (2002) *International Environmental Law and Policy*, 2nd edn. Foundation Press, New York, 1547 pp.

Kershen, D. (2002) Agricultural biotechnology: environmental benefits for identifiable environmental problems. *Environmental Law Reporter* 32, 11,312–11,316.

Kirwin, J. (2001) EU awaits sign from United States on trade complaint over GMO moratorium. *International Environment Reporter* 24, 158.

Krimsky, S. and Wrubel, R. (1996) *Agricultural Biotechnology and the Environment.* University of Illinois Press, Urbana, 294 pp.

Kyoto Protocol to the United Nations Framework Convention on Climate Change (1997) FCCC/CP/1997/L.7/Add.1. Reprinted in *International Legal Materials* 37, 32–42.

Lambrecht, B. (2001) *Dinner at the New Gene Café – How Genetic Engineering is Changing What We Eat, How We Live, and the Global Politics of Food.* Thomas Dunne Books, St Martin's Press, New York, 383 pp.

Liebenthal, A. (2002) *Promoting Environmental Sustainability in Development – An Evaluation of the World Bank's Performance.* World Bank, Washington, DC, 51 pp.

McGarity, T. (2002) Seeds of distrust: federal regulation of genetically modified foods. *University of Michigan Journal of Law Reform* 35, 403–510.

Mendelson, J., III (2002) Untested, unlabeled, and you're eating it – the health and environmental hazards of genetically engineered food. In: Kimbrell, A. (ed.) *The Fatal Harvest Reader – The Tragedy of Industrial Agriculture.* Foundation for Deep Ecology in collaboration with Island Press, Washington, DC, pp. 148–160.

Montreal Protocol on Substances that Deplete the Ozone Layer (1987) Reprinted in *International Legal Materials* 26, 1550–1561.

Nelson, A. (2002) Note: Legal liability in the wake of StarLink™: who pays in the end? *Drake Journal of Agricultural Law* 7, 241–266.

Pew Initiative on Food and Biotechnology (2002) *Three Years Later: Genetically Engineered Corn and the Monarch Butterfly Controversy.* Pew Charitable Trusts, Washington, DC, 19 pp.

Pomerance, R. (2000) The Biosafety Protocol: Cartagena and beyond. *New York University Environmental Law Journal* 8, 614–631.

Rotterdam Convention on the Prior Informed Consent (PIC) Procedure for Certain Hazardous Chemicals and Pesticides in International Trade (1998) UNEP/FAO/PIC/CONF/2. Reprinted in *International Legal Materials* 38, 1–23.

Safford, D. (2000) Risk-averse precautionary principle really ignores risk of inaction, panel says. *International Environment Reporter* 23, 798.

Safrin, S. (2002) Treaties in collision? The Biosafety Protocol and the World Trade Organization Agreements. *American Journal of International Law* 96, 606–628.

Schoenbaum, T.J. (1993) Agricultural trade wars: a threat to the GATT and global free trade. *St Mary's Law Journal* 24, 1165–1194.

Schweizer, G. (2000) The negotiation of the Cartagena Protocol on Biosafety. *Environmental Lawyer* 6, 577–602.

Speer, L. (2001) French court orders ministry to publish list, location of farms with biotech crops. *International Environment Reporter* 24, 205.

Staff (2000) Top administrative court in France rules farmers can grow, market GM corn seed. *International Environment Reporter* 23, 946.

Székely, A. (2001) Modified organisms and international law: an ethical perspective. *Transnational Lawyer* 14, 129–134.

Teitel, M. and Wilson, K. (2001) *Genetically Engineered Food – Changing the Nature of Nature,* 2nd edn. Park Street Press, Rochester, Vermont, 206 pp.

Thomas, V., Dailami, M., Dhaveshwar, A., Kaufmann, D., Kishnov, N., Lopez, R. and Wang, Y. (2000) *The Quality of Growth.* Oxford University Press, New York, 262 pp.

Tuna/Dolphin I, United States – Restrictions on imports of Tuna (1992) Reprinted in *International Legal Materials* 30, 1598–1623.

Tuna/Dolphin II, United States – Restrictions on Imports of Tuna (1994) Reprinted in *International Legal Materials* 33, 839–899.

Uchtmann, D.L. (2002) StarLink™– A case study of agricultural technology regulation. *Drake Journal of Agricultural Law* 7, 159–211.

United Nations Conference on Environment and Development (1992a) Convention on Biological Diversity. Reprinted in *International Legal Materials* 31, 818–841.

United Nations Conference on Environment and Development (1992b) Rio Declaration on Environment and Development, UNCED Doc A/CONF.151/5//rev.1. Reprinted in *International Legal Materials* 31, 874–880.

US Food and Drug Administration (FDA) (1992) Statement of policy: foods derived from new plant varieties. *Federal Register* 57, 22,984–23,005.

US Food and Drug Administration (FDA) (2001) Premarket notice concerning bioengineered foods. *Federal Register* 66, 4706–4738.

Vienna Convention on the Law of Treaties (1969) *United Nations Treaty Series* 1155, 331–512.

Weiskopf, J. (1999) Talks on biosafety protocol suspended: 'Miami Group' thwarts compromise accord. *International Environment Reporter* 22, 177.

Werner, K.L. (2000) Restrictions to prevent insect resistance announced for genetically modified corn. *International Environment Reporter* 23, 90.

World Trade Organization (WTO) (1998a) *EC Measures Concerning Meat and Meat Products (Hormones)*. Appellate Body Report, 13 February. World Trade Organization, Geneva, Switzerland.

World Trade Organization (WTO) (1998b) *Shrimp/Turtle, United States – Import Prohibition of Certain Shrimp and Shrimp Products*. Appellate Body Report, 6 November. World Trade Organization, Geneva, Switzerland.

World Trade Organization (WTO) (2001) *European Communities – Measures Affecting Asbestos and Asbestos-Containing Products*. Appellate Body Report, 5 April. World Trade Organization, Geneva, Switzerland.

Environmental Policy and the Reform of European Agriculture Law

11

Christopher Rodgers

Introduction

Since its inception, the Common Agricultural Policy (CAP) of the European Community (EC) has been based on the objectives of increasing agricultural productivity and increasing agricultural earnings through the promotion of stable internal markets for European produce. These remain its principal objectives, enshrined in Article 33.1 of the EC Treaty. The promotion of policies aimed at reducing the environmentally harmful effects of intensive agriculture is an issue that has been addressed only comparatively recently in the European Union (EU). The importance attached by European policy-makers to the integration of environmental-protection measures into agricultural law is now, however, quite considerable. This is reflected in the role which the EC has assumed in the renegotiation of the 1994 WTO Agreement on Agriculture ('URRA'), where it has been one of the principal proponents of the importance of the need to balance trade concerns (such as market access, export competition and domestic support) with 'non-trade concerns' that reflect wider societal goals. Although the concept encompasses a range of policy goals that are alleged to be non-trade-distorting, including animal welfare and food safety measures, the most wide ranging and potentially problematic category of 'non-trade concern' in the WTO negotiation is the protection of the environment.

The road from a strictly production oriented CAP to a position where sustainability and environmental protection are at the heart of European agriculture policy is one that the EC has travelled very quickly. In order to understand the position the EC has taken on this issue in the Millennium Round of WTO negotiations, it is necessary to consider the manner in which environmental concerns have emerged as a policy issue within the EC, and in particular the legal framework within which environmental policy has been incorporated into the law of the CAP. This chapter will consider the emergence of agri-environmental policy and its implementation within the legal framework of the CAP, and conclude with a consideration of the role which environmental protection may play as an important 'non-trade concern' in the future legal order for international agricultural trade.

The Origins of EC Agri-environmental Policy

Changes in the CAP to reorient it towards environmental objectives have been driven by two imperatives: one internal to the legal order of the EC itself and the other external, in the form of changes to pre-existing support arrangements to prepare for the renegotiation of the URAA. The reform measures initiated under the Agenda 2000 reform programme were aimed at putting the EC in a stronger negotiating position in the Millennium Round and were in part premised on a perceived need to avoid the problems surrounding the Uruguay Round of WTO negotiations in the early 1990s and the last substantial CAP reform in 1992.

The primary internal catalyst for a change in the legal order of the CAP and its environmental reorientation was the adoption of the environmental integration principle in EC law. The 1987 Single European Act added new articles to the Treaty to give environmental protection a legal basis as one of the overall objectives of the Community. These are now to be found in Arts 174–176 of the EC Treaty. After the ratification of the Maastricht Treaty, it became a legal obligation for the Community to integrate environmental protection requirements into the definition and implementation of all Community policies, in particular with a view to promoting sustainable development (see Art. 6 EC Treaty: the promotion of a common policy in the sphere of agriculture and fisheries is specified as a Community activity by Art. 3.1(e) of the EC Treaty). This obligation was reinforced in the Amsterdam Treaty, which also gave legal force to sustainable development as an objective of the EC. The CAP accounts for approximately 50% of the EC budget, and the effective integration of environmental protection requirements into its operations is therefore a key objective if the EC is to achieve the overall aim of the treaty provisions on sustainable development.

The primary external pressure for change in the CAP producer-support arrangements has come through the WTO and the need to prepare for the renegotiation of the 1994 Agreement on Agriculture. The 1992 CAP reform saw a partial decoupling of support for agricultural production. Under the 1992 'McSharry proposals', supply-control measures were introduced for some agriculture sectors, for example by the introduction of quotas on livestock subsidy payments paid on a headage basis to beef and sheep meat producers, and a mandatory requirement to set aside a proportion of land down to arable crops was introduced before arable payments could be paid. These measures were accompanied by reductions in guaranteed prices, aimed at bringing the price of EC products down to world market levels (e.g. for beans, beef and cereals).

To offset the economic impact of these measures on producers, a compensation package was introduced, which involved the payment of arable area payments to cereals farmers for land planted to crops annually and increases in beef and sheep headage payments to livestock producers payable on the number of animals they maintain. The switch away from product support and towards the increased use of direct support payments to producers was a key feature of the 1992 reform, although not entirely new. Direct support for beef producers is provided through the Beef Special Premium Scheme and Suckler Cow Premiums. The Beef Special Premium Scheme predates the 1992 reforms, having operated since 1989 (under Council Regulation (EC) 468/87 and Commission Regulation (EC) 859/87 and 714/89). Similarly, Suckler Cow Premiums have operated as a direct subsidy to farmers since 1980 (under Council Regulation (EC) 1357/80 as amended). Both schemes were, however, substantially reorganized under the 1992 McSharry reform package.

Importantly, the 'accompanying provisions' agreed as part of the reform package also introduced a greater environmental element into the reformed CAP. Of particular importance was the 'agri-environment' regulation (Council Regulation (EC) 2078/1992), under which the financing of agri-environmental policy was transferred to the Guarantee section of the European Agricultural Guarantee and Guidance Fund (EAGGF). This associated agri-environmental policy with the mainstream of CAP expenditure and meant that no financial limit was placed on initial expenditure commitments. However, designing and then implementing legal and economic instruments to achieve the required incorporation of environmental policy objectives into the operation of the reformed CAP have proved difficult. This is due in no small part to the close interrelationship between farming and the environment. The agricultural sector presents unique problems for the development of environmental policy and new regulatory instruments. Agricultural production involves shaping and using natural resources – it is an integral part of the natural environment, not external to it, and by its nature creates, shapes and preserves ecosystems. European policy-makers now recognize that agricultural production methods can have both beneficial and detrimental effects on local ecosystems and also on landscape values (see CEC, 2000a, esp. pp 10ff.).

Agriculture has shaped the European landscape, creating in the process an environment with a rich variety of animal and bird species, many of which are dependent on the continuation of farming. The market-management and price-support policies pursued within the CAP have, however, led to an intensification of farming practices. That this has led to problems for landscape preservation and biodiversity, and to problems of water, soil and air pollution, has only recently been recognized and addressed by European policy-makers. The EC's Fifth Action Programme on the Environment selected agriculture as one of the five target sectors for special attention, and established a range of objectives and target measures to be pursued in the period up to 2000 (CEC, 1993, esp. C138/38). This was accompanied by an explicit recognition that one of the effects of CAP expenditure across the Community had been an overemphasis in some areas on production levels, resulting in overintensification, with consequent degradation of the natural resources on which agriculture itself ultimately depends. The policy response at EC level has therefore been based on promoting the key concept of 'sustainable' agriculture. This raises special problems, unique to the law of the CAP. The establishment of indicators to evaluate sustainability is a complex and challenging task, raising complex issues of the interaction of agricultural practice with the environment. Moreover, the implementation of environmental policy within the CAP has required the adoption of new measures to give legal effect to sustainability principles in modern agriculture.

Since the adoption of the Single European Act in 1987, a number of legal instruments have been introduced aimed at 'greening' the CAP. Initially, agri-environmental measures were closely linked with market management within the CAP, and their environmental focus was, as a consequence, often blunted. The earliest agri-environment measures were introduced under the CAP farm structures legislation, most notably Council Regulation 797/85 on improving the efficiency of agricultural structures. This permitted Member States to establish zonal programmes to encourage the adoption of traditional farming methods in environmentally vulnerable areas. The impetus towards adopting an environmental agenda within the CAP was substantially strengthened by the 'accompanying measures' adopted under the McSharry reform package prior to the conclusion of the WTO Agreement on Agriculture in 1994. The environmental aspects of the 1992 CAP reform were to some extent peripheral to the central thrust of the overall reform package. This was aimed at reducing overproduction of many agricultural commodities within the CAP regime, and (at the same time) at responding to the needs of the (then) ongoing General Agreement on Tariffs and Trade (GATT) negotiations leading to the 1994 agreement. Nevertheless, the adoption of the 1992 Agri-environment Regulation marked a new departure. For the first time, European law required Member States to draw up agri-environmental programmes and submit them to the Commission to a set time-scale.

Three objectives for EU agri-environmental policy were laid down in Article 1 of the Agri-environment Regulation, namely to accompany the changes to be introduced under the commodity-management rules, to contribute to the Community's objectives regarding agriculture and the environment and to contribute to providing an adequate income for farmers. The focus was, therefore, at least in part socio-economic as well as environmental. Nevertheless, the Agri-environment Regulation provided Community cofinancing for a number of possible agri-environmental measures. It set out seven objectives that national schemes could aim to promote: the use of farming practices that reduce the polluting effects of agriculture, an environmentally favourable extensification of arable farming, ways of using farmland that are compatible with protection of the environment and natural resources, the upkeep of abandoned farmland and woodlands, long-term set-aside of land from production for environmental reasons, the management of land for public access and recreation and education/training for farmers in environmental protection requirements and countryside protection. The zonal programmes set up under earlier EC farm structures regulations were subsumed into the Agri-environment Regulation programmes, for which more generous community financing was available. Environmental requirements were also introduced into the operation of some of the common market organizations for produce within the CAP, e.g. the arable regime was reformed to require the management of land set aside from production for environmentally beneficial purposes. This was achieved by introducing a new technique of environmental regulation into the CAP – 'cross-compliance' – making the receipt of many subsidies for farming conditional upon the observance of basic environmental safeguards.

The 1992 reform process made a significant contribution towards increased transparency in the operation of the CAP by shifting public subsidy away from product support and towards the direct payment of support to farmers. The breadth of the programmes that were permitted, however, meant that there was uncertainty as to the primary objectives of the measure. While some Member States used the Agri-environment Regulation to introduce new environmental schemes with EC cofinancing, others (especially the Mediterranean states) viewed its primary focus in terms of providing an appropriate income for farmers (Buller, 1999, esp. pp. 200ff.). The very diversity of the stated objectives for the new agri-environmental policy therefore blunted its wider impact as a vehicle for introducing new environmental measures. The continuing link with commodity management also meant that agri-environmental measures were viewed as subsidiary to the market-support functions of the CAP and, moreover, led to tensions between market and agri-environmental policies. To take just one example, the environmental benefits achieved by the introduction of set-aside and the more rational use of fertilizers were largely offset by the encouragement of intensive crop production by the regionalization of direct payments for cereals, and by an intensification of livestock production encouraged, for example, by subsidizing silage crops (CEC, 1997a, p. 23; Winter, M., 1999). The continuing link with commodity management and market support is a particular problem, which, as we shall see, the subsequent Agenda 2000 reforms have not satisfactorily dealt with.

Commentators differ over the impact that the Uruguay Round had on the overall shape of the 1992 reform of the CAP. Some hold that the Uruguay Round had little impact on reforms that were primarily driven by problems of overproduction and spiralling budgetary costs (Paarlberg, 1997). The more persuasive view, however, is that the extent and shape of the 1992 CAP reform were conditioned by interpretations of what policies might successfully be enshrined in the WTO agreement following the Uruguay Round, and that the European Commission had to consider not only what policy instruments might be configured to resolve internal difficulties, but also whether possible sets of policy instruments would be acceptable to the international forum and lead to a successful conclusion of the WTO negotiations (Coleman and Tangermann, 1999). The overall level of subsidization of the agricultural sector was not reduced by the 1992 reform. Rather, the reform changed the form of producer support measures away from explicit export subsidies to compensatory payments in a variety of forms. In so doing, the EC exploited to the full the 'blue box' peace clause negotiated under the Blair House accord. This provides that payments to farmers for production control or as compensation for price cuts are exempt from the 20% reduction in the aggregate measure of support (AMS) otherwise required under the URAA This has, until now, exempted partially decoupled payments, such as those made under the arable area payments scheme, under which farmers are paid compensation for cuts in domestic price support on the basis of areas sown to arable crops, and the reformed arrangements for headage-based payments to livestock producers. And, of course, payments under environmental schemes established within the framework of the 1992 Agri-environment Regulation qualified for exemption from AMS undertakings under the 'green-box' exemption for policy instruments that are minimally trade-distorting.

Agenda 2000: a 'Greener' Direction for the CAP?

Because the overall level of subsidy administered by the CAP has remained undiminished following the 1992 reform, these arrangements were always likely to be a focus of particular controversy in the Millennium Round and possibly (if unreformed) subject to fierce challenge (Lowe and Brouwer, 1999). The subsequent Agenda 2000 CAP reform process was therefore strongly conditioned by the need to move towards a fuller decoupling of support and a switch towards a greater use of fully decoupled payments for environmental stewardship, qualifying for exemption from AMS reductions under the 'green box' arrangements in the WTO Agriculture Agreement. The Agenda 2000 measures were also, however, heavily conditioned by the shape of the pre-existing support arrangements that emerged from the Uruguay Round and the associated 1992 CAP reform. The constraint for radical change in the CAP since 1992 has therefore been minimized by the switch away from export subsidies, which are not allowed under WTO agreements, and towards compensatory payments which were allowed under the 1994 URAA and Blair House accord (Scott, 1996). The Agenda 2000 reform package retained compensatory payments as the principal agricultural support technique. The EC's stance in the 2001/2005 Millennium negotiating round has been to vigorously negotiate and exploit, as it did in the Uruguay Round, a wide variety of exemptions (i.e. green- and blue-box measures) which do not qualify for AMS reduction. The EC's Comprehensive Negotiating Proposal argues for the retention of the blue box, and for a revision of the green box to include qualifying criteria that ensure minimal trade distortion while at the same time allowing appropriate coverage of measures that meet 'important societal goals' – of which the protection of the environment is one of the most important (WTO, 2000b, paras 12 and 13). The development of a fuller range of agri-environmental policy instruments under the Agenda 2000 reform process and a reformulation of the legal order of the CAP to integrate environmental protection requirements are integral to this strategy.

Under Agenda 2000, the continuing expansion of agri-environmental measures has been set in the wider context of the development of rural development policy. When the European Commission published its original Agenda 2000 communication in 1997, it proposed an expansion of rural development policy to enable agriculture to adapt to changes in market evolution, market policy and trade rules, as well as the need to promote sustainability in land use (CEC, 1997b). The proposals envisaged deepening and extending the 1992 CAP reforms in a way which, if implemented, would produce a significant shift in agricultural policy towards the introduction of a multifaceted concept of agriculture and the development of an integrated rural development policy. In response, the Berlin European Council signalled that the content of the reform to be implemented from 1 January 2000 should be aimed at securing a multifunctional, sustainable and competitive agriculture throughout Europe.

The concept of 'multifunctionality' lies at the core of the Agenda 2000 reform process. The need to renegotiate the URAA to recognize its importance is also a key element in the EC position in the Millenium Round, where it has taken the lead in initiating the group of 'friends of multifunctionality', e.g. by hosting a meeting in July 2000 to promote the concept as a policy agenda to like-minded WTO Members. The legal basis for this wider negotiating agenda is, however, somewhat insecure. Article 20 of the URAA commits WTO Members to taking account of non-trade concerns in the renegotiation of the Agreement on Agriculture, but does not explicitly mention the concept of multifunctionality (Landau, 2001). Although proponents of multifunctionality stress that the management and enhancement of the rural environment and landscape are among the most important functions of a multifunctional agriculture sector, the concept takes in a number of other non-productive elements, including the protection of human, animal and plant health, food safety and food quality, and other consumer concerns relevant to agriculture (WTO, 2000a). Indeed, the concept is somewhat malleable and can encompass many non-productive aspects of modern agriculture (see Chapter 6 by Cardwell in this volume). This wider agenda may distract attention from the importance of environmental protection measures as a key non-trade concern at issue in the renegotiation.

Despite the new emphasis on 'multifunctionality' as a cornerstone of European farm policy, however, concerns that the inbuilt dynamics of the CAP following the 1992 reform would militate against fundamental change have to some extent proved justified. Instead of offering a radical new approach to sustainability and environmental issues, the Agenda 2000 reforms were in large measure an extension of the 1992 measures, and were premised on the need to continue the progress made since 1992 in reducing institutional surpluses and in introducing further agri-environmental measures on the model of the 'accompanying measures' adopted in 1992. The tenor of the Agenda 2000 reforms was therefore one of incremental change, not radical innovation.

An inevitable consequence of this approach has been that agri-environmental measures have remained linked to commodity management and hence are viewed, by many, as subsidiary to the principal market-support function of the CAP. Nevertheless, the reform of the financial structures and organization of the CAP introduced under Agenda 2000 moved agri-environmental policy into the broader framework of the new policy on rural development and, in so doing, strengthened its role within the framework of the CAP. Agenda 2000 also instituted a major reorientation in the administration and policy goals of agricultural policy, with rural development becoming the second pillar of the CAP alongside market management. Council Regulation (EC) 1257/99 (the 'Rural Development Regulation') brought together all previous rural development measures, including the 1992 accompanying measures on agri-environment, early retirement and forestry, into one composite framework regulation. The aim was to lay the foundation for what the European Commission claimed would be 'a comprehensive and consistent rural development policy whose task will be to supplement market management by ensuring that agricultural expenditure is devoted more than in the past to spatial development and nature conservancy, the establishment of young farmers etc.' and other development programmes in rural areas (CEC, 2000b).

The reorientation of policy towards rural development initiatives by Agenda 2000 was matched by a commitment to significantly higher expenditure on rural development and environmental measures. For each of the years 2000 to 2006, rural development expenditure was budgeted to run at between 4300 million and 4370 million Euros. Total CAP expenditure during the same period was budgeted to be between 40,920 million Euros (in 2000 itself) and a maximum of 43,900 million Euros (in 2002).

'Sustainable' Agriculture and the Legal Order for the CAP

From an environmental perspective, perhaps the most important issue underlying the Agenda 2000 reform process is a new emphasis on sustainability. The reorientation of agricultural policy towards the promotion of 'sustainable agriculture' is a key development underpinning the move to integrate environmental policy into the CAP, and encapsulates the desired relationship between agriculture and the environment that the Agenda 2000 reforms sought to achieve. The European Commission's original proposals for the introduction of sustainable agriculture called for the management of natural resources in a way that ensures that their benefits are also available in the future (CEC, 1999). Inasmuch as oversupply within the existing common market organizations implies a waste of resources, it can be claimed that the reformed market-management measures under Agenda 2000, whose principal aim was to bring demand and supply into closer equilibrium, were themselves a move towards sustainable development in the agriculture sector. Subsequent communications from the European Commission, however, recognized that a broader understanding of sustainability is required, extending to a larger set of features linked to land use, such as the protection of habitats, landscapes and biodiversity, as well as the prevention of pollution to drinking water and air (CEC, 2000a, esp. para. 1.3.1.2). Sustainability must, therefore, take account of the protection of the environment and the cultural heritage. Importantly, however, the Commission has also emphasized the continuing social role of agriculture and its importance to the sustainability of rural communities. Sustainability principles must also reflect society's concerns in regard to the social function of agriculture, especially the maintenance of viable rural communities and a balanced pattern of development.

The Commission's wide interpretation of sustainability principles, allocating a social role to agriculture, continues a policy strand within the CAP that underlie earlier measures, such as Directive 75/268/EEC on Farming in Less Favoured Areas and Title 7 of Council Regulation (EEC)

797/85 on farm structures, which sought to preserve the socio-economic fabric of those rural areas which are dependent upon agriculture as the dominant economic activity. To some extent, critics may argue that the new policy agenda is simply a repackaging of existing policy imperatives. Nevertheless, the establishment of sustainability as a key objective for European agriculture will undoubtedly involve a radical reorientation of the CAP, with much greater emphasis being placed on the importance of environmental stewardship and the protection of natural resources.

The effective implementation of the new policy by the Member States will generate a number of novel legal problems. Council Regulation (EC) 1259/1999 ('the Horizontal Regulation') laid down common rules to be applied by all Member States in the administration of direct support schemes under the CAP. For the first time, environmental safeguards must be mandatorily applied by Member States to all direct support payments, although a measure of discretion is given in the manner in which this is done. Article 3 of the Horizontal Regulation requires Member States to apply those environmental measures they consider appropriate in view of the land use and type of production concerned. Under the 1992 measures, the imposition of cross-compliance conditions was optional, and (not surprisingly) many Member States failed to impose environmental conditions on the receipt of support payments. Under the implementing regulations for Agenda 2000, this is not only a mandatory feature of direct support schemes, but Member States must apply cross-compliance conditions to direct support payments in all sectors, not just to livestock and (in relation to set-aside) arable producers. Moreover, the switch to an increased reliance on decoupled direct payments to producers, rather than product support, means that there is now a greater range of payment regimes to which cross-compliance can and must be applied.

Although attaching environmental compliance conditions to direct support payment regimes is an important step forward, this was not the only legal mechanism deployed to integrate environmental policy into the operation of the reformed CAP. It was coupled with the environmental reorientation of a number of existing policy instruments, for example the compensatory payments regime for livestock producers under the Less Favoured Areas (LFA) policy. And Member States are now empowered to use agri-environmental schemes to target problems or areas where special protection measures are required to safeguard natural habitat or landscape features. They can also attach specific environmental conditions to the granting of payments under a market regime where the environmental situation requires targeted extra measures, in addition to the mandatory cross-compliance measures applied to all producers under the direct-payment regime in question.

Although the introduction of cross-compliance within the common market organizations has an important role to play, therefore, the European Commission envisaged the development of a combined approach, using both cross-compliance techniques and more targeted measures aimed at particular habitats and landscapes. This is significant because cross-compliance is arguably an inappropriate tool with which to address some of the priorities for the reform of agri-environmental measures highlighted by the Commission. These include the improved targeting of environmental policy to site-specific problems on particular farms (or in particular localities) and the use of whole-farm plans to develop a more holistic approach to environmental protection (CEC, 1998, pp. 126–127). As currently applied, cross-compliance requirements apply across whole market sectors within the CAP in an unfocused and non-targeted manner. The development of agri-environmental measures within the context of the Rural Development Plan (the first of the options allowed by the Horizontal Regulation) may offer greater benefits. Moreover, if Member States choose to use an implementation strategy based on the use of contractual agreements, this will have the added merit of introducing greater flexibility and greater scope for the improved targeting of agri-environmental measures.

The measures in the 1999 Rural Development Regulation were premised on the need to ensure that agri-environmental schemes are WTO-compliant. The 1999 regulation lays down a number of important principles to be followed in scheme design, monitoring and evaluation and puts in place a rigorous regime for the monitoring of scheme performance and outcomes by the European Commission. The current WTO green-box exemption for payments under agri-environmental programmes requires eligibility for payments to be determined as part of a clearly defined government environmental or conservation programme and to be dependent upon the fulfilment of specific conditions under the government programme in question, including conditions relating to production inputs or methods. Importantly, the amount of payments under the scheme in question must be limited to the extra costs or loss of income involved in complying with the government programme (Annex 2, para. 12 of the URAA). The arrangements for EC agri-environmental schemes post-Agenda 2000 are aimed at coming within these conditions and exploiting the green-box exemption from AMS reduction that they offer.

A fundamental feature of the legal order for the CAP is that agri-environmental measures must adhere to the requirement that farmers be required to meet conditions going beyond what is required by the dictates of good agricultural practice.

> The philosophy underpinning the environmental aspects of the CAP reforms is that farmers should be expected to observe basic environmental standards without compensation. However, where society demands that farmers deliver an environmental service beyond the base line level, this service should be specifically purchased through the agri-environment measures.
>
> (CEC, 1999, para 3.2.1)

This philosophy underwrites the detailed measures contained in both the 1999 Horizontal Regulation and the Rural Development Regulation. It was applied, for example, to ensure that Member States' rural development plans only apply targeted measures with appropriate payment where the environmental goal to be achieved cannot be met by cross-compliance measures attached to direct producer payments, and subject to the requirement that minimum environmental standards should be observed or result from the action funded (for examples, see Council Regulation (EC) 1257/1999, Arts. 5 (investment), 26 (improving processing and marketing) and 31 (afforestation)).

For activities going beyond the baseline standard of good agricultural practice, agri-environmental measures with incentive payments will normally be applicable. But Article 23.2 of the Rural Development Regulation expressly requires that agri-environmental commitments shall involve more than the application of usual good farming practice. Support under agri-environmental schemes must be for commitments for at least 5 years, and support is to be granted annually on the basis of income forgone by participating farmers, additional costs and 'the need to provide an incentive' (Council Regulation (EC) 1257/1999, Art. 24). The reference point for calculating income forgone and the additional costs to the farmer of implementing the agri-environmental undertaking given must be the usual farming practice in the given area where the measure applies (Commission Regulation (EC) 445/2002 Art. 17). The level of incentive payment to be applied is left to the Member States to determine, but it must be determined on the basis of objective criteria, and a ceiling is applied limiting the incentive element to a maximum of 20% of income forgone and the additional cost of carrying out the undertaking given (Commission Regulation (EC) 445/2002, Art. 18). In the WTO context, the use of incentive payments could be problematic, as the current Agreement on Agriculture limits green-box exemption to environmental schemes with payments based on loss of income and compliance costs (Annex 2, para 12(b) of the URAA). The maximum aid that can currently be paid is 600 Euros/ha for annual crops, 900 Euros/ha for specialized crops and 450 Euros/ha for all other land uses.

Thus, the law on the CAP now establishes 'good agricultural practice' as a normative standard. This has property-rights implications, in that compensation payments for changes in land use will only be made if the land management concerned goes beyond what is considered to be good practice (see Rodgers, 2002). Landowners will be required to respect general requirements concerning environmental stewardship, without specific payment for doing so. In other words, landowners will be expected to bear environmental compliance costs up to a reference level of good agricultural practice reflected in property rights (CEC, 2000a). This is an attempt to apply the polluter pays principle to agriculture. It is likely to be problematic, however, not least because of the elasticity of the concept of 'good practice'. Clearly, good agricultural practice is a concept whose content will vary from one location to another, and it is entirely dependent on the nature of the husbandry practised in a locality and its historical interaction with wildlife, the landscape and the wider environment. It is not a homogeneous concept capable of any precise definition with universal validity. The implementing regulations for the Agenda 2000 reforms recognize this, and define 'usual good farming practice' as the standard of farming that a reasonable farmer would follow in the region concerned (Commission Regulation (EC) 445/2002, Art. 28). They require Member States to formulate verifiable standards in their rural development plans, with environmental indicators against which the impact of agricultural practice on the environment can be measured. The rigour with which Member States formulate environmental indicators will be fundamental to the success of the Agenda 2000 reform process and its impact on the environment, as will effective monitoring by the European Commission.

Implementing Environmental Policy

Essentially, the Agenda 2000 reform adopted a threefold approach to the integration of environmental protection requirements into the operation of the CAP. At a general level, attaching environmental compliance conditions to the various direct support payment regimes was an important step forward. This was supplemented in the new law of the CAP by amendments to the legal basis of a number of existing policies, aimed at integrating environmental issues into their market operation. And Agenda 2000 envisaged an extension in the use by the Member States of agri-environmental schemes to target special problems or areas where special protection measures are required to safeguard natural habitat or landscape features. The implementation of these new policy instruments at the level of the national legal order in the Member States reveals a number of issues of relevance to the debate over the WTO compatibility of existing EC subsidy arrangements.

Less Favoured Areas policy: a paradigm of change

A consideration of the impact of the Agenda 2000 reform on each of the various market organizations for agricultural products within CAP is beyond the scope of this chapter. A good example of the environmental reorientation of existing EC support schemes is, however, provided by the LFA policy, under which compensatory allowances are paid to livestock producers in disadvantaged areas to compensate for the handicaps of farming in difficult terrain and adverse weather/climatic conditions. As already noted above, this policy has a social dimension, in that Member States can also designate areas as 'Less Favoured' if they are in danger of depopulation, in circumstances where the predominance of infertile land and low productivity renders reliance on agriculture as the primary economic activity problematic. The policy was adopted by the EC in 1975 following the entry of the UK, which wanted the freedom to continue to give special help to hill farmers. Ensuring the continuation of hill farming, and thereby maintaining a minimum population level in sparsely populated rural areas, as well as conservation of the countryside, were key elements of the original LFA policy.

The environmental impact of the LFA policy, though considerable, was not initially an imperative. The scheme was formally established in 1975 under Council Directive 75/268, and facilitated the payment of compensatory allowances to livestock producers in LFA calculated by reference to the number of animals maintained on a holding for the claim period. The importance of this additional income for hill livestock producers cannot be overestimated. The bulk of LFA expenditure (80% of all allowances paid) is concentrated in four north-western EC members: France, Germany, the UK and Ireland. LFA payments in these countries have in recent years averaged approximately 2000 Euros per beneficiary holding (Dax and Hellegers, 1999). Inevitably, a system based on headage payments encouraged higher stocking levels of sheep and cattle in upland areas, and this has in turn resulted in an intensification of grazing pressures in sensitive landscapes and areas of seminatural habitat. This was partially addressed, prior to Agenda 2000, by the introduction of ceilings on stocking rates for the purposes of claiming hill livestock compensatory payments under the scheme – for example a maximum stocking rate of 6.6 ewes/ha was fixed for sheep payments. The ceilings for producer support were, however, often fixed at an artificially high level and bore no relation to the stocking capacity of upland holdings. Moreover, the livestock ceilings for support payments were often seen as 'targets' for livestock producers seeking to maximize income under the LFA arrangements. A criticism often made of the LFA policy was that it encouraged increased stocking levels on upland holdings by farmers wishing to maximize the compensatory allowances to which they were entitled – and that it thereby encouraged overgrazing in upland areas (Wathern, 1992). Because of the fragility and high nature value of many upland habitats and landscapes, it is alleged that this has had serious environmental consequences.

Under the Agenda 2000 reform, LFA policy has now been integrated into the rural development 'pillar' of the CAP, alongside the administration of agri-environmental policy. As a consequence, Member States must incorporate policy for hill livestock support in disadvantaged areas into their rural development plans. The legal basis for the operation of LFA policy and the administration of compensatory payments to livestock producers has also been fundamentally reformed. Article 14 of the 1999

Rural Development Regulation provides that all compensatory payments must in future be based on an area calculation and not paid on a headage basis (as was previously the case) according to the number of livestock on a holding.

The significance of the changes can be seen from the revised payment arrangements for livestock farmers in the UK. Before the Agenda 2000 reform, hill livestock producers received Hill Livestock Compensatory Allowances (HLCAs) based on the number of ewes or cattle maintained on the holding annually. The changes in the legal basis of the LFA policy have, however, resulted in the introduction of new schemes in the Rural Development Plan for England and its counterpart for Wales. Under the Hill Farm Allowance scheme, from 2001 LFA support in England has been paid on an area basis and made conditional on the use of sustainable farming practices. There are two elements to payments under the scheme. All eligible producers receive a basic area payment, defined by reference to the land classification and size of the holding. In addition there are top-up payments of either 10% or 20% that can be claimed by producers meeting environmental enhancement criteria. The latter include maintaining at least 5% of the holding under arable or woodland cover, converting land to organic production and maintaining mixed stocking of cattle and sheep to secure beneficial grazing patterns. Maintenance of very low stocking rates of below 1.0 livestock unit/ha is also rewarded by an environmental enhancement payment (see reg. 7(2),(3) of the Hill Farm Allowance Regulations 2002 (SI 2002, No. 271)). Because some producers will suffer under the change to an area-based payment system, the scheme was safety-netted for the first 3 years, to 2003.

Similar arrangements have been introduced under the *Rural Development Plan for Wales 2000–2006*. The arrangements for LFA payments in Wales under the Tir Mynydd scheme are, however, more complex than the parallel arrangements in England. Over 77% of Wales is designated as LFA, and livestock farming in the Principality is heavily dependent on the scheme. Hill-livestock payments under Tir Mynydd are made up of two elements: an area payment per hectare calculated by reference to the eligible forage area on each holding (differentiated according to disadvantaged and severely disadvantaged areas) (element 1) and an environmental incentive payment to reward compliance with a range of sustainable farming indicators (element 2). Depending on the number of points awarded to a producer under the latter, the environmental incentive payment (or 'top-up') will be an increase in payment of either 10% or 20%. If producers earn insufficient points to spend the budget allocated to the environmental enhancement fund, unused grant will be reallocated to producers *pro rata* under element 1, i.e. on an area basis according to the size of the producer's holding. Under this funding model, compliance with environmental enhancement indicators is essentially optional: moreover, if insufficient producers participate in the environmental-incentive element of the fund, the relevant funding will simply be reallocated to all LFA producers under element 1 on an area basis, according to the size of the holding (see generally The Tir Mynydd (Wales) Regulations 2001, as amended by SI 2002, No. 1806).

Clearly, the thrust of the changes to the LFA regime made by the 1999 Rural Development Regulation were intended to decouple livestock payments and to present the new arrangements as non-trade-distorting direct payments within the green box for the purposes of the WTO Agreement on Agriculture. LFA schemes approved by the European Commission under the new rural development policy will undoubtedly be closely monitored for green-box compliance in the Millennium Round. Neither the Hill Farming Scheme nor Tir Mynydd would appear to come within the current green-box exemption for decoupled income support (Annex 2, paras 1 and 6 of the 1994 URAA), as they require continued agricultural production as a precondition for the receipt of payments and as payment rates are linked to the type of livestock maintained on the holding (for example sheep and suckler cows in the case of Tir Mynydd). Neither do they fall within the exemption for payments under environmental programmes (Annex 2, para 12 of the URAA) – indeed, in the case of Tir Mynydd non-participation in environmental stewardship may even be 'rewarded' if funding is redistributed on an area basis. Even if they are accepted as minimally trade-distorting (Annex 2, para 1), therefore, the support arrangements would not appear to meet the additional policy-specific conditions for green-box exemption under the 1994 URAA. This may focus attention on the *de minimis* rule (currently Art. 6(4)), under which domestic support that does not exceed 5% of the total value of production of a specific product or, in the case of non-product-specific support, 5% of the value of total agricultural production, is not included in the AMS calculations. Support for British livestock producers under the Hill Farming and Tir Mynydd schemes would appear to be product-specific in that participation is limited to sheep and beef producers. Given the value of LFA support to hill farmers in the UK and the poor market return for sheep meat and beef in the current economic climate, there must be a question whether the current level of support comes within the limit for *de minimis* exemption from the current WTO undertakings.

Agri-environmental schemes: programme design and implementation

The implementation of the new agri-environmental measures in differing Member States may entail the use of a range of different legal instruments, depending on the legal tradition and structures of the national legal order. The 1999 'Horizontal' Regulation, as we have seen, gives Member States several options in the choice of legal mechanism for implementing agri-environmental policy within the framework of the new rural development regime. A key factor in shaping an appropriate implementation strategy concerns the interrelationship of these mechanisms both with the market itself and with the market-management instruments of the common organizations for agricultural produce under the CAP.

Many Member States currently implement agri-environmental measures using environmental contracts. The Fifth Action Programme on the Environment set a target of 15% of the utilized agricultural area (UAA) the EC to be under management contracts for the maintenance of natural habitats and minimizing natural risks (such as erosion) by 2000 (CEC, 1993, pp. C138/38). In statistical terms, progress towards this target across the Community has been good. Reviewing the implementation of agri-environmental programmes introduced under the 1992 reform, the European Commission found that, at the mid-point in the 1997 budget year, 1.35 million agreements had been concluded with farmers, covering 17% of the UAA of the EC at that time (CEC, 1997a, pp. 16–17). Behind the statistics, however, a closer examination reveals a more prosaic picture, with some Member States adopting agri-environmental measures across their whole territory with contracts imposing generalized and light obligations, while other Member States have adopted a targeted approach using agreements focused and available only in high nature- or landscape-value areas. The take-up of agreements also varies enormously from one Member State to another. Indeed, the Commission's 1997 review of agri-environmental programmes concluded that the divergences between high and low levels of take-up had resulted in an overall imbalance between Member States and between regions, and signalled the Commission's intention to encourage more active implementation in states where the take-up had been insufficient.

The use of management agreements has hitherto been the favoured mechanism for implementing agri-environmental measures in the UK, The Netherlands, Ireland and Finland, where it is used for zonal agri-environmental measures as well as contracts for the protection of rural landscapes (see CEC, 1998, Annex 3 of which gives a review of the implementation strategies in most EC Member States). An agri-environmental strategy based upon the use of contractual instruments has considerable flexibility. Member States currently use differing contractual models to implement agri-environmental strategy. Some have used standard agreements with prescribed terms to apply basic requirements for environmental land management and stewardship, usually either across their whole territory or in identified (but geographically large) zones. The French *prime à l'herbe* scheme and the Rural Environmental Protection Scheme (REPS) in Ireland are examples of this contractual model. The first generation of Environmentally Sensitive Area (ESA) agreements in England and Wales, though available only in designated ESA zones, also represents a variant of this model, with standardized and inflexible management prescriptions applicable to all participating farms in each ESA area. On the other hand, environmental contracts can also be used to deliver a basic level of environmental stewardship if made generally applicable, but with additional optional tiers of participation targeted at restoring or improving particular habitat types or providing public access to farmland. Second and third generation ESA agreements in the UK offer an example of this type of approach, combining participation in a basic tier of obligations aimed at preserving environmental features of the ESA concerned with optional additional (or higher) tiers of participation under which additional premiums could be paid for allowing public access to farmland or for additional environmental obligations targeted at particular habitat types. This is essentially a more sophisticated variant of the standardized or 'general' contractual model, in which prescriptions were targeted at particular ESA areas (many of which were geographically very large, such as the Cambrian Mountains ESA in Wales) rather than at individual farms and farm-based habitats (see Whitby, 1994).

More recently, however, a more sophisticated approach has developed, with the advent of schemes under which landowners tender for admittance into agri-environmental schemes, and participation is granted by reference to the environmental 'goods' offered in each case. This enables the targeting of aid to projects offering the best environmental potential for the enhancement and restoration of farmland habitats and wildlife species, e.g. by requiring farmers to submit bids for contracts according to the environmental 'goods' to be purchased. The Countryside Stewardship scheme in England and Tir Gofal in Wales are models of this type of environmental contract. This type of approach enables the 'screening out' of cases where an agreement would offer no tangible environmental advantage, limits participation to cases where a defined and measurable environmental 'good' is to be purchased by the taxpayer, and therefore represents a better use of public resources to fund agri-environmental agreements (Potter, 1998). This type of agreement also offers greater flexibility, in that management prescriptions can be tailored to the particular farm concerned, and the use of 'whole-farm' plans in order to develop a holistic conservation strategy for the entire holding is increasingly common. This type of contractual model is also more closely attuned to the priorities identified by the European Commission for programme design for agri-environmental measures and offers clear benefits for the delivery of the priorities identified by the Commission (CEC, 1998).

The implementing regulations for the Agenda 2000 reforms give Member States considerable discretion in programme design and implementation. Although the Commission has signalled its desire to see the introduction of more targeted measures, the use of generalized agri-environmental programmes across large geographical areas is retained as an option where it is appropriate to the environmental conditions in the Member State or region concerned. At the programme design level, there will inevitably be an important tension between the use of site-specific measures in designated areas and general measures implemented on a widespread geographical basis with light agri-environmental obligations. The Commission wants to see both types of measure feature in Member States' agri-environmental programmes (CEC, 1998). As we have seen, in the agricultural sector previous contractual measures have largely been based on the general model, coupled with the use of cross-compliance techniques within the common market organizations of the CAP to achieve generalized improvements in farming practices of benefit to the environment. Evaluation reports of the agri-environmental programmes established under Regulation 2078/92 show that these instruments are often successful in producing limited results, such as an improvement in water quality and the prevention of erosion. In some areas, however, especially those with fertile soil or high livestock stocking densities, there has been a low take-up of those agri-environmental measures which limit production capacity. Where used by Member States, management agreement strategies targeted at sensitive zones and aimed at promoting biodiversity in fact achieved greater positive effects under schemes approved under Council Regulation 2078/92 (CEC, 1997b, p.20).

Conclusion

Although the legal order for the CAP now puts agri-environmental policy at its heart, there remain doubts as to whether it will deliver long-term environmental benefits. Evaluation of the 1992 reform package indicates that environmental protection has not yet been subsumed to the traditional goals of the CAP and to its' basic objective of maintaining agricultural incomes. The Agenda 2000 reforms took the process a stage further, and their emphasis on the establishment of environmental indicators addresses a number of weaknesses in the arrangements introduced in 1992.

The key problems in establishing and giving normative legal effect to 'good agricultural practice' are central to the potential success of the Agenda 2000 reforms. The experience of the 1992 reform is not encouraging. Countryside management and environmental stewardship have proved to be highly elastic concepts, with environmental outputs that are hard to measure, and this has been exploited by agriculture departments in many EC Member States in order to establish measures that often do little more than subsidize existing good agricultural practice (Potter, 1998, p. 157). Some of the general agri-environmental schemes established under the 1992 reform, such as *prime à l'herbe* in France and REPS in Ireland, are particularly vulnerable to this criticism.

In principle, the Agenda 2000 reform addressed some of these issues. As this chapter has endeavoured to show, however, there remain questions as to the compatibility of some aspects of CAP policy on the environment with the undertakings contained in the WTO Agreement on Agriculture. In the case of green-box exemption, some of the current support arrangements linked to environmental or socio-economic policy initiatives are potentially problematic; some fail to meet the green-box criteria for payments under environmental programmes, while others fail to meet the strict criteria for decoupled income support.

Perhaps anticipating these problems, the Mid-term Review of the Agenda 2000 reform programme published by the Commission in July 2002 posited a number of further – and in some respects fundamental – changes in the orientation of policy on farming and the environment. These are explicitly targeted at securing green-box compatibility for CAP support arrangements in the Millenium Round (CEC, 2002). The proposals in the Review were more radical than many had expected and mark a fundamental change of approach.

Following ministerial agreement on the reform package in June 2003, the draft Council regulation implementing the Review posits a complete decoupling of support from production by the introduction of a single income payment per farm ('single farm payment', or "SFP"), integrating into one support payment all existing direct payments received by producers under the various CAP support schemes (CEC, 2003). The entitlement will be calculated by reference to subsidy payments received in the historical base period 2000–2002, and will be subject to cross-compliance requirements as to environmental, food safety and animal welfare obligations. There will be no requirement to continue production, but land will have to be maintained in 'good agricultural and environmental condition' if eligibility to receive payments is to be maintained (Art. 5, CEC, 2003). The new regulations also aim to simplify the administration of environmental regulation by providing for cross-compliance to be applied using a whole-farm approach. This will require the observance of environmental standards on both used and unused agricultural land within a holding.

These proposals have been carefully tailored to match the criteria for green-box exemption in the WTO Agreement on Agriculture. The green-box conditions for decoupled income support (currently Annex 2, para 6, of the URAA) provide that eligibility must be determined by reference to income, producer status or production level in a defined base period, and that the amount of any support payment must not be related to (or based on) the type or volume of production undertaken by a producer in any year after the base period. Payment entitlements can, however, be calculated by reference to production volume or income in the base period itself. Crucially, green-box compatibility is dependent on there being no requirement for continued production in order to receive payments.

Implementing the proposals in the Mid-term Review of the CAP would break the direct link with production for most forms of support, but an indirect link remains in that entitlement to SFPs is conditional on land being maintained in 'good agricultural and environmental condition' – whether it is actually farmed or not. This is a key concept in the proposals, and one that is likely to prove problematic. The implementing regulation leaves it to the Member States to lay down minimum requirements to ensure that a range of broadly defined agronomic and environmental conditions set out in annex IV of the regulation are met. The content of the obligations imposed will therefore be found in national measures introduced by some Member States.

Monitoring the standards imposed by Member States for WTO compliance will be difficult. This will be problematic, as a clear legal definition of when land can be deemed to be in 'good' agricultural condition will be fundamental to both the successful implementation of the Mid-term Review and to its potential compatibility with WTO obligations. There are clear parallels with the concept of 'good agricultural practice', which, as explained above, was adopted as a normative standard for assessing land use practices under the reforms initiated by Agenda 2000. Clearly, what constitutes 'good' agricultural condition will depend upon the nature of the farming system historically practised on the land concerned and/or any future agricultural use to which it might be put. Similarly, it will vary from one geographical area to another and be dependent upon such factors as soil composition, climate and the husbandry customarily practised in the locality. In other words, it is likely to be a highly mobile concept incapable of precise or universally valid definition. Moreover, this aspect of the proposals might actually inhibit the establishment of environmental schemes targeted at marginal agricultural land. The proposed entitlement changes would, for instance, appear to rule out the use of previously productive land for the establishment of nature reserves if this were to involve taking the land out of production long-term, other than under set-aside or one of the approved agri-environmental programmes funded through the rural development budget. In some areas (for example, the uplands) where there is heavy dependence upon farm support, it is conceivable that dereliction could become a problem unless grazing is permitted in order to maintain semi-natural grassland in good agricultural condition. But herein lies the paradox at the heart of the proposals in the Mid-term Review. In formal legal terms, continuing with production may well become optional, but if grazing by livestock is necessary in order to keep land in good agricultural condition – for example, to prevent the re-establishment of dwarf shrub populations on hill land – the artificiality of the 'break' with production will be readily apparent. Continued farming will in practice be a necessity in order to claim the decoupled single farm payment. The decoupling of support from production will, in cases like this, be arguably notional.

The EC's comprehensive negotiating position in the Millennium Round posited that measures to protect the environment should be accommodated in the Agreement on Agriculture, but that they should be well targeted, transparent and implemented in no more than minimally trade-distorting ways (WTO, 2000b, para. 16). If this laudable objective is to be achieved, a number of issues will require clarification and resolution before the conclusion of the Millenium Round. If environmental protection is to fulfil a meaningful role as a non-trade concern in the legal order for international trade, the qualifying criteria for those measures which will qualify as 'environmental' for the purposes of green-box exemption must be clarified and more closely defined. The criteria for environmental schemes coming within the green box must also be transparent and workable. Furthermore, as the EC's own position paper on non-trade concerns points out, where there is joint production of both marketable agricultural produce and environmental services, this should not be used to conceal domestic economic subsidies. It follows that, where society asks farmers to produce a legitimate environmental objective, then government should only recompense them for the additional costs and income forgone which they incur in doing so, taking full account of the farmers' income from selling commodities in the market (WTO, 2000a, p. 37).

This is a sound position from which to consider the relevance of environmental policy within the agriculture debate, but its incorporation into the legal framework for international trade would require a clear formulation of what constitute legitimate 'environmental' goods. A related issue concerns green-box flexibility. The need for flexibility in the conditions for green-box eligibility has been illustrated in the course of this chapter by reference to the technical legal problems concerning the compatibility of several CAP support regimes with the 1994 URAA. Finally, the need for transparency in support arrangements was a criticism of CAP support regimes frequently made in earlier WTO rounds. The adoption of the Mid-term Review, and the full decoupling of support by the introduction of single farm payments, effectively finessed this argument. Nevertheless, the EC's position on the need for transparency in arrangements that involve both support for environmental or cultural services and for agricultural output may prove difficult to sustain.

References

Buller, H. (1999) The agri-environment measures. In: Lowe, P. and Brouwer, F. (eds) *CAP Regimes and the European Countryside*. CAB International, Wallingford, UK, pp. 199–219.

Coleman, W.D. and Tangermann, S. (1999) The 1992 CAP reform, the Uruguay Round and the Commission: conceptualised linked policy games. *Journal of Common Market Studies* 37, 385–405.

Commission of the European Community (CEC) (1993) *Towards Sustainability: A European Community Programme of Policy and Action in Relation to the Environment and Sustainable Development*. OJ C 138/5.

Commission of the European Community (CEC) (1997a) *Report from the Commission to the Council and the European Parliament on the application of Council Regulation (EEC) 2078/92.* COM (97) 620 Final.

Commission of the European Community (CEC) (1997b) *Agenda 2000: for a Wider and Stronger Union.* COM (97) 2000 Final.

Commission of the European Community (CEC) (1998) *Evaluation of Agri-Environment Programmes.* DGVI Working Document VI/7655/98.

Commission of the European Community (CEC) (1999) *Directions Towards Sustainable Agriculture.* COM (1999) 22 Final.

Commission of the European Community (CEC) (2000a) *Indicators for the Integration of Environmental Concerns into the Common Agricultural Policy.* CEC COM (2000) 20 Final.

Commission of the European Community (CEC) (2000b) *Agenda 2000: Reform of the Common Agricultural Policy.* Available at: http://www.europa.eu.int/scadplus/leg/en/lvb/160002.htm

Commission of the European Community (CEC) (2002) *Mid-term Review of the Common Agricultural Policy.* Communication from the Commission to the Council and the European Parliament, (CEC) COM (2002) 394.

Commission of the European Community (CEC) (2003) Proposal for a Council Regulation establishing common rules for direct support schemes under the common agricultural policy and esblishing certain support schemes for farmers (draft 4.7.2003 Final).

Dax, T. and Hellegers, P.(1999) Policies for Less Favoured Areas. In: Lowe, P. and Brouwer, F. (eds) *CAP Regimes and the European Countryside.* CAB International, Wallingford, UK, pp. 179–197.

Landau, A. (2001) The agriculture negotiations in the WTO. *Journal of Common Market Studies* 39, 913–925.

Lowe, P. and Brouwer, F. (1999) Agenda 2000: a wasted opportunity? In: Lowe, P. and Brouwer, F. (eds) *CAP Regimes and the European Countryside.* CAB International, Wallingford, UK, pp. 321–334.

Paarlberg, R. (1997) Agricultural policy reform in the EC: a three level game analysis. *International Organisation* 51, 413–444.

Potter, C. (1998) *Against the Grain: Agri Environmental Reform in the United States and the European Union.* CAB International, Wallingford, UK.

Rodgers, C. (2002) Agenda 2000, land use and the environment: towards a new theory of 'environmental property rights'? In: Holder, J. and Harrison, C. (eds) *Law and Geography.* Oxford University Press, Oxford, pp. 239–258.

Scott, J. (1996) Tragic triumph: agricultural trade, the Common Agricultural Policy and the Uruguay Round. In: Emiliou, N. and O'Keefe, D. (eds) *The European Union and World Trade Law: After the GATT Uruguay Round.* Wiley, London.

Wathern, P. (1992) Less Favoured and Environmentally Sensitive Areas: a European dimension to the rural environment. In: Howarth, W. and Rodgers, C.P. (eds) *Agriculture Conservation and Land Use: Law and Policy for Rural Areas.* University of Wales Press, Cardiff.

Whitby, M. (ed.) (1994) *Incentives for Countryside Management: the Case of Environmentally Sensitive Areas.* CAB International, Wallingford, UK (see Chapters 10 (Colman, D.) and 11 (Whitby, M.)).

Winter, M. (1999) The arable crops regime and the countryside implications. In: Lowe, P. and Brouwer, F. (eds) *CAP Regimes and the European Countryside.* CAB International, Wallingford, pp. 117–136.

World Trade Organisation (2000a) *Note on Non-Trade Concerns.* G/AG/NG/W/36/Rev. 1, 9 November.

World Trade Organisation (2000b) *EC Comprehensive Negotiating Proposal.* G/AG/NG/W/90, 14 December.

Reforming United States Environmental Regulations for Agriculture: Impediments and Opportunities

12

David E. Adelman*

Introduction

One of the most engaging aspects of studying agricultural policy is that, after thousands of years of development, much remains to be learned and many opportunities for improvement exist. What is most dismaying is that after so much time we continue to practise agriculture using brute-force techniques with relatively little appreciation for or understanding of their impacts on the environment. This chapter explores some of the reasons for this impasse, focusing largely on crop production, and suggests some promising areas for significant improvement.

Agricultural production implicates many fundamental environmental, socio-economic and regulatory issues. In the USA, heated debates continue over the risks associated with pesticides, the challenges of tracking and mitigating polluted runoff from farms and agriculture's impacts on biodiversity. The sheer breadth of environmental issues raised by agriculture and the inherent uncertainties in our scientific understanding of them pose significant challenges to evaluating potential options for reducing agriculture's impacts on the environment.

Socio-economic forces provide both impediments against and impetus for change. In the USA, the powerful political muscle of the

---

* David Adelman, BA Reed 1988, PhD Stanford 1993, JD Stanford 1996, is Associate Professor of Law at the University of Arizona, James E. Rodgers College of Law. This chapter builds on and draws from a paper co-authored with John H. Barton (Adelman and Barton, 2002).

agricultural community has almost uniformly opposed new regulatory standards and controls. The significant economic pressures under which farmers operate, particularly with spectacularly low commodity prices and the intense global competition for export markets, are of critical importance. These same economic pressures, however, have motivated farmers to enrol in a variety of incentives programmes designed to promote environmental preservation and thus are not necessarily adverse to improved environmental protection.

Global socio-economic forces add another dimension to the problem. Continued population growth, combined with deteriorating or diminishing resources, such as soil quality and irrigation water, puts farmers in the position of having to maximize agricultural production, on the one hand, while reducing its negative impacts on sustainability, on the other. Perverse national agricultural subsidies, however, have eclipsed the need for more sustainable practices in favour of increased production. As resource limits become increasingly important, some kind of balance will have to be struck between increasing yields and sustainability, but significant reforms in agricultural trade will have to occur before this can be resolved.

The, development of effective environmental regulations also faces major challenges. Agricultural production is both highly diffuse and conducted in an enormous range of local environments. Even the USA, where agricultural production is highly mechanized, has over 2 million farms. This natural variability makes it difficult, if not impossible, to establish national standards and, even at the local level, the complexity of tracking agricultural pollutants and understanding their environmental impacts is both technically challenging and expensive. Thus, even if it were acknowledged that stronger regulation was necessary, it would take substantial effort to determine what form it should take.

This chapter seeks to elucidate, at a general level, how each of these factors – environmental, socio-economic and regulatory – influence the USA's approach to regulating agriculture and to identify how they interact, negatively or positively, to affect environmental policies. The first three sections discuss the environmental and socio-economic factors, the fourth section outlines and analyses environmental regulations for agriculture in the USA and the fifth section discusses several promising approaches derived from these evaluations.

Agriculture's Environmental Impacts

At a global level, agricultural production, including crops and grazing, utilizes 37% of the earth's land surface (WRI, 1994).[1] Agriculture is the largest consumer of freshwater resources (WRI, 1994), the largest contributor to a doubling of the earth's level of fixed nitrogen (Vitousek *et al.*, 1997a) and a significant source of greenhouse gases and tropospheric-ozone-generating chemicals (Matson *et al.*, 1997). Species are threatened by agriculture indirectly, through habitat destruction, and directly, through agrochemical runoff and introductions of exotic species. According to the World Conservation Monitoring Center, approximately 39% of the known losses of animal species since 1600 are attributable to species introductions and 36% to habitat destruction (WRI, 1994). Currently, land transformation, particularly in the tropics, is the most significant factor in loss of biodiversity globally (Vitousek *et al.*, 1997b). Agriculture contributes significantly to each of these effects.

In the USA, agricultural practices are implicated in almost a third of the species listings under the Endangered Species Act (ESA) and have been the most significant cause of habitat destruction (OTA, 1996). Agricultural production is believed to be the primary contributor to surface-water deterioration in the USA, predominantly through sedimentation and agrochemical runoff (NRC, 1989; OTA, 1996). A 1990 study by the US Environmental Protection Agency (EPA) ranking the severity of environmental threats in the USA concluded that pesticides, non-point-source water pollution, physical degradation of terrestrial ecosystems and physical degradation of water and wetlands were among the top five most urgent environmental threats (EPA, 1990). Each of these is linked to agriculture.

While crop cultivation is a significant source of three greenhouse gases (i.e. carbon dioxide, methane and nitrous oxide) (IPCC, 1996), the leading effluent flow from agriculture is into surface and groundwater (NRC, 1989). The existing data demonstrate that agriculture is the greatest contributor to water pollution in the USA. In a June 2000 EPA report, agriculture was identified as the leading cause of impairment to US rivers (siltation, nutrients and pesticides), lakes (nutrients and siltation) and wetlands (nutrients and siltation) and one of the leading causes of groundwater contamination (fertilizers and pesticides) (EPA, 2000). The EPA study also found that agriculture continues to be the leading cause of wetland destruction.

Seventy-one per cent of US cropland is located in watersheds in which at least one agricultural contaminant exceeds EPA guidelines (Smith *et al.*, 1994; OTA, 1996). Agricultural activities are responsible for 64% of the degradation of river-water quality and 57% of the degradation of lake-water quality. Atrazine and other herbicides are virtually always detectable in surface waters in regions where they are used, sometimes at levels far in excess of EPA standards. In several regions, agricultural non-point sources account for 70–94% of the nitrogen loading in rivers (Knopman and Smith, 1993). Agrochemicals also reach coastal regions in significant quantities. Herbicides and nitrogen fertilizers applied on farms within the Mississippi basin, for example, create a 'dead zone' extending over 16,000 km^2 in the Gulf of Mexico (OTA, 1996; Malakoff 1998).

Wetlands are of particular ecological importance because they provide important habitat for about 40% of the species listed as endangered or threatened under the ESA (Novitski, 1997). Until recently, when other factors such as urban growth became a significant factor, agriculture was the major cause of wetland losses in the USA, accounting for 87% of all conversions between the mid-1950s and mid-1970s (Smith, 1995), and agriculture continues to be the single largest cause of wetland destruction in the USA.

Agriculture is therefore a leading cause of pollution in the USA, particularly in respect of water pollution, wetlands destruction and threats to biodiversity.

Constraints Imposed by Limits on Global Natural Resources

In the brief period between 1961 and 1993, global population increased by 80%; total cropland increased by 8%; fertilizer use increased by 287%; and irrigated crops expanded from 10% to 17% of the total land area cultivated, but by 1993 produced 40% of the food harvested globally (Goklany, 1998). Over essentially the same period, the global per capita food supply increased by 20% and the number of malnourished people decreased from 917 million to 839 million. As these statistics imply, the large increases in production, particularly over the last 40 years, were achieved by dramatic increases in crop yields, irrigation and agrochemical use.

These impacts were (and are) offset to some degree by the significant reduction in land required to meet the increased food needs of the growing global population. Agricultural experts estimate that, without the technical gains made after 1961, feeding the 1993 global population would have required an increase in agricultural land of 80% over that cultivated in 1961 (Goklany, 1998). This would have required an additional 970 million hectares of cropland, or almost seven times the net global loss of forest and woodlands over that period. Put another way, by 1995 improvements in grain yields since 1960 saved a land area estimated to be equal to that of the entire Amazon rain forest (Goklany, 1998).

These impressive gains will not be sufficient to meet future global food requirements. Demographers predict that the population will grow to between 8 and 10 billion people during the 21st century. Global demand for food could easily double between 1990 and 2030, with a two-and-a-half to threefold increase occurring in developing countries (Daily *et al.*, 1998). Moreover, while malnutrition is now caused more by poverty and inequitable food access, many regions of the world, particularly Africa, are not self-sufficient in food production. As the global population grows and expands into undeveloped regions, it will be all the more important to increase food production by improving yields from land already in production. The unavoidable implication of these observations is that the necessary increases in food production cannot be met without expanding agrochemical usage and irrigated lands and – most importantly – improving the efficiency of agricultural production.

The environmental impacts of agrochemicals and limited global water resources add particular urgency to improving the efficiency of agricultural production. But there are important limits to low-input crop management. Some experts believe that nutrients removed from soil by modern high-yield crops cannot be adequately replenished using low-input methods, such as rotations of legumes and complete recycling of organic wastes from crops and domestic animals (Matson *et al.*, 1997; Smil, 2000). According to one of these estimates, feeding the world's projected population using such methods would require a doubling, or even tripling, of the land area currently cultivated – which would entail eliminating all tropical rain forests and converting large portions of tropical and subtropical grasslands. At the same time, there are limits to the efficacy of agrochemicals, as evidenced by the rise in resistance to pesticides and the diminished quality and increased erosion of intensively farmed soils, and yields of many crops have hit a plateau. These constraints are even more severe in the tropics, where soil fertility is often low.

Existing water withdrawals for irrigation are also straining supplies in many regions. For example, in Asia, where most of the population growth is projected, many rivers are completely depleted during the drier periods of the year (Postel, 1998). Further, almost half of the world's irrigated land area is located in China, India and Pakistan, each of which contends with limited water-supplies and intense population pressures (Smil, 2000). Thirty-four countries in Africa, Asia and the Middle East currently have annual runoff volumes below that needed to ensure food self-sufficiency – all but two are net grain importers (Smil, 2000). In the next 25 years, population growth will add ten more African nations, India, Pakistan and several other Asian countries to the list of countries with per capita water resources below that necessary for food self-sufficiency, and China will be struggling on the margins (Goklany, 1998). For these regions, long-term food self-sufficiency is probably not an option, and it is questionable whether food surpluses in other regions will be sufficient to meet their needs or be affordable.

Economic Barriers to Agricultural Reform

After a short-lived decline between 1988 and 1997, agricultural subsidies have risen to record levels (OECD, 2000). In 1999, an estimate by the Organization for Economic Cooperation and Development (OECD) found that agricultural subsidies among its member countries totalled $361 billion (OECD, 2000), and the 2000 harvest was projected to break subsidy-level records yet again. Agricultural subsidies now account, on average, for about 40% of total receipts among OECD members (17% for the USA), with some countries supplying more than 70% (USDA, 2000).

Positive steps away from reliance on trade-distorting and often environmentally harmful subsidies have been made through the Uruguay Round Agreement on Agriculture (URAA) and subsequent World Trade Organization (WTO) negotiations, recent efforts by the United States and, to a lesser extent, changes within the European Union (EU) (URAA, 1994). The URAA and WTO negotiations have initiated a process for nations to reduce their agricultural subsidies and tariffs and to establish an implementation schedule. These efforts have faced strong public opposition and the hard realities of depressed commodity prices – brought about, in part, because of these same agricultural policies. Under the current socio-economic constraints, it will be difficult for individual countries to make headway, but signs of potential progress exist.

The USA has made a serious attempt to revise its agricultural trade policies, both at the national level, with legislation passed in 1996, and internationally, through the WTO negotiations. The 1996 Federal Agricultural Improvement and Reform Act was intended ultimately to eliminate federal agricultural subsidies through a 7-year phase-out period and relaxation of restrictions on crops that could receive subsidies.[2] Four years later, however, low commodity prices have driven subsidies, in the form of 'loan deficiency payments' and emergency appropriations, to levels that are, on average, 70% higher than before the 1996 reforms.[3] Even in the absence of large export subsidies, federal commodity-support programmes supplied more than half of farmers' net income in 1998 and about two-thirds in 1999.

The 1996 Act's (partial) lifting of restrictions on eligible crops was expected to promote more environmentally sustainable cropping regimes, while the reduction in subsidy levels would reduce the incentive to maximize yields rather than net return in the marketplace. This too has failed. Farmers are continuing to monoculture major commodity crops and the incentive to maximize crop yields above all else remains. The reforms failed because loan deficiency payments – now the largest crop subsidy – were retained without lifting the restrictions on eligible crops or basing payments on something other than the amount produced. These restrictions effectively preclude the use of more sustainable practices, such as crop rotations to maintain soil fertility, control pests and reduce erosion, and promote aggressive overuse of agrochemicals and water to maximize yields. Further reform is necessary to eliminate these incentives and the bias against alternative agricultural methods under US policies.

The other OECD countries lag behind the USA, with the EU, Japan, Korea and Switzerland being among the most regressive in their policies. The EU plays a dominant role, expending over $46 billion on agricultural subsidies in 1997 and accounting for 83% of the total agricultural export subsidies in 1996 (USDA, 1999). Because of fiscal constraints and growing commodity surpluses, the EU is moving towards reducing its agricultural subsidies, which consume about 50% of its annual budget (USDA, 1999). The EU's Agenda 2000 targets subsidies under the Common Agricultural Policy as a key issue to be addressed. Yet significant points of dispute remain under the WTO negotiations, particularly recognition of the social aspects of agriculture (referred to as 'multifunctionality'), which the USA believes the EU could use, among other things, to exclude genetically modified crops (as to which see further Chapters 5 and 6 by Grossman and Cardwell respectively, in this book). These, and other, differences will have to be resolved before further advances can be made.

In response to agricultural subsidies levied by developed countries, developing countries have also instituted extensive subsidy programmes, and these have compounded economic distortions in agricultural trade. Developing countries are harmed by these distortions in several ways: first, they lose trade revenue because subsidies depress international commodity prices; secondly, consumers in developing countries rarely benefit from buying subsidized goods because their governments limit imports and levy significant tariffs; and, thirdly, developed-country efforts to buffer domestic price variability magnify world price fluctuations of agricultural goods. Further, because farmers generally are politically marginal in developing countries, crop prices are kept low by government-imposed price restrictions and monopolization of agricultural trade. Making matters still worse, to offset the negative impacts of the price limits and to improve international competitiveness, developing countries heavily subsidize agrochemicals and irrigation water (Knudsen *et al.*,1990; Baffes and Meerman, 1998).

These policies have reduced the productivity of small rural farmers and promoted the inefficient usage of agrochemicals and irrigation water, which has caused widespread salinization of fields and substantial environmental damage from agrochemical and sediment runoff. Thus, far from benefiting poor communities in developing countries, agricultural subsidies have allowed governments to justify wasteful agricultural programmes, increased agricultural production costs and promoted unsustainable, environmentally damaging practices. Fortunately, developing countries are also beginning to appreciate the need to eliminate agricultural subsidies and, under the WTO negotiations, have made initial commitments to reduce their subsidy levels.

The evolution of the WTO negotiations and domestic efforts to remove perverse subsidies illustrate the complexity of these issues and the obstacles to reform. Thus, while the WTO negotiations have had mixed results – particularly in the wake of the aborted WTO meeting in Seattle – they represent a potentially significant change towards instituting trade liberalization and the lifting of environmentally damaging subsidies. As such, the WTO negotiations could lead to much broader changes in agricultural trade. These reforms are critical because open agricultural trade will be essential to global food security – and environmental preservation – as food imports become more important to feeding the rapidly growing populations in Asia and Africa.

The United States Regulation of Agriculture

Despite agriculture's extensive impacts on the environment, US programmes regulating agriculture are generally far weaker than those for other industries. They are also unnecessarily complex – involving at least four federal agencies and ten major laws, many of which were designed to promote or protect agriculture rather than regulate it. Moreover, several of the most important US environmental laws, such as the Clean Water Act (CWA), have exempted agricultural activities from regulation, which has profound environmental impacts.

Controlling agricultural impacts on water resources

The 1972 CWA (33 USC §§ 1251–1387) and the 1974 Safe Drinking Water Act (SDWA) (42 USC §§ 300f-300j-26) are the principal environmental laws that protect water resources. The CWA's provisions addressing regulation of 'non-point sources', however, remained moribund until recently, thereby creating a *de facto* exemption for 'return flows from irrigated agriculture' and voiding the Act's regulation of agri-environmental threats. The SDWA has suffered from a laborious standard-setting process and limited funding, which have restricted the range of chemicals it regulates.

In 1987, the CWA was amended to promote regulation of non-point sources through federal funding and oversight, but these reforms had little effect because EPA continued to rely on voluntary arrangements to reduce pollution. EPA began to address non-point-source pollution – i.e. pesticide and fertilizer runoff – only after being successfully sued by several environmental groups in the mid-1990s.[4] These suits prompted EPA ultimately to issue a final rule in July 2000 imposing deadlines for states to set 'total maximum daily loads' limiting non-point-source releases – invoking strident opposition by many members of Congress.[5]

The 25-year delay in implementing these regulations provides a case-study on the complexity – both political and technical – of regulating agricultural effluents. Because the regulations unavoidably have an impact on a broad range of interests (local and state governments, federal agencies, farmers and developers), they have generated enormous political opposition. The technical and fiscal obstacles are equally serious. Data on the concentrations of pollutants and their fate are limited, and contaminant monitoring and health data are at a very 'primitive' stage

(Houck, 1999). Further, the costs of adequately evaluating the environmental effects of pollutants and the control costs will be high, if not prohibitive – \$4 billion per state by one account (Houck, 1999). This experience has proved sobering for everyone involved.

The SDWA authorizes EPA to identify and limit contaminants in public drinking-water-supply systems and thus indirectly affects agricultural sources. The regulations are both feasibility-limited and health-based, with the primary focus on carcinogenicity to humans. As of August 2000, EPA had listed 85 substances – only 16 of which are actively used pesticides – and had identified another 60 candidates for listing. To put this in perspective, of the hundreds of pesticides on the market, all but 16 remain unregulated under the Act – including four listed among the ten highest-volume pesticides. Moreover, progress is likely to remain limited because Congress revised the Act in 1996 to require that only five substances be listed every 5 years.

The USA has also established a special regulatory system for approving pesticides, the Federal Insecticide, Fungicide and Rodenticide Act (FIFRA) (7 USC §§ 136–136y), which limits the types of agrochemicals that can be used. EPA is responsible for registering pesticides under FIFRA and, of great operational importance, for reregistering older pesticides that were reviewed prior to the establishment of stricter standards in 1972. Under FIFRA, registration is contingent on the pesticide 'perform[ing] its intended function without unreasonable adverse effects on the environment . . . when used in accordance with widespread and commonly recognized practice[s]'. The regulations interpreting this language focus almost exclusively on human carcinogenicity, but also require limited consideration of the off-site environmental effects of the candidate pesticide. About 890 pesticides are registered with EPA.

Even with this relatively limited scope of analysis and a standard strongly favouring approval, registration of pesticides has historically taken an average of 4–6 years (although the review period has decreased markedly) and can cost in excess of \$10 million per pesticide (NRC, 1996b). This process is further complicated by the fact that EPA must cope with a huge number of products – more than 20,000 in all. As a result, 30% of the registered agricultural pesticides are not properly approved because they were registered prior to newer standards and have not been 'reregistered' under this stricter system (EPA, 1999b). In a number of instances involving pesticides used in limited markets, these delays have discouraged registration of new, less harmful pesticides.

FIFRA regulations have had little effect on the overall use or sale of agricultural pesticides. In the USA, total pesticide usage doubled between the mid-1960s and a peak in the mid-1980s, and has remained a little below that peak level ever since (EPA, 1999a). This slight decrease, however, is attributable to reductions in acreage farmed, rising pesticide prices and the introduction of more potent, low-volume compounds, rather than to a response to regulation. These trends suggest that there are limits to health-based standards – in both cost and technical complexity – which will take time to resolve. Indeed, this overview strongly suggests that agricultural effluents, particularly those derived from fertilizers and pesticides, are under-regulated in the USA – the actual levels of pollution are high and have not responded to existing regulations.

In response, the USA has increasingly adopted incentive programmes to safeguard water resources and wetlands. Initiated in 1990, the Water Quality Incentives Program (WQIP) (16 USC §§ 3838ff.), is a voluntary programme that provides incentives, in the form of crop subsidies, for farmers to adopt environmentally and economically sound management practices to prevent soil erosion, protect wildlife habitat and conserve water resources. The potential benefits of the programme have been minimal, though, because of Congress's failure to fund it adequately (Faeth, 1995). The 1996 Farm Bill, which merged the WQIP with the US Department of Agriculture (USDA)'s Environmental Quality Incentives Program (EQIP), fortunately boosted funding for the programme substantially.

Protecting biodiversity from high-input agricultural practices

No simple methodology exists for striking a balance between human activities and species loss. The kinds of issues involved – property rights, economic trade regulations, cultural values, highly complex scientific analyses, etc. – and the dramatic variation in the issues themselves from region to region preclude a mechanistic approach. In the discussion that follows, two critical issues are addressed: (i) introductions of harmful exotic species; and (ii) destruction of species' habitats.

Regulatory control of exotic species introductions

Most countries regulate deliberate species introductions, but the regulations are typically designed to protect agriculture itself, not biodiversity. In the USA, among a panoply of agencies, the USDA's Animal and Plant Health Inspection Service (APHIS) is the lead agency. The recently passed Plant Protection Act (7 USC. §§ 7701–7772) authorizes APHIS to prevent the import, export or interstate transfer of 'noxious weeds' and 'plant pests'. 'Noxious weed' is defined broadly to include any plant that can 'cause damage to [agriculture] . . . the natural resources of the United States, the public health, or the environment'. The new Act, like its predecessor, applies only to listed noxious weeds. The listing process, however, has been extremely slow. Despite strong evidence that 750 known weeds met the definition of noxious weed, it took APHIS 8 years to list a total of 93 species – currently, only 96 species are listed (OTA, 1996). The likelihood is thus low that USDA will intercept an invasive species before it is released.

The Plant Protection Act also controls the transfer of plant pests, which include pathogens, animals and parasitic plants. The statute does not require USDA to regulate or evaluate imported plants themselves, unless they are either a plant parasite or a listed noxious weed. Consequently, apart from those specifically regulated, exotic plants receive minimal regulatory review – neither APHIS nor the Agricultural Research Service performs regular tests to evaluate foreign plants for invasiveness or weediness. This regulatory gap persists despite $28.8 billion in losses to agriculture annually from invasive species (NRC, 1996a) and prior introductions of 'beneficial' plants (e.g. kudzu and multiflora rose) causing profound ecological damage. A 1999 presidential executive order mandating that federal agencies coordinate their efforts on invasive species could prompt more proactive government action, but it is unlikely to achieve much without greater public pressure.[6]

International restrictions on species movements would seem to be a straightforward – though non-trivial – approach to protecting biological diversity against unwise introductions, especially given that the 1992 United Nations Convention on Biological Diversity incorporates provisions that address the issue.[7] Despite the magnitude of the problem, public awareness and the scientific understanding needed to evaluate these risks remain limited. Moreover, international negotiations, which overcame a long-standing deadlock recently, have instead focused on regulation of genetically modified organisms.[8] By failing to address exotic species, though, the agreement's value is substantially diminished.

Restrictions on the destruction of wildlife habitat

Species-protection regulations face difficult obstacles, including constitutional property-rights issues and the complexity of assessing the ecological significance to a species' survival of specific private or public properties. The USA's regulatory approach to preventing habitat destruction has relied largely on the ESA (16 USC §§ 1531–1544), which was designed, at least initially, as a stopgap measure to save seriously threatened species, rather than as an affirmative mechanism to safeguard biodiversity. The ESA creates procedures for listing species as either threatened or endangered, designating specific areas as critical habitat and defining recovery plans for endangered species.[9] Enforcement actions occur pursuant to Section 1538, which provides a legal basis for preventing 'takings' of endangered species on private lands or in waters affected by activities on private lands. This section can be used to limit agricultural pesticide use, conversion of grasslands or wetlands to cropland and other development options.

The ESA allows flexibility by permitting 'incidental taking' of a species provided there is an overall habitat conservation plan. While the US Department of Interior has attempted to assuage farmers' concerns by utilizing habitat-conservation plans in creative ways, land-use disputes are likely to increase already volatile political tensions. Allocation of scarce water resources in western states between endangered fish species, municipalities and irrigation for agriculture will, in particular, result in continuing and probably escalating disputes (Postel, 1998).[10] The inevitable increase in the number of listed species will further augment these tensions, which makes the ESA's failure to protect species prospectively before they are endangered all the more apparent.

In place of a traditional regulatory scheme, the US government has established two incentive programmes to protect wetlands, the Wetland Reserve and Swampbuster programmes. Under the Wetlands Reserve programme, (16 USC §§ 3837–3837f.), USDA compensates landowners for restoring wetlands through either permanent or 30-year easements. Despite high farmer interest, low funding has seriously limited restoration efforts. It will take more than 10 years for the programme to enrol 975,000 acres (about 10% of the current agricultural wetlands). The Swampbuster programme, (16 USC § 3821(a)), creates a disincentive to alter wetlands by denying USDA benefits to farmers if they produce a commodity crop on converted wetlands. The benefits of this programme have also been limited because the restrictions have been severely diluted by an array of exceptions. In spite of the enormous resources for other programmes, USDA implementation and support of these incentive programmes has, as a general rule, been meagre; this parsimony is reflected in their limited success.

Protecting agricultural environments and resources

Maintaining agricultural environments requires that appropriate incentives exist to protect soil quality and promote sustainable water use, both of which are critical to maintaining and increasing crop yields. Although estimating the severity of soil erosion in the USA remains controversial (Pimentel *et al.*, 1995; Ganz, 1999; Trimble and Crosson, 2000), a recent study found that crop yields from severely eroded plots were 21% less than those for slightly eroded plots in a US maize-belt state (Mokma and Seitz, 1992). Maintaining soil nutrients and carbon levels, both of which are linked to erosion, is also critical to sustaining yields, as well as reducing agrochemical runoff. Freshwater supplies for irrigation, however, remain among the most binding environmental constraints (Postel, 1997). In the USA, this factor will be most critical in the midwestern and western states, where water is limited and groundwater supplies are being depleted. Careful water use is also important because inefficient irrigation methods deposit salts that harm the soil and reduce crop yields.[11]

Neither water nor fertilizer use is governed by incentives designed to promote its sustainable use. Irrigation water is both overused and used highly inefficiently and, without proper pricing, investments in efficient technologies will not be made. Similarly, inexpensive, poorly regulated fertilizers give farmers little incentive to adopt practices to minimize reductions in soil quality.

The USA has instead implemented several modestly successful conservation programmes. Efforts to reduce soil erosion, for example, achieved about a 33% reduction in erosion levels nationwide over the past decade. There are two primary incentive programmes, the Conservation Compliance Program and the Conservation Reserve Program (CRP). The Conservation Compliance Program (16 USC §§ 3831–3836 (2000)) fosters sustainable soil-management practices, using education programmes and, more importantly, conditioning certain agricultural benefits on the adoption of farming practices that meet designated conservation standards. USDA has not, however, uniformly enforced these requirements and has often failed to promote erosion control effectively.

The most important US programme is the CRP (16 USC §§ 3831–3836 (2000)), under which the government 'rents' agricultural land for a term of years (usually 10), during which time the owner must take the land out of production and plant it with grasses or trees. Land is selected using an 'environmental benefits index', which ranks factors such as benefits to wildlife, water quality, air quality, and erosion reduction. The CRP's impact, however, derives more from its size – 33 million acres (10% of US farmland) and average budget of $1.7 billion annually – than from its structure and effectiveness (OTA, 1996; Gardner, 1999). Similar to other programmes, the USDA needs to improve programme prioritization and programme oversight.

Several findings emerge from this overview of these statutes and regulations. First, there are a number of programmes that are being seriously underutilized because of poor funding. In the case of the incentives programmes, such as the Swampbuster or Conservation Compliance programmes, opposition from the agricultural community is not the controlling factor. Secondly, federal agencies, particularly the USDA, often undermine the potential environmental benefits of a programme by inconsistent or weak implementation of the governing requirements. Thirdly, US regulations are predominantly reactive, as opposed to forward-looking – obvious examples include the ESA and the US government's poor record of promoting new, environmentally friendly technologies.

Conclusions

The preceding discussion provides a general framework for understanding the major factors that influence the environmental regulation of agriculture in the USA. The findings distil down to a few central observations for each section:

1. Environmental impacts from agriculture predominantly concern water quality, wetland destruction and threats to biodiversity.
2. Resource constraints on agricultural production include limited irrigation water, soil nutrient levels and low land availability.
3. The major economic drivers are agricultural subsidies, low commodity prices and international trade regulations.
4. Traditional regulatory approaches are severely hampered by poor information, high costs and the diverse settings in which crops are produced, whereas incentive-based programmes directed at agricultural practices have shown modest success but have been limited by funding.

Comparing these findings suggests several overarching themes. First, for the foreseeable future critical trade-offs will have to be made between limiting the expansion of agriculture to new (often marginal) land, preserving water resources and minimizing the volume and environmental impacts of agrochemicals. Secondly, direct regulation of agrochemical usage – by either economic incentives or prescribing methods – shows promise as a viable regulatory approach. Thirdly, fundamental changes in agricultural subsidies are essential to any effort to reform agricultural methods and to protect the environment.

There is no simple answer to achieving a fair balance between preserving water resources and natural habitats and minimizing agrochemical inputs. In the short run, the USA does not have to contend with significant tensions between protecting habitat and maintaining or increasing agricultural production; existing surpluses eliminate this dynamic. This, however, will not always be the case, particularly as the global population increases and urban growth continues to expand into agriculturally valuable areas.

The more immediate issue concerns protecting water resources from agrochemicals, which existing agricultural subsidies seriously undermine. The significance of these impacts to agricultural sustainability, local ecology and human health should spur governments to renew their commitment to agricultural research and the development of alternative methods. This threat is of particular importance because the range of conditions under which agriculture is practised will require a plurality of approaches. Thus far, there is little evidence that the US government appreciates the need for new, more sustainable methods.

Traditional end-of-the-pipe regulatory approaches have not worked for agriculture. The recent efforts to formulate regulations under the CWA, while important, stretch such regulatory mechanisms – and the existing science – to their breaking point. Significant gains, however, may be achievable by promoting more sustainable cultivation methods. Promising techniques include precise fertilizer methods that reduce application levels and costs, combined fertilization–crop-rotation methods that increase soil nutrients and diverse crop-variety regimes that eliminate the need for pesticides. Hybrid approaches using precise agrochemical treatments, biological controls and integrated pest management have proved very successful as well (NRC, 1989, 1996a). To this end, US incentive programmes should be expanded substantially and leveraged by conditioning receipt of other agricultural subsidies (e.g. loan deficiency payments) on compliance with their requirements.

This approach is not a panacea. Where critical resources are at risk (e.g. wetlands or other unique ecosystems), specific limitations should be placed on agricultural practices, using the precautionary principle. Specifically, where potential threats are substantial and alternatives exist, regulations should be imposed to require the use of more sustainable methods. Similarly, the enormous ecological damage caused by exotic species introductions must be addressed by strict regulatory standards and protocols. Again, where the potential risks are significant – or highly uncertain – strict limits should be placed on species introductions or the import of certain goods. The obvious transnational implications of such regulations must be addressed internationally through the Convention on Biological Diversity.

The WTO negotiations are likely to offer the best forum available to promote the elimination of national agricultural price supports that severely undermine agricultural sustainability. Agricultural subsidies are extremely damaging because they discourage investment in and the adoption of alternative technologies, even where the technologies are economically viable (NRC, 1989; Benbrook, 1996; Curtis, 1998). By promoting yield maximization or directly subsidizing agrochemicals and water, the USA continues to encourage their inefficient overuse and implicitly creates an economic disincentive against efficient, environmentally sustainable methods. The WTO's goal of placing agricultural trade on a more liberal economic basis necessarily entails reforming national subsidies. Accordingly, if trade liberalization were enacted thoughtfully – i.e. by requiring signatories to introduce mechanisms to internalize environmental costs – substantial environmental gains could be made.

The WTO's 'Green-Box' subsidies, which are exempt from aggregate measure of support (AMS) reduction and restrictions, are an example of a mechanism that could lead to important reforms.[12] Early experience with the WTO rules suggests that countries wishing to protect their farmers will refashion price-support subsidies as exempt programmes, such as payments for land set-asides, which have significant environmental benefits. The WTO restrictions could thus be used to redirect negative agricultural subsidies into programmes that promote agricultural sustainability (as to which, in the European context, see Chapter 11 by Rodgers in this book). The existing green-box programme, however, may need to be broadened, at least initially, to enable countries to wean themselves from their massive subsidy programmes. The WTO negotiations therefore have the potential to achieve more than economic harmonization; they could reform an entrenched economic order that has profound implications for agriculture's impacts on the environment.

Notes

1 In the USA, almost half of the coterminous land area is dedicated to agriculture (OTA, 1996).

2 See 7 USC §§ 7202 (5), 7312, 7235 (2000) (the 'commodity crops' to which subsidies were restricted include maize, cotton, wheat, grain sorghum, barley, oats and rice).

3 Average annual farm subsidies since 1996 were $16 billion per year, compared with $9.4 billion previously (Stuteville, 2000). More recently, Congress passed a 10-year Farm Bill with $180 billion in agricultural subsidies, which has received condemnation from the international community (Becker, 2002).

4 See, for example, *Idaho Sportsmen's Coalition* v. *Browner*, 951 F. Supp. 962 (WD Wash. 1996) (Federal court directed EPA, which must cooperate with the states, to develop 'total maximum daily limits' for deposits of substances into Idaho waters on a 5-year schedule, rather than the 25-year schedule EPA had sought).

5 See 65 Federal Register 43585 (2000) (codified at 40 CFR Part 9) (the central provisions include (i) requiring states to identify polluted waters and establish prioritized schedules and (ii) defining elements of a total maximum daily limit (TMDL) and state implementation plans). Congress subsequently attached a legislative rider to an emergency supplemental assistance bill blocking implementation of the regulation for a year.

6 Executive Order 13112, 3 CFR Exec. Order 13112, reprinted in 42 USC § 4321 (2000).

7 Convention on Biological Diversity, 5 June 1992, Article 8 (h), UN Doc. [ST/]DPI/1307.

8 The Cartagena Protocol on Biosafety to the Convention on Biological Diversity, 29 January 2000, Articles 6–20. http://www.biodiv.org/biosafe/BIOSAFETY-PROTOCOL.htm.

9 See 16 USC § 1533 (2000). Currently, 515 animals and 743 plants are listed under the ESA (USFWS, 2002); the total number of species potentially imperilled or more seriously endangered is estimated to be 3740 (NC, 1997).

10 A recent 17-county assessment of western states determined that agricultural activities were a factor in the decline of 50 endangered fish species. At a global level, human activities already appropriate approximately 54% of the available water runoff; by 2025, human water withdrawals could exceed 70% (Postel, 1998).

11 According to one estimate, 63% of the irrigated area of the lower Colorado Basin is impaired by salinity (Lee and Howitt, 1996).

12 'Green-box' subsidies appear in Art. 6, Annex 2(12) of the URAA (URAA, 1994).

References

Adelman, D.E. and Barton, J.H. (2002) Environmental regulation for agriculture. *Stanford Environmental Law Journal* 21, 3.

Baffes, J. and Meerman, J. (1998) From prices to incomes: agricultural subsidization without protection? *World Bank Research Observer* 13, 191, 193.

Becker, E. (2002) Raising farm subsidies, US widens international rift. *New York Times*, 15 June, A3.

Benbrook, C.M. (1996) *Pest Management at the Crossroads.* Consumers Union, New York.

Curtis, J. (1998) *Fields of Change: a New Crop of American Farmers Finds Alternatives to Pesticides.* Natural Resources Defense Council, New York.

Daily, G., Dasgupta, P., Bolin, B., Crosson, P., du Guerny,J., Ehrlich, P., Folke, C., Jansson, A.M., Jansson, B.-O.,Kautsky,N., Kinzig, A., Levin, S., Mäler, K.-G., Pinstrup-Andersen, P., Siniscalco, D. and Walker, B. (1998) Food production, population growth, and the environment. *Science* 281, 1291.

Environmental Protection Agency (EPA) (1990) *Reducing Risk: Setting Priorities and Strategies for Environmental Protection.* US Government Printing Office, Washington, DC, pp. 36, 55–56, 66–67.

Environmental Protection Agency (EPA) (1999a) *Pesticide Industry Sales and Usage 1996 and 1997.* US Government Printing Office, Washington, DC, 3 pp.

Environmental Protection Agency (EPA) (1999b) *Office of Pesticide Programs Biennial Report: Fiscal Year 1998 and 1999.* US Government Printing Office, Washington, DC, 44 pp.

Environmental Protection Agency (EPA) (2000) *National Water Quality Inventory: 1998 Report to Congress.* US Government Printing Office, Washington, DC, pp. 61–62, 65, 87–88, 91, 149–50, 164.

Faeth, P. (1995) *Growing Green: Enhancing the Economic and Environmental Performance of US Agriculture.* World Resources Institute, Washington, DC, 28 pp.

Ganz, J. (1999) Sharp drop seen in soil erosion rates. *Science* 285, 1187.

Gardner, B. (1999) Agricultural relief legislation in 1998: the bell tolls for reform. *Regulation* 22, 31.

Goklany, I.M. (1998) Saving habitat and conserving biodiversity on a crowded planet. *Bioscience* 48, 941, 943.

Houck, O. (1999) TMDLs IV: the final frontier. *Environmental Law Review* 29, 10,469, 10,475–76.

Intergovernmental Panel on Climate Change (IPCC) (1996) *Climate Change 1995: Impacts, Adaptations and Mitigation of Climate Change.* Cambridge University Press, New York, 748 pp.

Knopman, D. and Smith, R. (1993) 20 years of the Clean Water Act: Federal Water Pollution Control Act Amendments of 1972. *Environment* 35, 16, 17.

Knudsen, O., Nash, J., Bovard, J., Gardner, B. and Winters, L.A. (1990) Redefining the role of government in agriculture for the 1990s. *World Bank Discussion Papers* 105, 49–57.

Lee, D. and Howitt, R. (1996) Modeling regional agricultural production and salinity control alternatives for water quality policy analysis. *American Journal of Agricultural Economics* 78, 41.

Malakoff, D. (1998) Death by suffocation in the Gulf of Mexico. *Science* 281, 190.
Matson, P.A., Parton, W.J., Power, A.G.and Swift, M.J. (1997) Agricultural Intensification and Ecosystem Properties. *Science* 277: 504, 507–508.
Mokma, D. and Seitz, M. (1992) Effects of soil erosion on corn yields on Marlette soils in south-central Michigan. *Journal of Soil and Water Conservation* 47, 325.
National Research Council (NRC) (1989) *Alternative Agriculture.* National Academy Press, Washington, DC, pp. 89, 176–188.
National Research Council (NRC) (1996a) *Ecologically Based Pest Management: New Solutions for a New Century.* National Academy Press, Washington, DC, pp. 31–2, 44–47, 50–51, 65–67.
National Research Council (NRC) (1996b) *Linking Science and Technology to Society's Environmental Goals.* National Academy Press, Washington, DC, 55 pp.
Nature Conservancy (NC) (1997) *1997 Species Report Card: State of US Plants and Animals.* http://www.natureserve.org/publications/97reportcard/contents.html
Novitzki, R.P. (1997) *Restoration, Creation, Recovery of Wetlands: Wetlands Functions, Values and Assessment.* Water Supply Paper 2425, US Geological Survey Washington, D.C. http://water.usgs.gov/nwsum/WSP2425/functions.html.
Office of Technology Assessment (OTA) (1996) *Agriculture, Trade, and the Environment.* US Government Printing Office, Washington, DC, pp. 69, 73–74, 75–78, 80, 92, 98.
Organization for Economic Cooperation and Development (OECD) (2000) *Agricultural Policies in OECD Countries: Monitoring and Evaluation 2000.* OECD Publications, Paris, pp. 2, 19.
Pimentel, D., Harvey, C., Resosudarmo, P., Sinclair, K., Kurz, D., McNair, M., Crist, S., Shpritz, L., Fitton, L., Saffouri, R. and Blair, R. (1995) Environmental and economic costs of soil erosion and conservation benefits. *Science* 267, 1117.
Postel, S.L. (1997) *Lost Oasis: Facing Water Scarcity.* Norton, New York.
Postel, S.L. (1998) Water for food production: will there be enough in 2025? *Bioscience* 48, 629–631.
Smil, V. (2000) *Feeding the World.* MIT Press, Cambridge, pp. 46–47, 93–95, 129.
Smith, K. (1995) Time to 'Green' US farm policy. *Issues in Science and Technology* 11, 71, 73.
Smith, R.A., Schwartz, G.E. and Alexander, R.B. (1994) *Regional Estimates of the Amount of US Agricultural Land Located in Watersheds with Poor Water Quality.* Open-File Report, US Geological Survey, US Government Printing Office, Washington, DC, 399 pp.
Stuteville, G. (2000) Crop subsidies keep growing: 'freedom to farm' Bill actually costs taxpayers a hefty chunk of change. *The Indianapolis Star*, 16 July, A1 pp.
Trimble, S.W. and Crosson, P. (2000) US soil erosion rates – myth and reality. *Science* 289, 248.
Uruguay Round Agreement on Agriculture (URAA) (1994) Marrakesh Agreement Establishing the World Trade Organization, Annex 1A, Multilateral Agreements on Trade in Goods (15 April). http://www.wto. org/english/docs_e/legal_e/final_e.htm
US Department of Agriculture (USDA) (1999) *The European Union's Common Agricultural Policy: Pressures for Change.* US Government Printing Office, Washington, DC, pp. 5, 6–7.

US Department of Agriculture (USDA) (2000) *Farm Income Forecast.* http://ers.usda.gov/briefing/ farmincome/fore/fore.htm
US Fish and Wildlife Service (USFWS) (2002) *Summary of Listed Species Listing and Recovery Plans 8 October.* http://ecos.fws.gov/tess/html/boxscore.html
Vitousek, P., Aber, J.D., Howarth, R.W., Likens, G.E., Matson, P.A., Schindler, D.W., Schlesinger, W.H. and Tilman, D.G. (1997) Human alteration of the global nitrogen cycle: sources and consequences. *Ecological Applications* 7, 737.
Vitousek, P., Mooney, H.A., Lubchenco, J. and Melillo, J.M. (1997b) Human domination of earth's ecosystems. *Science* 277, 494, 495.
World Resources Institute (WRI) (1994) *World Resources 1994–95.* World Resources Institute, Washington, DC, pp. 149, 182.

Appendix
General Web Addresses

European Community: http://europa.eu.int/rapid
OECD: www.oecd.gov
US Department of Agriculture: www.usda.gov
World Trade Organization: www.wto.org

Index

Italics are used to indicate legal cases as well as titles of documents. A page number followed by *n* indicates that the reference is to a note on that page.

Printed and bound by CPI Group (UK) Ltd, Croydon, CR0 4YY
16/09/2026
14963431-0003